Advances in

BOTANICAL RESEARCH

VOLUME 50

Advances in

BOTANICAL RESEARCH

Advances in

BOTANICAL RESEARCH

Series Editors

JEAN-CLAUDE KADER

Laboratoire Physiologie Cellulaire et Moléculaire des Plantes,
CNRS, Université de Paris,
Paris, France

MICHEL DELSENY

Laboratoire Génome et Développement des Plantes,
CNRS IRD UP, Université de Perpignan,
Perpignan, France

VOLUME 50

ELSEVIER

AMSTERDAM • BOSTON • HEIDELBERG • LONDON
NEW YORK • OXFORD • PARIS • SAN DIEGO
SAN FRANCISCO • SINGAPORE • SYDNEY • TOKYO
Academic Press is an imprint of Elsevier

Academic Press is an imprint of Elsevier
32 Jamestown Road, London NW17BY, UK
30 Corporate Drive, Suite 400, Burlington, MA 01803, USA
525 B Street, Suite 1900, San Diego, CA 92101-4495, USA
Radarweg 29, PO Box 211, 1000 AE Amsterdam, The Netherlands

First edition 2009

ISBN: 978-0-12-374835-5
ISSN: 0065-2296

For information on all Academic Press publications
visit our Web site at www.elsevierdirect.com

Printed and bound in USA
09 10 10 9 8 7 6 5 4 3 2 1

CONTENTS

Low Oxygen Signaling and Tolerance in Plants

FRANCESCO LICAUSI AND PIERDOMENICO PERATA

Roles of Circadian Clock and Histone Methylation in the Control of Floral Repressors

RYM FEKIH, RIM NEFISSI, KANA MIYATA, HIROSHI EZURA AND TSUYOSHI MIZOGUCHI

CONTRIBUTORS TO VOLUME 50

DUNCAN D. CAMERON *Department of Animal and Plant Sciences, University of Sheffield, Alfred Denny Building, Western Bank, Sheffield S10 2TN, United Kingdom*

NICOLAS CARELS *Fundação Oswaldo Cruz (FIOCRUZ), Instituto Oswaldo Cruz (IOC), Laboratório de Genômica Funcional e Bioinformática, Rio de Janeiro, RJ, Brazil, and Universidade Estadual de Santa Cruz (UESC), Núcleo de Biologia Computacional e Gestão de Informações Biotecnológicas, Ilhéus, BA, Brazil*

BRUNO G. DEFILIPPI *Instituto de Investigaciones Agropecuarias (CRI La Platina), Santa Rosa 11610, La Pintana and The Plant Cell Biotechnology Millennium Nucleus, Santiago, Chile*

HIROSHI EZURA *Gene Research Center, University of Tsukuba, Tennodai 1-1-1, Tsukuba, Ibaraki 305-8572, Japan*

RYM FEKIH *Gene Research Center, University of Tsukuba, Tennodai 1-1-1, Tsukuba, Ibaraki 305-8572, Japan*

MAURICIO GONZÁLEZ-AGÜERO *Instituto de Investigaciones Agropecuarias (CRI La Platina), Santa Rosa 11610, La Pintana and The Plant Cell Biotechnology Millennium Nucleus, Santiago, Chile*

LOUIS J. IRVING *Graduate School of Agricultural Science, Tohoku University, 1-1, Amamiyamachi Tsutsumidori, Aoba-ku, Sendai 981-8555, Japan*

FRANCESCO LICAUSI *PlantLab, Scuola Superiore Sant'Anna, Piazza Martiri della Libertà 33, 56127 Pisa, Italy*

KIETSUDA LUENGWILAI *Department of Plant Sciences, University of California, Davis, CA 95616, USA*

DANIEL MANRÍQUEZ *Research and Development, AgroFresh Inc. Chile, Isidora Goyenechea 3477, Oficina 221, Las Condes, Santiago, Chile*

KANA MIYATA *Gene Research Center, University of Tsukuba, Tennodai 1-1-1, Tsukuba, Ibaraki 305-8572, Japan*

TSUYOSHI MIZOGUCHI *Gene Research Center, University of Tsukuba, Tennodai 1-1-1, Tsukuba, Ibaraki 305-8572, Japan*

RIM NEFISSI *Gene Research Center, University of Tsukuba, Tennodai 1-1-1, Tsukuba, Ibaraki 305-8572, Japan*

PIERDOMENICO PERATA *PlantLab, Scuola Superiore Sant'Anna, Piazza Martiri della Libertà 33, 56127 Pisa, Italy*

CONTENTS OF VOLUMES 35–49

Series Editor (Volumes 35–44)

J.A. CALLOW
School of Biosciences, University of Birmingham,
Birmingham, United Kingdom

Contents of Volume 35

Contents of Volume 36

PLANT VIRUS VECTOR INTERACTIONS

Edited by **R. Plumb**

Contents of Volume 37

ANTHOCYANINS IN LEAVES

Edited by **K. S. Gould and D. W. Lee**

Contents of Volume 38

Contents of Volume 39

Contents of Volume 40

Contents of Volume 41

Contents of Volume 42

Contents of Volume 43

Contents of Volume 44

Series Editors (Volume 45–)

JEAN-CLAUDE KADER
Laboratoire Physiologie Cellulaire et Moléculaire des Plantes, CNRS, Université de Paris, Paris, France
MICHEL DELSENY
Laboratoire Génome et Développement des Plantes, CNRS IRD UP, Université de Perpignan, Perpignan, France

Contents of Volume 45

RAPESEED BREEDING

Contents of Volume 46

Contents of Volume 47

Contents of Volume 48

Contents of Volume 49

Aroma Volatiles: Biosynthesis and Mechanisms of Modulation During Fruit Ripening

BRUNO G. DEFILIPPI,*,†,[1] DANIEL MANRÍQUEZ,[‡]
KIETSUDA LUENGWILAI[§]
AND MAURICIO GONZÁLEZ-AGÜERO*,†

*Instituto de Investigaciones Agropecuarias (CRI La Platina), Santa Rosa 11610, La Pintana, Santiago, Chile
†The Plant Cell Biotechnology Millennium Nucleus, Santiago, Chile
‡Research and Development, AgroFresh Inc. Chile, Isidora Goyenechea 3477, Oficina 221, Las Condes, Santiago, Chile
§Department of Plant Sciences, University of California, Davis, CA 95616, USA

[1]Corresponding author: Email: bdefilip@inia.cl

Advances in Botanical Research, Vol. 50 0065-2296/09 $35.00

DOI: 10.1016/S0065-2296(08)00801-X

ABSTRACT

Flavor composition has been defined as a complex attribute of quality, in which the mix of sugars, acids, and volatiles play a primary role. In addition to the four basic flavors (sweet, sour, salty, and bitter) that humans can recognize in fruits and vegetables, aroma has an important influence on the final consumer acceptance of the commodity. Fruit aroma is determined by a complex mixture of a large number of volatile compounds including alcohols, aldehydes, and esters. During fruit development, especially at ripening, there are many changes of these metabolites caused by their synthesis, transport or degradation. In terms of volatile biosynthesis, several studies have been performed identifying and characterizing the most important genes and encoded enzymes involved in aroma-related volatiles; however, research in the mechanisms of regulation or modulation is still limited. To have an updated overview about aroma biosynthesis in fruit species, the main objective of this manuscript was to review the recent advances in this topic, mainly in terms of the new insights in volatile characterization, gene identification, and regulation of aroma during fruit ripening.

I. INTRODUCTION

In the plant kingdom, volatile compounds play an important role in coordinating many processes during a plant's interaction with its environment (Dudareva *et al.*, 2006; Dudareva and Pichersky, 2008). For example, these metabolites are involved in plant defense (against microorganism and herbivore attack), as well as communication with others plants. Another important function of aroma-related volatile compounds occurs during seed dispersal (Borges *et al.*, 2008). In this respect, this role is similar to those compounds responsible for changes in the color and palatability of fruit, increasing their appeal to different organisms that can spread the seeds contained therein.

Worldwide, the fruit market is quite important, and the quality of organoleptic attributes represents a key issue at the consumer level. This quality is related to many attributes, such as sweetness, acidity, aroma, color, and firmness, all of which are associated with specific metabolic pathways that are typically coordinated during the ripening process. The development of these qualities depends on many factors, such as variety, growing conditions, stage of harvest, and storage conditions. During ripening, many changes in fruit composition occur; these include the synthesis and degradation of pigments, changes in the concentrations of organic acids and sugars, and the accumulation of volatile compounds. These changes contribute to the consumer's overall perception of quality, which includes appearance, texture, and flavor.

In climacteric fruit, such as apple (*Malus × domestica*), melon (*Cucumis melo*), and banana (*Musa sp*), the typical aroma develops during ripening, with a maximum endogenous concentration occurring at the climacteric peak (Dixon and Hewett 2000; El-Sharkawy *et al.*, 2005; Fellman *et al.*, 2000). For climacteric fruit, the gaseous plant hormone ethylene plays a key role in the ripening process, initiating and enhancing softening, increasing the proportion of soluble solids, and facilitating development of the characteristic flavor (Abeles *et al.*, 1992; Giovannoni, 2004; Lelièvre *et al.*, 1997; Mir *et al.*, 1999; Theologis, 1992). Therefore, ethylene is considered as a critical factor in determining fruit quality, post-harvest life, and sensory impact at the consumer level.

Flavor composition has been defined as a complex attribute of quality, in which the mixture of sugars, acids, and volatiles plays a primary role (Baldwin, 2002). In addition to the four basic flavors (sweet, sour, salty, and bitter) recognized by humans in fruits and vegetables, aroma has an important influence on final consumer acceptance of the commodity (Lewinsohn *et al.*, 2001). Furthermore, volatile compounds include a broad group of metabolites that function in a biologically important manner to determine the interactions of plants with other organisms. These metabolites are also a key component of the overall flavor of fruits and vegetables (Pichersky and Gershenzon, 2002). The aroma properties of fruits depend upon the combination of volatiles produced, as well as on the concentration and potency of the individual volatile compounds.

During the last few decades, many researchers have tried to identify volatile compounds present in fruit aromas in the attempt to elucidate some of the biosynthetic pathways using bioconversion techniques or precursor tracing (D'Auria *et al.*, 2002; Dudareva *et al.*, 2004; Sanz *et al.*, 1997). Recently, research efforts have been directed toward the isolation of genes involved in the production of fruit volatile aromas (Aharoni *et al.*, 2000a; Beekwilder *et al.*, 2004; El-Sharkawy *et al.*, 2005) or flower scents (Dudareva and Pichersky, 2000; Shalit *et al.*, 2003). Although fruit aroma is generally a complex mixture of a wide range of compounds, volatile esters often represent the major contribution in apple and pear (*Pyrus communis*) (Paillard, 1990), banana (Shiota, 1993), melon (Beaulieu and Grimm, 2001), pineapple (*Ananas comosus*) (Elss *et al.*, 2005), and strawberry (*Fragaria × ananassa*) (Zabetakis and Holden, 1997).

To achieve an updated overview regarding the volatile aromas associated with these and other species, the main objective of this manuscript was to review recent advances in this topic, mainly in terms of new insights into volatile characterization, gene identification, aroma regulation during fruit ripening, and the environmental factors affecting this complex trait. In addition, we wished to provide the scientific community with information that complements that described in recently published valuable reviews (Dudareva and Pichersky, 2008; Dudareva *et al.*, 2006; Schwab *et al.*, 2008; Song and Forney, 2008; among others).

II. AROMA COMPOSITION IN FRUITS

Research on the broad range of volatile compounds involved in fruit aroma revealed that there are three major chemical groups that appear to be common to several fruits: alcohols, aldehydes, and esters. The volatile profiles of fruit are complex and vary depending on the cultivar, ripeness, pre- and post-harvest environmental conditions and analytical methods utilized.

Due to the large number of studies characterizing the aroma profiles of fruit, we selected a few specific species to provide an overall view of volatile compound composition with regard to common chemical groups. These species were selected based on economic importance and the amount of relevant information published in the literature.

In apple, melon, banana, mango (*Mangifera indica*), and papaya (*Carica papaya* L.), the major volatile-related aromas present in ripening fruit are alcohols, aldehydes and, particularly, esters (Beaulieu and Grimm, 2001; Dixon and Hewett, 2000; Zabetakis and Holden, 1997).

In apple, the aroma profile changes during fruit development from an abundance of aldehyde volatiles, for example, Z-2-hexenal, to a profile dominated by esters (Fellman *et al.*, 2000). A small number of these compounds, with different concentrations and thresholds, finally determine the characteristic aroma of a particular cultivar. Some volatiles are found at low concentrations but generate a significant contribution to the final aroma, for example, 2-methylbutyl butanoate. Dixon and Hewett (2000) summarized the important apple volatile compounds found in different cultivars. Among fruit tissues, it has been shown that epidermal tissue produces a greater amount of volatiles than internal tissues, including hypanthial and carpellary tissue (Rudell *et al.*, 2002). This higher capacity for aroma production by the peel has been attributed to either the abundance of fatty acid substrates (Guadagni *et al.*, 1971) or the higher metabolic activity (Defilippi *et al.*, 2005a; Rudell *et al.*, 2000). In melon, approximately 200 aromatic components have been identified in different melon varieties. Among those, volatile esters, mainly acetate derivatives such as ethyl 2-methylpropyl acetate, ethyl butyrate, and 2-methylbutyl acetate, are dominant with a 37% of the total volatile profile (Aubert and Bourger, 2004). In addition, lower amounts of lactones, sulfur compounds (such as [methylthio] acetate, 2-[methylthio] ethyl acetate, and 3-[methylthio] propyl acetate), short-chain alcohols, and aldehydes compose the complex mixture of volatile compounds (Aubert and Bourger, 2004; Aubert and Pitrat, 2006; Beaulieu and Grimm, 2001; Ibdah *et al.*, 2006; Manríquez *et al.*, 2006; Shalit *et al.*, 2001; Yahyaoui *et al.*, 2002). Non-climacteric cultivars often exhibit much lower levels of total volatiles and lack volatile esters, but they nonetheless demonstrate high levels of volatile aldehydes and alcohols (Shalit *et al.*, 2001). The biochemical and molecular characteristics of the enzymes involved in aroma production of melon have been recently extensively studied (El-Sharkawy *et al.*, 2005; Lucchetta *et al.*, 2006; Manríquez *et al.*, 2006; Yahyaoui *et al.*, 2002).

Similarly to apple and melon, more than 300 compounds have been identified that could contribute to the volatile profile of strawberries. The major components of strawberry flavor include esters, acids, aldehydes, alcohols, and terpenes (Zabetakis and Holden, 1997). Other contributing groups include sulfur compounds, acetals, furans, phenols, epoxides, and hydrocarbons. Among these, methyl, ethyl ester, furanones, C_6-compound aldehydes, and C_6-derivative compounds are considered to be key flavor compounds responsible for strawberry aroma (Pelayo *et al.*, 2003; Schieberle and Hofmann, 1997; Zabetakis and Holden, 1997; Zabetakis *et al.*, 1999). In Cigaline and Chandler varieties, C_6 aldehydes and alcohols, products of the enzymatic breakdown of unsaturated fatty acids, are major contributors to the flavor of immature fruits in the absence of furanones and esters. During fruit ripening, C_6 compound levels decrease drastically with increased furanone, acid, lactone, and ester production (Menager *et al.*, 2004; Pérez *et al.*, 1992, 1996).

Within the above chemical groups mentioned above we found terpenoid compounds. Terpenoids are an important family of secondary metabolites comprising close to 40,000 compounds, many of which are not volatiles and are involved in different plant processes such as membrane structure, photosynthesis, redox reactions, and plant regulation (Croteau *et al.*, 2000; Dudareva and Pichersky, 2000; McGarvey and Croteau, 1995; Schwab *et al.*, 2008). In the context of fruit aroma, many C_{10} monoterpenes and C_{15} sesquiterpenes compose the most abundant group of compounds present in the aroma profile. In some cases, these are also the key compounds determining the characteristic aroma. For example, the terpenoids *S*-linalool, limonene, valencene, and β-pinene are key compounds in the aroma profile of tomato, strawberry, koubo (*Cereus peruvianus* L.), citrus (*Citrus* sp.), and mango (Akakabe *et al.*, 2008; Buttery *et al.*, 1990; MacLeond and Gonzalez de Troconis, 1982; Moshonas and Shaw, 1994; Ninio *et al.*, 2003; Zabetakis and Holden, 1997). In strawberry, differences have been observed between cultivated and wild-type varieties, with the monoterpene *S*-linalool and the sesquiterpene nerolidol being the most abundant in cultivated varieties

(Aharoni *et al.*, 2004; Hampel *et al.*, 2006; Zabetakis and Holden, 1997). In contrast, oleafinic monoterpenes and myrtenyl acetate are more important in the wild-type varieties (Aharoni *et al.*, 2004). Koubo, another non-climacteric fruit, represents a unique case, since *S*-linalool and its derivates are responsible for 99% of the total volatile profile at ripening (Ninio *et al.*, 2003; Sitrit *et al.*, 2004). Volatile compounds in citrus fruits accumulate in the oil glands of flavedo and in the oil bodies of the juice sacs. In oranges (*Citrus sinensis* L.), the role of terpenoids in the aroma profile has been elucidated, with limonene being the most abundant compound (Obenland *et al.*, 2008; Maccarone *et al.*, 1998; Moshonas and Shaw, 1994). In other citrus species, such as *Citrus natsudaidai* (a natural hybrid of pummelo and *Citrus nagato-yuzukichi* [the Japanese sour citrus]), other terpenoids are less abundant, but they exert a profound effect on aroma. These include γ-terpinene, β-phellandrene, mycerene, and α-pinene (Akakabe *et al.*, 2008; Phi *et al.*, 2006).

In climacteric fruit, such as mango, terpenes are also key compounds of the overall fruit aroma (MacLeond and Gonzalez de Troconis, 1982; Singh *et al.*, 2004). Compounds such as α-pinene, car-3-ene, limonene, γ-terpinene, α-humulene, and β-selinene are part of the major group determining aroma (MacLeond and Gonzalez de Troconis, 1982). In tomato, a sesquiterpene (eugenol) and a monoterpene (*S*-linalool), are the most important factors, with *S*-linalool causing the sweet, floral, and alcoholic note observed for the aroma bouquet (Buttery *et al.*, 1990).

In other fruits, different terpenoids play an important role in the flavor profile. For example, limonene is one of the most important compounds in acerola (*Malpighia glabra*) (Pino and Marbot, 2001). In guava (*Psidium guajava* L.), there are at least nine key compounds, including (*E*)-β-caryophyllene, α-terpineol, α-pinene, α-selinene, β-selinene, δ-cadinene, 4,11-selinadiene, and α-copaene (Pino *et al.*, 2002). Terpenes, C_{13}-norisoprenoids, benzene derivatives, and aliphatic alcohols, mainly present in the skin, are responsible for grape (*Vitis vinifera*) volatility.

III. ETHYLENE AS MODULATOR OF VOLATILE BIOSYNTHESIS DURING RIPENING

A. ETHYLENE AND FRUIT RIPENING, CLIMACTERIC AND NON-CLIMACTERIC FRUITS

During fruit development, many changes in flavor metabolites are caused by their synthesis, transport, or degradation. In climacteric fruits, ethylene plays an important role as a modulator of ripening. All of these fruit quality-related metabolites may be directly regulated by ethylene (ethylene-dependent processes) or by other signals (ethylene-independent processes) (Flores *et al.*, 2001). Fruit ripening corresponds to biochemical, physiological, and structural changes that give the fruit its organoleptic qualities and make it consumable. Although these processes vary from one type of fruit to another, fruit can be classified in two categories: climacteric and non-climacteric (Biale and Young, 1981). The sharp increase in ethylene synthesis observed in climacteric fruit is considered to induce the changes in color, aroma, texture, flavor, and other biochemical and physiological parameters. In contrast, among non-climacteric fruit, the changes that occur during maturation are thought to be ethylene-independent, with unknown regulatory factors. There is no dominant model for non-climacteric fruit, probably because ethylene does not play a regulatory role in the ripening processes that occur in these fruit. It should be noted that efforts have been made to answer this question in strawberries (Aharoni *et al.*, 2000a) and grapes (Waters *et al.*, 2005). For example, genetic studies in citrus led to the isolation of genes encoding a chlorophyllase (Trebitsh *et al.*, 1993) and the discovery of chlorophyllase regulation by ethylene (Jacob-Wilk *et al.*, 1999), showing that the maturation of non-climacteric fruit includes events that are regulated by this plant hormone. The discovery of transcription factors type MADS box involved in the development of maturation in tomato will undoubtedly have repercussions in other fruits, especially non-climacterics (Ezura and Owino, 2008; Giovannoni, 2007; Pech *et al.*, 2008).

Increasing post-harvest life has been a key goal for breeders in recent decades, and this practice has been generally associated with a loss in flavor. In general, there is an inverse relationship between ethylene production and post-harvest life (Gussman *et al.*, 1993; Zheng and Wolff, 2000). For example, melons producing large amounts of ethylene, such as Charentais cantaloupe (cantalupensis group), have a shorter post-harvest life and stronger aroma than the American cantaloupes or American rockmelons (reticulates group), which produce much less ethylene and, consequently, exhibit decreased aroma. On the other hand, non-climacteric melons such as cassabas and *piel de sapo* (inodorus group) have a long post-harvest life and produce less aroma. Therefore, the particular climacteric and its unique aroma intensity profile are critical characteristics affecting the post-harvest life and sensory quality of fruit (Obando-Ulloa *et al.*, 2008).

Over the past two decades, with the advances in molecular biology, fruit ripening has emerged as a genetically programmed phenomenon during development, and it involves the expression of specific genes (Argueso *et al.*, 2007; Grierson *et al.*, 1986). The ripening mechanisms of climacteric fruit are, by far, the most well studied due to the key role of ethylene and the substantial progress made in understanding both the biosynthesis (Hamilton *et al.*, 1991; Pech *et al.*, 2004; Sato and Theologis, 1989; Van der Straeten *et al.*, 1991) and activity of this plant hormone (Ecker, 1995; Hall *et al.*, 2007; Kendrick and Chang, 2008; Stepanova and Ecker, 2000; Underwood *et al.*, 2005). Consequently, many scientific manuscripts have sought to explain the mechanisms of fruit ripening in view of biochemical, physiological, and molecular perspectives (Alexander and Grierson, 2002; Brady, 1987; Giovannoni, 2001, 2004; Pech *et al.*, 2004; Lelièvre *et al.*, 1997).

B. ETHYLENE AND AROMA BIOSYNTHESIS

As stated above, fruit ripening is a complex process in which ethylene plays an important role, in combination with other hormones and developmental factors. This process involves mechanisms that are both dependent on and independent of ethylene (Pech *et al.*, 2004, 2008).

Evidence for a relationship between ethylene and aroma production is based on the observation that the concentration of aroma-related volatiles increases significantly as ripening progresses, and that application of ethylene inhibitors or enhancers results in changes in volatile production. However, many aspects of this relationship remain unclear, and the steps responsive to ethylene remain unknown. Moreover, it is still not known whether the onset of ethylene production during ripening is concurrent with the onset of volatile biosynthesis, or rather precedes and plays a significant role in the initiation of ester production (Dixon and Hewett, 2000; Fellman *et al.*, 2000). In early studies performed during the ripening stage, Paillard (1986) determined that, upon onset of the respiratory climacteric period, ester biosynthesis directly followed advancement of the climacteric period. However, whether the volatile biosynthetic enzymes are constitutive or induced during the climacteric period remains unclear (Dixon and Hewett, 2000).

Inhibitors of ethylene biosynthesis or ethylene activity have been extensively used as experimental tools for identifying ethylene-related processes. In studies evaluating the effects of the ethylene inhibitor 1-methylcyclopropene (1-MCP) on many fruits, including bananas (Golding *et al.*, 1998), plum (*Prunus salicina* Lindl) (Abdi *et al.*, 1998), and apple (Defilippi *et al.*, 2004; Fan and Mattheis, 1999; Lurie *et al.*, 2002), a decrease in volatile ester production was noted, indicating that a high rate of ester production requires continuous ethylene activity. Aldehyde and alcohol compounds were also affected by inhibition of ethylene activity (Lurie *et al.*, 2002). Similarly, the use of L-2-amino-4-(1-aminoethoxy)-*trans*-3-butenoic acid (AVG), an ethylene biosynthesis inhibitor, delays maturity and ripening in apple, with a reduction in ester production exceeding 20%. This reduction was restored after external ethylene application (Fan *et al.*, 1998; Mir *et al.*, 1999). Moreover, the use of AVG only has shown that many volatile alcohols (ester substrate) were reduced after its application (Mir *et al.*, 1999), suggesting that overall aroma biosynthesis is regulated by ethylene. Using a biotechnological approach based on the suppression of ethylene biosynthesis, ethylene has been clearly demonstrated to regulate aroma production. Oeller *et al.* (1991) demonstrated that suppression of the climacteric ethylene peak affected the characteristic in transformed tomato plants. State-of-the-art tools such as antisense RNA have been used to further elucidate the role of ethylene during fruit ripening. In climacteric 1-aminocyclopropane-1-carboxylate oxidase (ACO) antisense transgenic melon, ripening parameters including rind color and aroma production (esters) were strongly reduced in response to low ethylene levels (Bauchot *et al.*, 1998; Flores *et al.*, 2001), suggesting that these parameters are physiologically regulated by ethylene during fruit development.

Yahyaoui *et al.* (2002) isolated two AAT genes from melon fruit; the authors observed that both genes were differentially regulated by ethylene during ester formation. The effect of ethylene suppression in transgenic apple lines is dramatic, resulting in a remarkable reduction or delay in the accumulation of ester compounds of 85–88% in transgenic lines compared to non-transformed apples (Dandekar *et al.*, 2004; Defilippi *et al.*, 2004). Similar levels of ester inhibition were previously observed in transgenic ACO antisense melon lines, as discussed below (Bauchot *et al.*, 1998), as well as in apple treated with the ethylene inhibitors AVG and 1-MCP (Fan *et al.*, 1998; Lurie *et al.*, 2002).

In cantaloupe melons, esters represent the major group of aroma volatiles emitted and are likely to be the key contributors to the unique aroma of ripe melon (Beaulieu and Grimm, 2001; Homatidou *et al.*, 1992). Among cantaloupe melons, the Charentais type is highly aromatic, and the synthesis of aroma volatiles is regulated by the plant hormone ethylene (Flores *et al.*, 2002). Extension of the shelf life either by breeding or genetic engineering led to a strong reduction in aroma volatiles production (Aubert and Bourger, 2004; Bauchot *et al.*, 1998). In order to better understand the effects of ethylene on fruit maturation, Charentais cantaloupe melons were transformed with an antisense DNA construct encoding ACO oxidase under control of the 35S promoter (Ayub *et al.*, 1996). Similarly to the results described for apple, a melon line expressing antisense ACO exhibited a reduction in ethylene production close to 99.5%, which led to inhibition of a number of maturation phenomena, including volatile production. Bauchot *et al.* (1998) demonstrated that inhibition of ethylene synthesis due to antisense ACO resulted in a 60–80% reduction in aroma-related volatiles compared to wild-type. Inhibition of ethylene synthesis primarily affects esters. The metabolic steps of biosynthesis controlled by ethylene were determined using the discs of fruit witnesses or antisense ACO transcribed in the presence of various esters biosynthetic pathway precursors (Flores *et al.*, 2002). In the ACO transgenic melon lines the last step in ester biosynthesis was partially affected by ethylene suppression, suggesting that there are signal other than ethylene involved in the modulation of aroma biosynthesis (ethylene-independent components) (Flores *et al.*, 2002; Pech *et al.*, 2008).

In the case of non-climacteric fruits such as koubo, studies showed that there is an increase in the activity of *S*-linalool synthase during ripening; this activity is related to the rise in *S*-linalool that occurs during fruit ripening (Ninio *et al.*, 2003; Sitrit *et al.*, 2004). In the case of citrus, one study shows that the accumulation of sesquiterpene valencene in citrus occurs in response to ethylene. This study provides evidence that ethylene regulates the late ripening stages of non-climacteric fruits (Sharon-Asa *et al.*, 2003). In raspberry (*Rubus idaeus* L.), the terpenenoids comprise the major compounds and increase during storage in air. This increase could be related to terpene synthase activity (Harb *et al.*, 2008). Evidence obtained for mango (Herianus *et al.*, 2003), a climacteric fruit, shows that monoterpenes increase during mango ripening, followed by a decrease during later ripening stages. The observed increase runs parallel to ethylene production and may be affected by it: some sesquiterpenes follow the same trend of ethylene production observed during mango ripening.

IV. VOLATILE BIOSYNTHESIS IN FRUITS

Volatiles are important for aroma and flavor and are synthesized from amino acids, membrane lipids, and carbohydrates (Sanz *et al.*, 1997). These compounds are composed of different chemical classes, and several pathways are involved in their biosynthesis. Fatty acids are major precursors of aroma volatiles in several fruits, and the biosynthetic pathway includes β-oxidation (in intact fruit) and lipoxygenase action (in disrupted tissue), ultimately forming aldehydes, acids, alcohols, and esters from lipids (Sanz *et al.*, 1997; Schreier, 1984; Yahia, 1994).

An important step in the biosynthetic pathway of aroma compounds is the availability of primary precursor substrates, including fatty acids and amino acids, which are highly regulated during fruit development in terms of amount and composition (Ackermann *et al.*, 1992; Song and Bangerth, 2003). Several studies have demonstrated the significance of fatty acids as precursor of esters. In particular, the oxidation of linoleic and linolenic acids generates many of the alcohols, aldehydes, acids, and esters typically found in fruit (Bartley *et al.*, 1985; Rowan *et al.*, 1999). Notably, transgenic modification of fatty acid biosynthesis in plant tissues resulted in significant changes in aroma compounds (Wang *et al.*, 2001). Fatty acid levels are highly regulated during fruit development, where they accumulate during apple ripening, especially during the climacteric peak, followed by a decline due to changes in lipid metabolism (Meigh and Hulme, 1965). Recently, Song and Bangerth (2003) showed that free fatty acids increase at least four times during the climacteric period, coinciding with the increase in aroma production.

Esters produced in several fruits (i.e., apple, melon, and pear) are enzymatically synthesized by coupling the respective acid (C_2–C_8 acids) and alcohol. Most of the esters involved in the apple aroma profile are biosynthesized from either lipids or amino acid precursors through a series of enzyme-mediated steps (Sanz *et al.*, 1997). In addition, ester biosynthesis is considered to be limited by the alcohol concentration, as it has been demonstrated that the availability of alcohol can modify the final aroma profile within a specific cultivar (Berger and Drawert, 1984). Alcohol acyltransferase (AAT), which catalyzes linkage of the acetyl moiety from acetyl CoA to the appropriate alcohol, is the major aroma-related enzyme that has been studied in some detail (e.g., in ripe apple) (Fellman *et al.*, 2000).

Alcohol dehydrogenase (ADH) is another enzyme that functions upstream of AAT in the biosynthetic pathway. In fruits, ADH has been related to the interconversion of aldehyde and alcohol forms of flavor volatiles, and to their accumulation in the fruit during ripening (Chen and Chase, 1993; Speirs *et al.*, 1998). Prestage *et al.* (1999) determined that mature green tomato contained lower levels of ADH transcripts than ripened fruits. This trend was correlated with lower alcohol and higher aldehyde levels. The upstream enzymes involved in the ester biosynthetic pathway, excluding ADH, may play an important role in the production of C_6 volatile compounds in plants. Lipoxygenase (LOX) may aid in determining the composition of precursors for ester production in fruits (Bate *et al.*, 1998; Fellman *et al.*, 2000). For example, in tomato transformed with two antisense LOX genes, Griffiths *et al.* (1999a) determined that very low levels of enzyme activity are adequate for the production of C_6 aldehydes and alcohols, demonstrating levels comparable to those present in non-transformed plants. Further experiments performed by Griffiths *et al.* (1999b) showed that ethylene was differentially involved in the regulation of gene expression for three LOX genes during fruit ripening; however, they concluded that additional work is required to elucidate the role of ethylene and other developmental factors.

Amino acids are also involved in aroma biosynthesis in fruit, and their metabolism is responsible for the production of a broad number of compounds, including alcohols, carbonyls, acids, and esters (Sanz *et al.*, 1997). The most important amino acids responsible for generating aroma compounds as direct precursors are alanine, valine, leucine, isoleucine, phenylalanine, and aspartic acid (Baldwin, 2002; Sanz *et al.*, 1997). The addition of amino acid precursors to apple tissue slices increased total aroma volatile compound production. For example, experiments using deuterium-labeled substrates showed that isoleucine is the precursor for 2-methylbutanoate in apple (Rowan *et al.*, 1996). Another important compound identified in some apple varieties, 4-methoxyallylbenzene, appears to be generated from phenylalanine (Hansen and Poll, 1993). However, the importance of individual amino acids as aroma precursors depends on the fruit species; for example, isoleucine is an important precursor of branched-chain-derived esters in strawberries, but not in bananas (Sanz *et al.*, 1997). These differences may be explained by the variation in relative individual amino acid content during ripening. In apple, aspartic acid is the most abundant free amino acid, followed distantly by glutamic acid, serine, and phenylalanine (Ackermann *et al.*, 1992). All of these amino acids decrease during fruit development, reaching more or less constant levels during ripening. Despite the importance of amino acids as potential substrates for volatile production, their levels cannot always be related to the formation of a specific aroma compound. This suggests that enzymes upstream of AAT play an important role in the formation of volatiles (Defilippi *et al.*, 2005a; Dixon and Hewett, 2000; Wyllie and Fellman, 2000).

In the case of other key compounds determining aroma, terpenoids also comprise a large group of different compounds. All terpenoids are based on the five-carbon molecule isopentenyl diphosphate (IPP) and its allilyc isomer dimethylallyl diphosphate (DMAPP). Both molecules result from the conversion of acetyl-Coenzyme A (CoA), pyruvate and glyceraldehyde-3-phosphate (Fig. 1) (Bohlmann *et al.*, 1998; Croteau *et al.*, 2000; Dudareva *et al.*, 2004, 2006; Pichersky *et al.*, 2006). The next step in the biosynthesis of terpenoids is the formation of direct precursors; this occurs in different cell compartments via two parallel pathways, the first, the mevalonic acid (MVA) pathway, in the cytosol and the second, the methyl-erythriol-phosphate pathway (MEP), in plastids. Both pathways work independently, but there is some "cross-talk" (Bick and Lange, 2003; Dudareva *et al.*, 2004; Lichtenthaler *et al.*, 1997; Rodríguez-Concepción and Boronat, 2002).

The direct precursors of terpenoids are linear geranyl diphosphate (GDP, C_{10}), farnesyl diphosphate (FDP, C_{15}), and geranylgeranyl diphosphate (GGDP, C_{20}). The reaction is catalyzed by a group of enzymes called prenyl transferases, which use the common C_5 units IPP and DAMP as substrates (Dudareva *et al.*, 2004, 2006; McGarvey and Croteau, 1995;

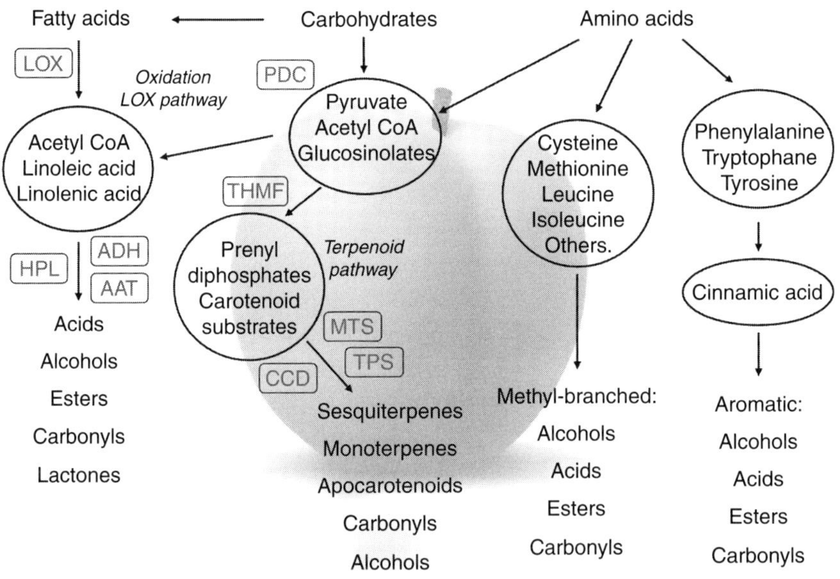

Fig. 1. *Summarized scheme of the biosynthetic pathways leading to the formation of major volatile compounds in fruits.* Common pathway names are italicized, enzymes are boxed, intermediary compounds are in circles, and volatile compounds are in bold in the final steps. Abbreviations: LOX, lipoxygenase; HPL, fatty acid hydroperoxide lyase; ADH, alcohol dehydrogenase; AAT, alcohol acyl transferase; PDC, pyruvate decarboxylase; THMF, 3-ketoacyl-CoA thiolase; CCD, carotenoid cleavage dioxygenase; MTS, monoterpene synthase; TPS, terpene synthase; Acetyl CoA, acetyl coenzyme A.

Pichersky *et al.*, 2006; Schwab *et al.*, 2008). Sesquiterpenes are synthesized in the cytosol, whereas hemiterpens, monoterpenes, diterpens, and carotenoids are synthesized in plastids (Dudareva *et al.*, 2006; Pichersky *et al.*, 2006; Schwab *et al.*, 2008). The last step in the biosynthetic pathway is the formation of the final terpenoids: hemiterpenes (C_5), monoterpenes (C_{10}), sesquiterpenes (C_{15}), or diterpenes (C_{20}). This reaction is catalyzed by terpene synthases (Dudareva *et al.*, 2004; Pichersky *et al.*, 2006; Schwab *et al.*, 2008). One important characteristic of some of these enzymes is that they can form different compounds utilizing a single substrate (Dudareva *et al.*, 2004).

The apocarotenoid compounds, also called norisoprenoids, derive from carotenoids (Lewinsohn *et al.*, 2005a; Schwab *et al.*, 2008). The carotenoids are pigments and correspond to tetraterpenes; they are present in many parts of the plant. The carotenoid cleavage dioxygenases (CCD) are enzymes that catalyze the oxidative cleavage of carotenoids, producing apocarotenoids (Ibdah *et al.*, 2006; Lewinsohn *et al.*, 2005a,b). Studies in tomato and watermelon show that carotenoids not only influence flavor through color perception, but also have a direct impact on aroma via apocarotenoids synthesis (Lewinsohn *et al.*, 2005a).

V. GENE DISCOVERY

The biogenesis and production rate of key volatile compounds depends on the activity and substrate-specificity of relevant enzymes implicated in the biosynthetic pathway, as well as on substrate availability. The enzymes have not been fully described, but appear to be common to different fruits.

A. ALCOHOL ACYL TRANSFERASE

Volatile esters, a major class of compounds contributing to the aroma of many fruits, are synthesized by alcohol acyltransferases (AAT, EC 2.3.1.84). As mentioned above, the AAT enzyme is the only enzyme that has been studied in some detail in ripe fruit, including apple (Echeverría *et al.*, 2004; Fellman *et al.*, 2000), apricot (*Prunus armeniaca* L.) (González-Agüero *et al.*, in press), banana (Harada *et al.*, 1985), melon (El-Sharkawy *et al.*, 2005; Lucchetta *et al.*, 2006; Yahyaoui *et al.*, 2002), and strawberry (Pérez *et al.*, 1993). Strawberry AAT is the only such enzyme that has been purified, characterized, and genetically cloned (Aharoni *et al.*, 2000a; Pérez *et al.*, 1993). Experiments performed using bananas and strawberries indicate a correlation between substrate specificity and the volatile esters present in each fruit's aroma, suggesting a determining role for AAT in flavor biogenesis in these species (Pérez *et al.*, 1996; Dixon and Hewett, 2000).

The ability to synthesize esters was evaluated for several fruits (whole fruit or discs) in response to different storage conditions or treatments: banana (Wyllie and Fellman, 2000), apple (Rowan *et al.*, 1996, 1999), strawberry (Hamilton-Kemp *et al.*, 1996), and melon (Ueda *et al.*, 1997). The first studies examining AAT in plants showed that this enzyme is located in the soluble fraction of banana pulp (Harada *et al.*, 1985). Similar work has been performed using strawberry (Pérez *et al.*, 1993) and melon (Ueda *et al.*, 1997) AAT. The first plant AAT encoded gene (Benzyl Alcohol Acetyl Transferase, *BEAT*) was isolated and described in *Clarkia breweri* flowers (Dudareva *et al.*, 1998). The AAT genes described in the literature are quite different in terms of amino acid sequences, substrates, and molecular weight (i.e., apricot, 50 kDa; strawberry, 70 kDa; melon, 400 kDa). A large number of acyltransferase genes are present in plants, around 70 in *Arabidopsis* (Pichersky and Gang, 2000), and more than 10 fruit species (see Table I). In melon, AATs are encoded by a family of four putative genes (El-Sharkawy *et al.*, 2005; Yahyaoui *et al.*, 2002), with amino acid identity ranging from 84% (between *CmAAT1* and *CmAAT2*) and 58% (*CmAAT1*–*CmAAT3*) to only 22% (*CmAAT1*–*CmAAT4*). All encoded proteins, except for *CmAAT2*, show AAT activity upon expression in yeast, as well as differential substrate preferences (El-Sharkawy *et al.*, 2005).

Despite performing the same reaction, AAT proteins from different fruit species may be highly divergent. For example, *P. armeniaca* AAT (*PaAAT*) shares a maximum of 58% identity with AAT from *P. communis* (*PAAT*) and *M. domestica* (*MdAAT2*), despite being from a closely related species in the *Rosaceae* family (González-Agüero *et al.*, in press). This low sequence identity is not uncommon for the enzyme, and has already been described for apple (Li *et al.*, 2006; Souleyre *et al.*, 2005), melon (Yahyaoui *et al.*, 2002), and strawberry (Aharoni *et al.*, 2000a) AATs, notwithstanding a similar preference for substrates across species. AATs produce a wide range of short- and long-chain acyl esters depending on the species (Table I) and substrate utilized. For instance, the *CmAAT1* protein has a strong preference for the formation of (*E*)-2-hexenyl acetate and hexyl hexanoate (Yahyaoui *et al.*, 2002), *MdAAT2* preferentially produces pentyl acetate and hexyl acetate (Li *et al.*, 2006), and *SAAT* typically yields methyl hexanoate, hexyl acetate, hexyl butyrate, and octyl acetate (Aharoni *et al.*, 2000a).

The activities of AAT proteins measured with the preferred substrates sharply increase during fruit ripening. The expression of *AAT* genes is regulated during ripening and was found to be regulated by ethylene, while others enzymes such as LOX and ADH were unaffected by ethylene modulation (Defilippi *et al.*, 2005b). In strawberries, *SAAT* is one of the most studied genes in volatile compound biosynthesis; this gene is exclusively expressed in fruit tissue and its expression increases significantly (more than 15-fold) during the transition from the pink to the full red stage during ripening (Aharoni *et al.*, 2000a).

B. ALCOHOL DEHYDROGENASE

Fruits produce acetaldehyde and ethanol during maturation and ripening. Pyruvate decarboxylase (PDC; EC 4.1.1.1) and alcohol dehydrogenase (ADH; EC 1.1.1.1) are two important enzymes responsible for acetaldehyde and ethanol production, respectively. ADH is an oxidoreductase involved in the reversible conversion of aldehydes to their corresponding alcohols. ADH has been implicated in the stress response of plants, and is responsible for ethanol production under anaerobic conditions. ADHs are also involved in a wide range of responses to other stresses, elicitors, and to abscisic acid (Matton *et al.*, 1990; Peters and Frenkel, 2004). However, *ADH* gene expression has been shown to be tissue-specific and developmentally regulated, particularly during fruit ripening (Echeverría *et al.*, 2004; Van der Straeten *et al.*, 1991).

TABLE I

Genes Putatively Encoding Major Volatiles-Forming Enzymes from Different Fruit Species

Enzymes	Genes[a]	Fruit species	Accession number[b]	Major volatiles reported[c]	References
Alcohol acyl transferase (AAT) [EC 2.3.1.84]	*TomAAT*	*Solanum lycopersicum*	AAS48091	Isoamyl acetate; isoamyl 3-methyl butyrate; hexyl acetate	Petro-Turza (1987)
	PAAT	*Pyrus communis*	AAS48090	Hexyl acetate; butyl acetate; ethyl (2*E*,4*Z*)-2,4-decadienoate; methyl (2*E*,4*Z*)-2,4-decadienoate	Morton and McLeod (1990)
	CmAAT1	*Cucumis melo*	CAA94432	(*E*)-2-Hexenyl acetate; hexyl hexanoate	Yahyaoui *et al.* (2002)
	CmAAT2	*Cucumis melo*	AAL77060	NR	Yahyaoui *et al.* (2002)
	CmAAT3	*Cucumis melo*	AAW51125	Benzyl acetate	El-Sharkawy *et al.* (2005)
	CmAAT4	*Cucumis melo*	AAW51126	Cinnamoyl acetate	El-Sharkawy *et al.* (2005)
	PaAAT	*Prunus armeniaca*	ACF07921	Hexyl acetate	González-Agüero *et al.* (in press)
	MdAAT1	*Malus × domestica*	AAU14879	Hexyl acetate; butyl acetate; 2-methylbutyl acetate	Souleyre *et al.* (2005)
	MdAAT2	*Malus × domestica*	AAS79797	Pentyl acetate; hexyl acetate	Li *et al.* (2006)
	SAAT	*Fragaria × ananassa*	AAG13130	Methyl hexanoate; hexyl acetate; hexyl butyrate; octyl acetate	Aharoni *et al.* (2000a)
	VAAT	*Fragaria vesca*	CAC09062	Ethyl acetate; ethyl butanoate; ethyl hexanoate; octyl acetate	Pyysalo *et al.* (1979)
	ManAAT	*Mangifera indica*	CAC09378	*cis*-3-Hexenyl butanoate; *cis*-3-hexenyl acetate; butyl butanoate	Morton and McLeod (1990)
	BanAAT	*Musa sp*	CAC09063	Isoamyl acetate; isobutyl acetate; 2-pentanol acetate; hexyl acetate; isoamyl butyrate	Morton and McLeod (1990)
	LAAT1	*Citrus limon*	CAC09049	Butyl acetate; geranyl acetate; neryl acetate; citronellyl acetate	Morton and McLeod (1990)

Pyruvate decarboxylase (PDC) [EC 4.1.1.1]	*FaPDC1*	*Fragaria × ananassa*	AAL37492	NR	Moyano *et al.* (2004)
	FaPDC2	*Fragaria × ananassa*	AAG13131	NR	Aharoni *et al.* (2000a)
	VvPDC1	*Vitis vinifera*	AAG22488	NR	Or *et al.* (2000)
	PaPDC	*Prunus armeniaca*	ABZ79223	NR	González-Agüero *et al.* (in press)
	StPDC	*Solanum lycopersicum*	BAC23043	NR	Nakane *et al.* (2003)
	CsPDC	*Citrus sinensis*	AAZ05069	NR	Kanellis and Pasentsis (Unpublished)
Alcohol dehydrogenase (ADH) [EC 1.1.1.1]	*pAADH*	*Malus × domestica*	CAA88271	Hexanol	Reid *et al.* (1996)
	PaADH	*Prunus armeniaca*	ABZ79222	1-Hexanol	González-Agüero *et al.* (in press)
	CmADH1	*Cucumis melo*	ABC02081	NR	Manríquez *et al.* (2006)
	CmADH2	*Cucumis melo*	ABC02082	NR	Manríquez *et al.* (2006)
	pTADH2	*Solanum lycopersicum*	CAA54450	Hexanol; (*Z*)-3-hexenol	Longhurst *et al.* (1994) and Speirs *et al.* (1998)
	VvADH1	*Vitis vinifera*	AAG01381	NR	Tesniere and Verries (2000)
	VvADH2	*Vitis vinifera*	AAG01382	NR	Tesniere and Verries (2000)
	VvADH3	*Vitis vinifera*	AAG01383	NR	Tesniere and Verries (2000)
	PcADH3	*Pyrus communis*	AAB86868	Ethanol; methanol	Chervin and Truett (1999)
	PcADH4	*Pyrus communis*	AAB86869	Ethanol; methanol	Chervin and Truett (1999)
	FaADH	*Fragaria × ananassa*	CAA33613	NR	Wolyn and Jelenkovic (1990)

(*continues*)

TABLE I (*continued*)

Enzymes	Genes[a]	Fruit species	Accession number[b]	Major volatiles reported[c]	References
Lipoxygenase (LOX) [EC 1.13.11.12]	*TomLOXA*	*Solanum lycopersicum*	AAA53184	NR	Ferrie *et al.* (1994)
	TomLOXB	*Solanum lycopersicum*	AAA53183	NR	Ferrie *et al.* (1994)
	TomLOXC	*Solanum lycopersicum*	AAB65766	NR	Heitz *et al.* (1997)
	PaLOX	*Prunus armeniaca*	ABZ05753	Hexanal; (*E*)-2-hexenal	González-Agüero *et al.* (in press)
	PdLOX	*Prunus dulcis*	CAB94852	NR	Mita *et al.* (2001)
	AdLOX1	*Actinidia deliciosa*	ABF59997	NR	Zhang *et al.* (2006)
	AdLOX5	*Actinidia deliciosa*	ABF60001	NR	Zhang *et al.* (2006)
	FaLOX	*Fragaria* × *ananassa*	CAE17327	NR	Leone *et al.* (Unpublished)
	MdLOX1	*Malus* × *domestica*	AAV50006	NR	Goulao and Oliveira (2007)
Fatty acid hydroperoxide lyase (HPL) [EC 4.1.2.92]	*LeHPL*	*Solanum lycopersicum*	AAF67142	*cis*-3-Hexenal	Howe *et al.* (2000)
	CmHPL	*Cucumis melo*	AAK54282	3(*Z*)-Nonenal	Tijet *et al.* (2001)
	PgHPL	*Psidium guajava*	AAK15070	3(*Z*)-Hexenal	Tijet *et al.* (2000)
	CsHPL	*Cucumis sativus*	AAF64041	*n*-Hexanal; 3(*Z*)-hexenal	Matsui *et al.* (2000)
	BanHPL	*Musa sp*	CAB39331	NR	Haeusler *et al.* (1997)
	CsHPL	*Citrus sinensis*	AAO72740	NR	Wu and Burns (Unpublished)
3-ketoacyl-CoA thiolase (THMF) [EC 2.3.1.16]	*pTHMF1*	*Mangifera indica*	CAA53078	NR	Bojorquez and Gomez-Lim (1995)
	CurTHMF	*Cucurbita cv. K. Amakuri*	BAA11117	NR	Kato *et al.* (1996)
	CsTHMF	*Cucumis sativus*	CAA47926	NR	Preisig-Muller and Kindl (1993)
	FaTHMF	*Fragaria* × *ananassa*	CAC09051	NR	Aharoni *et al.* (2000b)

Terpene synthase [EC 4.2.3.20]	*LeMTS1*	*Solanum lycopersicum*	AAX69063	(*R*)-Linalool; (*E*)-nerolidol	van Schie *et al.* (2007)
	LeMTS2	*Solanum lycopersicum*	AAX69064	β-Phellandrene; β-myrcene; sabinene	van Schie *et al.* (2007)
	FaLIS	*Fragaria × ananassa*	CAD57106	NR	Aharoni *et al.* (2000b)
	VvTPS	*Vitis vinifera*	AAS79352	Linalool; geraniol; nerol; citronellol; α-terpineol	Martin and Bohlmann (2004)
	VvVAL	*Vitis vinifera*	AAS66358	(+)-Valencene; (−)-7-epi-α-selinene	Lücker *et al.* (2004)
	MdEAFAR	*Malus × domestica*	AAX19772	Linalool; (*Z*)- and (*E*)-β-ocimene; β-myrcene	Green *et al.* (2007)
	CsTPS1	*Citrus sinensis*	AAQ04608	Valencene	Sharon-Asa *et al.* (2003)
	ClLIMS1	*Citrus limon*	AAM53944	Limonene	Lücker *et al.* (2002)
	ClβPINS	*Citrus limon*	AAM53945	β-Pinene	Lücker *et al.* (2002)
	ClγTS	*Citrus limon*	AAM53943	γ-Terpinene	Lücker *et al.* (2002)
	PcPFS	*Pyrus communis*	AAT70237	NR	Li *et al.* (Unpublished)
Carotenoid cleavage dioxygenase [EC 1.13.11.51]	*LeCCD1A*	*Solanum lycopersicum*	AAT68187	β-Ionone; pseudoionone; geranylacetone	Simkin *et al.* (2004a)
	LeCCD1B	*Solanum lycopersicum*	AAT68188	β-Ionone; pseudoionone; geranylacetone	Simkin *et al.* (2004a)
	CmCCD1	*Cucumis melo*	ABB82946	Geranylacetone; pseudoionone; β-ionone	Ibdah *et al.* (2006)
	PaNCED2	*Persea americana*	AAK00622	NR	Chernys and Zeevaart (2000)
	MdCCD4	*Malus × domestica*	ABY47995	NR	Huang and Schwab (Unpublished)
	VvCCD1	*Vitis vinifera*	AAX48772	3-Hydroxy-β-ionone	Mathieu *et al.* (2005)
	CitCCD1	*Citrus sinensis*	BAE92958	3-Hydroxy-β-ionone; C14 Dialdehyde; β-ionone	Kato *et al.* (2006)
	FaCCD1	*Fragaria × ananassa*	ACA13522	NR	Munoz-Blanco *et al.* (Unpublished)

[a]Name reported in the associated references or assigned in this review for better understanding.
[b]GenBank access for codified protein.
[c]Compounds reported in the literature associated to activity of recombinant enzymes and/or related to volatile production in each fruit species. NR, not reported in the literature.

The ADH gene has been identified and characterized in a number of fruit species (see Table I), including apple (Reid *et al.*, 1996), apricot (González-Agüero *et al.*, in press), melon (Manríquez *et al.*, 2006), tomato (Longhurst *et al.*, 1994), and grape (Tesniere and Verries, 2000); enzyme expression and/or activity have been reported to increase during ripening (Chen and Chase, 1993; Longhurst *et al.*, 1994). In apple, *pAADH* expression during apple ripening indicated that fruit ripening was associated with a decrease in ADH mRNA levels (Reid *et al.*, 1996). This result contrasts with observations in tomato, for which *pTADH2* mRNA is strongly upregulated in ripening fruit (Longhurst *et al.*, 1994). Overexpression of *pTADH2* led to improved flavor by increasing alcohol levels, particularly (*Z*)-3-hexenol (Speirs *et al.*, 1998). In grapes, three *ADH* genes (*VvADH1* to *3*) are expressed during fruit development. *VvADH1* and *VvADH3* transcripts accumulate transiently in the young developing berry, while *VvADH2* transcripts strongly increase at the onset of ripening (denoted by the term *véraison*) (Tesniere and Verries, 2000).

Manríquez *et al.* (2006) reported two highly divergent *ADH* genes (*CmADH1* and *CmADH2*) in melon that exhibited 15% shared identity at the amino acid level. *CmADH1* belongs to the medium-chain zinc-binding and *CmADH2* to the short-chain type of ADHs; both genes are expressed specifically in fruit and are upregulated during ripening. Apricot ADH (*PaADH*), a member of the short-chain ADH subfamily, has a variable percentage of shared identity with other ADHs, ranging from 57 to 99% (González-Agüero *et al.*, in press). The amino acid sequence diversity of the short-chain ADH may be related to the broad range of biological functions in which it is involved and the wide range of substrates used by this enzyme in higher plants (Jörnvall *et al.*, 1995).

C. LIPOXYGENASE

Fatty acids serve as ester precursors and are catabolized via two major pathways, β-oxidation and the lipoxygenase system (LOX; EC 1.13.11.12). LOXs are non-heme iron-containing deoxygenases that are widely distributed in the plant kingdom and possess diverse functions (Porta and Rocha-Sosa, 2002). Among the roles associated with fruit ripening, LOX is involved in the generation of C_6 alcohols and aldehydes, which constitute major volatile flavor components in ripening fruits such as tomato (Chen *et al.*, 2004), providing the fruit with a green aroma character. The green notes: hexanal, hexanol, (*E*)-2-hexenal, 3(*Z*)-hexenal, (*E*)-2-hexenol, and 3(*Z*)-hexenol are used widely in flavors to impart a fresh green character (Tijet *et al.*, 2000). The low capacity for fatty acid precursor biosynthesis in apple could be a major limiting factor for ester production in immature fruit (Song and Bangerth, 1994).

LOX activity has also been associated with the development of apple (Defilippi *et al.*, 2005b; Echeverría *et al.*, 2004), strawberry (Pérez *et al.*, 1999), pear (Lara *et al.*, 2003), and kiwifruit (Zhang *et al.*, 2006); furthermore, specific *LOX* genes have been identified in several fruit species (Table I). Three *LOX* genes have been identified in tomato (*TomLOXA* to *C*) and shown to be regulated by different processes in different tissues. Heitz *et al.* (1997) found that *TomLOXC* mRNA is not wound-inducible in tomato leaves, but accumulates in fruit upon ripening. Two other *LOX* genes, *TomLOXA* and *TomLOXB*, were found to be expressed in these organs (Ferrie *et al.*, 1994). Six kiwifruit *LOX* genes (*AdLOX1* to *6*) have been identified and characterized according to their differential expression in ripening fruit. According to the classification of plant genes belonging to the *LOX* family (Feussner and Wasternack, 2002), *AdLOX5* is grouped in the 9-LOX family, and *AdLOX1* is proposed to have 13-LOX activity (Zhang *et al.*, 2006). Remarkably, expression of *AdLOX1* and *AdLOX5* markedly increased during the developmental progression of fruit to the climacteric stage, and were upregulated by ethylene treatment, following a similar pattern as that observed for LOX enzyme activity (Zhang *et al.*, 2006).

D. FATTY ACID HYDROPEROXIDE LYASE

The metabolism of fatty acid hydroperoxides (LOX products) involves conversion to aldehydes, alcohols, and other derivatives, and these reactions are often catalyzed by cytochrome P450 enzymes (Tijet *et al.*, 2000). In plant tissues, the fatty acid hydroperoxide lyase (HPL; EC 4.1.2.92), a member of the cytochrome P450-family, including CYP74, and an enzyme in the LOX pathway, catalyzes the cleavage of 13- and 9-hydroperoxides of linoleic and linolenic acid into volatile C_6- or C_9-aldehydes and C_{12}- or C_9-oxoacids, respectively (Hatanaka *et al.*, 1987). HPL was first cloned from green bell peppers and is designated as *CYP74B* (Matsui *et al.*, 1996). HPL-codified enzymes have been cloned (Table I) and characterized in several species, including tomato (Howe *et al.*, 2000), guava fruit (*Psidium guajava* L.) (Tijet *et al.*, 2000), cucumber (*Cucumis sativus* L.) (Matsui *et al.*, 2000), and melon (Tijet *et al.*, 2001).

In guava fruit and cucumber, the major aldehyde produced for HPL is 3(*Z*)-hexenal; in comparison, 3(*Z*)-nonenal was preferentially produced in melon. The primary aldehyde product, 3(Z)-hexenal ("leaf aldehyde"), formed from the 13*S*-hydroperoxide of linolenic acid, is described to have distinct physiological functions in plant wound response and pathogen attack.

Furthermore, the family inclusive of this C_6 aldehyde and alcohols derived from 3(*Z*)-hexenal comprises important compounds in the flavor industry (Tijet *et al.*, 2001).

E. 3-KETOACYL-COA THIOLASE

Changes in fatty acids and triglycerides have been associated with changes in aroma, flavor, and the production of volatiles during fruit ripening (Bojorquez and Gomez-Lim, 1995). Therefore, fatty acid-related enzymes may be important for the production of aroma volatiles compounds; this might explain their induction during ripening. Thiolase (THMF; EC 2.3.1.16) is the last enzyme in the fatty acid β-oxidation pathway. THMF catalyzes the thiolytic cleavage of the β-ketoacyl-CoA substrate carbon chain by CoA-SH, yielding acetyl-CoA and a saturated acyl-CoA ester that is shorter by two carbon atoms (Bojorquez and Gomez-Lim, 1995). The acyl-CoA formed in the cleavage reaction may be utilized at the final stage of the biosynthetic pathway for ester formation in fruit.

To date, this enzyme has been reported in few species (see Table I) and its role in aroma biosynthesis is not fully understood. A peroxisomal 3-ketoacyl CoA thiolase (*pTHMF1*) has been identified and characterized in mango and demonstrates high homology with cucumber THMF (*CsTHMF*); notably, the protein was upregulated during mango ripening (Bojorquez and Gomez-Lim, 1995).

F. TERPENE SYNTHASE

The flavor and aroma of certain fruits, such as particular grape varieties, is dominated by small volatile aldehydes and volatile monoterpenes (Martin and Bohlmann, 2004). Monoterpenes comprise the C_{10} branch of the terpene family and consist of two head-to-tail coupled isoprene units (C_5). Monoterpenes are beneficial to plants as they function in defense against herbivores and plant pathogens or as attractants for pollinators. Furthermore, monoterpenes contribute to the final grape and wine aroma and flavor, in the form of free volatiles and as glycoside conjugates of monoterpene alcohols. Typical monoterpenol components of aroma-rich grape varieties are *S*-linalool, geraniol, nerol, citronellol, and α-terpineol. Tomato also emits a blend of volatile organic compounds, which mainly consist of terpenes. The advances in these species, especially tomato, have been largely favored by the used of transgenic approaches (Davidovich-Rikanati *et al.*, 2008; Lewinsohn *et al.*, 2001).

The enzymes that synthesize these compounds are varied and depend on the substrate and volatile compounds produced. To facilitate this review, we grouped terpene synthase (MTS; EC 4.2.3.20) within a single group (Table I). The first two monoterpene synthases have been identified and characterized in tomato (*LeMTS1* and *LeMTS2*). Although these proteins are highly homologous, recombinant *LeMTS1* protein produces (*R*)-linalool from GPP and (*E*)-nerolidol from FPP, while *LeMTS2* produces β-phellandrene, β-myrcene, and sabinene from GPP (van Schie *et al.*, 2007). The profile of terpenoid volatiles in various citrus species and their importance as aroma compounds has been studied in detail. However, much is still lacking in our understanding of the physiological, biochemical, and genetic regulation of their production. The sequences of several monoterpene synthases were identified in lemon: ClγTS produces c-terpinene, Cl(+)LIMS1 produces limonene, and Cl(–)βPINS produces β-pinene. These were divided into two separate groups. One group comprises C1γTS and Cl(–)βPINS, revealing 84% identity. The other group consists of Cl(+)LIMS1 and other MTSs, which demonstrate 97% identity. Between groups, the sequence identity does not exceed 51% (Lücker *et al.*, 2002). In another citrus species, sweet orange, researchers identified a gene (*CsTPS1*) encoding a sesquiterpene synthase that converts farnesyl diphosphate to a single product, valencene. Phylogenetic analysis revealed that this gene belongs to the group containing sesquiterpene synthase (Sharon-Asa *et al.*, 2003)

In grapes, the sesquiterpene valencene is a key aroma compound present in different organs and at different stages. The full-length cDNA *VvVAL* was expressed in *E. coli,* and the recombinant protein was shown to be necessary for valencene production. In expression studies performed during berry ripening, *VvVAL* transcripts were not detected in the mesocarp and exocarp during early stages of fruit development, but increased during the final stages of berry ripening (Lücker *et al.*, 2004). Recently, with the analysis of the grapevine genome sequence, at least four monoterpene synthase genes related to aroma biosynthesis have been identified (Velasco *et al.*, 2007).

G. CAROTENOID CLEAVAGE DIOXYGENASE

Volatile terpenoid compounds, potentially derived from carotenoids, are important components of flavor and aroma in many fruits, vegetables, and ornamentals. Despite their importance, little is known about the enzymes responsible for generating these volatiles. Recently, families of carotenoid cleavage dioxygenases (CCD; EC 1.13.11.51) that cleave carotenoid substrates at a variety of double bonds have been identified (Table I). The first member of the family to be identified was *VP14*, a 9-*cis*-epoxycarotenoid deoxygenase from *Zea mays* involved in the synthesis of the phytohormone abscisic acid (Tan *et al.*, 1997). Recombinant CCDs cleave carotenoids symmetrically at the 9, 10 bonds, resulting in the formation of C_{13}- and C_{14}-apocarotenoids; such enzymes have been reported in *A. thaliana* (Schwartz *et al.*, 2001) and *Crocus sativus* (Bouvier *et al.*, 2003). Two closely related genes potentially encoding carotenoid cleavage deoxygenases (*LeCCD1A* and *LeCCD1B*) were identified in tomato (Simkin *et al.*, 2004a) and petunia (Simkin *et al.*, 2004b). *LeCCD1A* and *LeCCD1B* silencing resulted in a significant decrease in the β-ionone content of ripe fruits, implicating a role for these genes in C_{13}-norisoprenoid synthesis *in vivo* (Simkin *et al.*, 2004a). C13-norisoprenoids are terpenoids that are commonly found in the flowers, fruits, and leaves of many plants (Winterhalter and Rouseff, 2002) and possess interesting flavor/aroma properties, in conjunction with low aroma thresholds.

In melon, *CmCCD* (a carotenoid cleavage deoxygenase) was isolated and characterized by Ibdah *et al.* (2006). They showed that the expression of this gene is upregulated during fruit development in different melon varieties. Interestingly, the heterologous protein can cleave carotenoids in positions 9, 10, 9′, and 10′, generating geranylacetone, pseudoionone, β-ionone, and α-ionone (Ibdah *et al.*, 2006).

VI. CONCLUSIONS

Fruit consumers are not only looking for traditional quality attributes such as sugar, acidity, firmness, and color. They also value other attributes, including nutrients availability, antioxidant, and aroma. Therefore, a major goal for growing fruits should emphasize on a good balance among the quality attributes already mentioned. In terms of aroma production, several studies have been focused mainly in identifying the volatile profile and the impact of individual compounds in overall aroma. Due to the complexity of the mechanism involved in determining fruit aroma, several issues remain to be studied, such as (i) the pre-harvest factors affecting aroma volatile production, (ii) the specific process under ethylene modulation affecting aroma production in climacteric fruit, (iii) the key player affecting aroma evolution in non-climacteric fruit, (iv) the actual involvement and role of genes and encoded enzymes involved in aroma-related volatiles, in the broad range of fruit available in the market. Within this final issue, an important step would be to provide new tools, as molecular markers, for example, for breeders in order to support breeding programs focused in varieties with more flavor and aroma.

ACKNOWLEDGMENTS

BGD thanks Fondecyt grant N°1060179 for funding studies on apricot aroma. DM thanks CONICYT Chile for a doctoral fellowship, as well as the *Laboratoire de Biologie Moléculaire et Physiologie de la Maturation des Fruits* INP-ENSAT in Toulouse, France for supporting the research work performed in aroma biosynthesis of melon. MGA gratefully acknowledges the PBCT-Conicyt (PSD03) project for financial support for a postdoctoral fellowship.

REFERENCES

Abdi, N., McGlasson, W. B., Holford, P., Williams, M. and Mizrahi, Y. (1998). Responses of climacteric and suppressed-climacteric plums to treatment with propylene and 1-methylcyclopropene. *Postharvest Biology and Technology* **14**, 29–39.

Abeles, F. B., Morgan, P. W. and Saltveit, M. E. (1992). "Ethylene in plant biology," p. 414. Academic Press, San Diego, CA.

Ackermann, J., Fischer, M. and Amado, R. (1992). Changes in sugars, acids, and amino acids during ripening and storage of apples (cv. Glockenapfel). *Journal of Agricultural and Food Chemistry* **40**, 1131–1134.

Aharoni, A., Keizer, L. C. P., Bouwmeester, H. J., Sun, Z., Alvarez-Huerta, M., Verhoeven, H. A., Blaas, J. L., van Houwelingen, M. L., De Vos, R. C. H., van der Voet, H., Jansem, R. C. Guis, M. *et al.* (2000a). Identification of the SAAT gene involved in strawberry flavor biogenesis by use of DNA microarray. *The Plant Cell* **12**, 647–661.

Aharoni, A., Verhoeven, H. A., Luecker, J., O'Connell, A. P. and Van Tunen, A. J. (2000b). Fruit flavour related genes and use thereof Patent: WO 0032789-A 08-JUN-2000.

Aharoni, A., Giri, A. P., Verstappen, F. W. A., Bertea, C. M., Sevenier, R., Sun, Z., Jongsma, M. A., Schwab, W. and Bouwmeester, H. J. (2004). Gain and loss of fruit flavor compounds produced by wild and cultivated strawberry species. *The Plant Cell* **16**, 3110–3131.

Akakabe, Y., Sakamoto, M., Ikeda, Y. and Tanaka, M. (2008). Identification and characterization of volatile components of Japanese sour citrus fruit *Citrus nagato-yuzukichi* Tanaka. *Bioscience, Biotechnology, and Biochemistry* **72**, 1965–1968.

Alexander, L. and Grierson, D. (2002). Ethylene biosynthesis and action in tomato: A model for climacteric fruit ripening. *Journal of Experimental Botany* **53**, 2039–2055.

Argueso, C. T., Hansen, M. and Kieber, J. J. (2007). Regulation of ethylene biosynthesis. *Journal of Plant Growth Regulation* **26**, 92–105.

Aubert, C. and Bourger, N. (2004). Investigation of volatiles in Charentais cantaloupe melons (*Cucumis melo* var. cantalupensis). Characterization of aroma constituents in some cultivars. *Journal of Agricultural and Food Chemistry* **52**, 4522–4528.

Aubert, C. and Pitrat, M. (2006). Volatile compounds in the skin and pulp of Queen Anne's pocket melon. *Journal of Agricultural and Food Chemistry* **54**, 8177–8182.

Ayub, R., Guis, M., Ben Amor, M., Gillot, L., Roustan, J. P., Latché, A., Bouzayen, M. and Pech, J. C. (1996). Expression of ACC oxidase antisense gene inhibits ripening of cantaloupe melon fruits. *Nature Biotechnology* **14**, 862–866.

Baldwin, E. A. (2002). Fruit flavor: Volatile metabolism and consumer perceptions. *In* "Fruit quality and its biological basis" (M. Knee, ed.), pp. 89–106. Sheffield Academic Press, Sheffield.

Bartley, I. M., Stoker, P. G., Martin, A. D. E., Hatfield, S. G. S. and Knee, M. (1985). Synthesis of aroma compounds by apples supplied with alcohols and methyl esters of fatty acids. *Journal of the Science of Food and Agriculture* **36**, 567–574.

Bate, N. J., Riley, J. C. M., Thompson, J. E. and Rothstein, S. J. (1998). Quantitative and qualitative differences in C-6-volatile production from the lipoxygenase pathway in an alcohol dehydrogenase mutant of *Arabidopsis thaliana*. *Physiologia Plantarum* **104**, 97–104.

Bauchot, A. D., Mottram, D. S., Dodson, A. T. and John, P. (1998). Effect of aminocyclopropane-1-carboxylic acid oxidase antisense gene on the formation of volatile esters in cantaloupe Charentais melon (cv. Vedrandais). *Journal of Agricultural and Food Chemistry* **46**, 4787–4792.

Beaulieu, J. C. and Grimm, C. C. (2001). Identification of volatile compounds in cantaloupe at various developmental stages using solid phase microextraction. *Journal of Agricultural and Food Chemistry* **49**, 1345–1352.

Beekwilder, J., Alvarez-Huerta, M., Neef, E., Verstappen, F. W., Bouwmeester, H. J. and Aharoni, A. (2004). Functional characterization of enzymes forming volatile esters from strawberry and banana. *Plant Physiology* **135**, 1865–1878.

Berger, R. G. and Drawert, F. (1984). Changes in the composition of volatiles by postharvest application of alcohols to Red Delicious apples. *Journal of the Science of Food and Agriculture* **35**, 1318–1325.

Biale, J. B. and Young, R. E. (1981). Respiration and ripening in fruits-retrospective and prospect. *In* "Recent advances in the biochemistry of fruits and vegetables" (J. Friend and M. J. C. Rhodes, eds.), pp. 1–39. Academic Press, New York.

Bick, J. A. and Lange, M. (2003). Metabolic cross talk between cytosolic and plastidial pathways of isoprenoid biosynthesis: Unidirectional transport of intermediates across the chloroplast envelope membrane. *Archives of Biochemistry and Biophysics* **415**, 146–154.

Bohlmann, J., Meyer-Gauen, G. and Croteau, R. (1998). Plant terpenoid synthases: Molecular biology and phylogenetic analysis. *Proceedings of the National Academy of Sciences* **95**, 4126–4133.

Bojorquez, G. and Gomez-Lim, M. A. (1995). Peroxisomal thiolase mRNA is induced during mango fruit ripening. *Plant Molecular Biology* **28**, 811–820.

Borges, R. M., Bessière, J.-M. and Hossaert-Mckey, M. (2008). The chemical ecology of seed dispersal in monoecious and dioecious figs. *Functional Ecology* **3**, 484–493.

Bouvier, F., Suire, C., Mutterer, J. and Camara, B. (2003). Oxidative remodeling of chromoplast carotenoids: Identification of the carotenoid dioxygenase CsCCD and CsZCD genes involved in crocus secondary metabolites biogenesis. *The Plant Cell* **15**, 47–62.

Brady, C. J. (1987). Fruit ripening. *Annual Review of Plant Physiology* **38**, 155–178.

Buttery, R. G., Teranishi, R., Ling, L. C. and Turnbaugh, J. G. (1990). Quantitative and sensory studies on tomato paste volatile. *Journal of Agricultural and Food Chemistry* **38**, 336–340.

Chen, A. R. S. and Chase, T. (1993). Alcohol dehydrogenase 2 and pyruvate decarboxylase induction in ripening and hypoxic tomato fruit. *Plant Physiology and Biochemistry* **31**, 875–885.

Chen, G. P., Hackett, R., Walker, D., Taylor, A., Lin, Z. F. and Grierson, D. (2004). Identification of a specific isoform of tomato lipoxygenase (TomloxC) involved in the generation of fatty acid-derived flavor compounds. *Plant Physiology* **136**, 2641–2651.

Chernys, J. T. and Zeevaart, J. A. (2000). Characterization of the 9-*cis*-epoxycarotenoid dioxygenase gene family and the regulation of abscisic acid biosynthesis in avocado. *Plant Physiology* **124**, 343–353.

Chervin, C. and Truett, J. (1999). Alcohol dehydrogenase expression and alcohol production during pear ripening. *Journal of the American Society for Horticultural Science* **124**, 71–75.

Croteau, R., Kutchan, T. M. and Lewis, N. G. (2000). Natural products (secondary metabolites). *In* "Biochemistry and molecular biology of plants" (B. Buchanan, W. Gruisem and R. Jones, eds.), pp. 1250–1318. American Society of Plant Physiologist, Rockville, MD.

Dandekar, A. M., Teo, G., Defilippi, B. G., Uratsu, S. L., Passey, A. J., Kader, A. A., Stow, J. R., Colgan, R. J. and James, D. J. (2004). Effect of down-regulation of ethylene biosynthesis on fruit flavor complex in apple fruit. *Transgenic Research* **13**, 373–384.

D'Auria, J., Chen, F. and Pichersky, E. (2002). Characterization of an acyltransferase capable of synthesizing benzylbenzoate and other volatile esters in flowers and damaged leaves of *Clarkia breweri*. *Plant Physiology* **130**, 466–476.

Davidovich-Rikanati, R., Lewinsohn, E., Bar, E., Lijima, Y., Pichersky, E. and Sitrit, Y. (2008). Overexpression of the lemon basil α-zingiberene singiberene synthase gene increases both mono- and sesquiterpene contents in tomato fruit. *The Plant Journal* **56**, 228–238.

Defilippi, B. G., Dandekar, A. M. and Kader, A. A. (2004). Impact of suppression of ethylene action or biosynthesis on flavor metabolites in apple (*Malus* $\times$ *domestica* Borkh) fruits. *Journal of Agricultural and Food Chemistry* **52**, 5694–5701.

Defilippi, B. G., Dandekar, A. M. and Kader, A. A. (2005a). Relationship of ethylene biosynthesis to volatile production, related enzymes and precursor availability in apple peel and flesh tissues. *Journal of Agricultural and Food Chemistry* **53**, 3133–3141.

Defilippi, B. G., Kader, A. A. and Dandekar, A. M. (2005b). Apple aroma: Alcohol acyltransferase, a rate limiting step for ester biosynthesis, is regulated by ethylene. *Plant Science* **168**, 1199–1210.

Dixon, J. and Hewett, E. W. (2000). Factors affecting apple aroma/flavour volatile concentration: A review. *New Zealand Journal of Crop and Horticultural Science* **28**, 155–173.

Dudareva, N. and Pichersky, E. (2000). Biochemical and molecular genetic aspects of floral scents. *Plant Physiology* **122**, 627–633.

Dudareva, N. and Pichersky, E. (2008). Metabolic engineering of plant volatiles. *Current Opinion in Biotechnology* **19**, 181–189.

Dudareva, N., D'Auria, J. C., Nam, K. H., Raguso, R. A. and Pichersky, E. (1998). Acetyl-CoA: Benzylalcohol acetyltransferase – an enzyme involved in floral scent production in *Clarkia breweri*. *Plant Journal* **14**, 297–304.

Dudareva, N., Pichersky, E. and Gershenzon, J. (2004). Update on biochemistry of plant volatiles. *Plant Physiology* **135**, 1893–1902.

Dudareva, N., Negre, F., Nagegowda, D. A. and Orlova, I. (2006). Plant volatiles: Recent advances and future perspectives. *Critical Reviews in Plant Sciences* **25**, 417–440.

Echeverría, G., Graell, J., López, M. L. and Lara, I. (2004). Volatile production, quality and aroma-related enzyme activities during maturation of 'Fuji' apples. *Postharvest Biology and Technology* **31**, 217–227.

Ecker, J. R. (1995). The ethylene signal transduction pathway in plants. *Science* **268**, 667–675.

El-Sharkawy, I., Manríquez, D., Flores, F. B., Regard, F., Bouzayen, M., Latché, A. and Pech, J. C. (2005). Functional characterization of a melon alcohol acyltransferase gene family involved in the biosynthesis of ester volatiles. Identification of the crucial role of a threonine residue for enzyme activity. *Plant Molecular Biology* **59**, 345–362.

Elss, S., Preston, C., Hertzig, C., Heckel, F., Richling, E. and Schreier, P. (2005). Aroma profiles of pineapple fruit (*Ananas comosus* L. Merr.) and pineapple products. *Lebensm-Wissu-Technology* **38**, 263–274.

Ezura, H. and Owino, W. O. (2008). Melon, an alternative model plant for elucidating fruit ripening. *Plant Science* **175**, 121–129.

Fan, X. and Mattheis, J. P. (1999). Impact of 1-methylcyclopropene and methyl jasmonate on apple volatile production. *Journal of Agricultural and Food Chemistry* **47**, 2847–2853.

Fan, X., Mattheis, J. P. and Buchanan, D. (1998). Continuous requirement of ethylene for apple fruit volatile synthesis. *Journal of Agricultural and Food Chemistry* **46**, 1959–1963.

Fellman, J. K., Miller, T. W., Mattinson, D. S. and Mattheis, J. P. (2000). Factors that influence biosynthesis of volatile flavor compound in apple fruits. *HortScience* **35**, 1026–1033.

Ferrie, B. J., Beaudoin, N., Burkhart, W., Bowsher, C. G. and Rothstein, S. J. (1994). The cloning of two tomato lipoxygenase genes and their differential expression during fruit ripening. *Plant Physiology* **106**, 109–118.

Feussner, I. and Wasternack, C. (2002). The lipoxygenase pathway. *Annual Review of Plant Biology* **53**, 275–297.

Flores, F., Ben Amor, M., Jones, B., Pech, J. C., Bouzayen, M., Latché, A. and Romojaro, F. (2001). The use of ethylene-suppressed lines to assess differential sensitivity to ethylene of the various ripening pathways in cantaloupe melons. *Physiologia Plantarum* **113**, 128–133.

Flores, F., El Yahyaoui, F., De Billerbeck, G., Romojaro, F., Latché, A., Bouzayen, M., Pech, J. C. and Ambid, C. (2002). Role of ethylene in the biosynthetic pathway of aliphatic ester aroma volatiles in Charentais Cantaloupe melons. *Journal of Experimental Botany* **53**, 201–206.

Giovannoni, J. (2001). Molecular biology of fruit maturation and ripening. *Annual Review of Plant Physiology and Plant Molecular Biology* **52**, 725–749.

Giovannoni, J. (2004). Genetic regulation of fruit development and ripening. *Plant Cell Supplement* **16**, 170–180.

Giovannoni, J. J. (2007). Fruit ripening mutants yield insights into ripening control. *Current Opinion in Plant Biology* **10**, 283–289.

Golding, J. B., Shearer, D., Wyllie, S. G. and McGlasson, W. B. (1998). Application of 1-MCP and propylene to identify ethylene-dependent ripening processes in mature banana fruit. *Postharvest Biology and Technology* **12**, 87–98.

González-Agüero, M., Troncoso, S., Gudenschwager, O., Campos-Vargas, R., Moya-Leon, A. and Defilippi, B. G. (in press). Differential expression levels of aroma related genes during ripening of apricot (*Prunus armeniaca* L.). *Plant Physiology and Biochemistry*.

Goulao, L. F. and Oliveira, C. M. (2007). Molecular identification of novel differentially expressed mRNAs up-regulated during ripening of apples. *Plant Science* **172**, 306–318.

Green, S., Friel, E. N., Matich, A., Beuning, L. L., Cooney, J. M., Rowan, D. D. and MacRae, E. (2007). Unusual features of a recombinant apple alpha-farnesene synthase. *Phytochemistry* **68**, 176–188.

Grierson, D., Tucker, G. A., Keen, J., Ray, J., Bird, C. R. and Schuch, W. (1986). Sequencing and identification of a cDNA clone for tomato polygalacturonase. *Nucleic Acids Research* **14**, 8595–8603.

Griffiths, A., Prestage, S., Linforth, R., Zhang, J., Taylor, A. and Grierson, D. (1999a). Fruit-specific lipoxygenase suppression in antisense-transgenic tomatoes. *Postharvest Biology and Technology* **17**, 163–173.

Griffiths, A., Barry, C., Alpuche-Solis, A. G. and Grierson, D. (1999b). Ethylene and developmental signals regulate expression of lipoxygenase genes during tomato fruit ripening. *Journal of Experimental Botany* **50**, 793–798.

Guadagni, D. G., Bomben, J. L. and Hudson, J. S. (1971). Factors influencing the development of aroma in apple peel. *Journal of the Science of Food and Agriculture* **22**, 110–115.

Gussman, C., Goffredz, J. and Gianfagna, T. (1993). Ethylene production and fruit-softening rates in several apple fruit ripening variants. *HortScience* **28**, 135–137.

Haeusler, A., Lerch, K., Muheim, A. and Silke, N. (1997). Hydroperoxide lyases Patent: EP 0801133-A 15-OCT-1997.

Hall, B. P., Shakeel, S. N. and Schaller, G. E. (2007). Ethylene receptors: Ethylene perception and signal transduction. *Journal of Plant Growth Regulation* **26**, 118–130.

Hamilton, A. J., Bouzayen, M. and Grierson, D. (1991). Identification of a tomato gene for the ethylene-forming enzyme by expression in yeast. *Proceedings of the National Academy of Sciences USA* **88**, 7434–7437.

Hamilton-Kemp, T. R., Archbold, D. D., Loughrin, J. H., Collins, R. W. and Byers, M. (1996). Metabolism of natural volatile compounds by strawberry fruit. *Journal of Agricultural and Food Chemistry* **44**, 2802–2805.

Hampel, D., Mosandl, A. and Wüst, M. (2006). Biosynthesis of mono- and sesquiterpenes in strawberry fruits and foliage: 2H labeling studies. *Journal of Agricultural Food and Chemistry* **54**, 1473–1478.

Hansen, K. and Poll, L. (1993). Conversion of L-isoleucine into 2-methylbut-2-enyl esters in apples. *Lebensmittel Wissenschaft und Technologie* **26**, 178–180.

Harada, M., Ueda, Y. and Iwata, T. (1985). Purification and some properties of alcohol acyltransferase from banana fruit. *Plant Cell Physiology* **26**, 1067–1074.

Harb, J., Bisharat, R. and Streif, J. (2008). Changes in volatile constituents of blackcurrants (*Ribes nigrum* L. cv. 'Titania') following controlled atmosphere storage. *Postharvest Biology and Technology* **47**, 271–279.

Hatanaka, A., Kajiwara, T. and Sekija, J. (1987). Biosynthetic pathway for C6-Aldehydes formation from linolenic acid in green leaves. *Chemistry and Physics of Lipids* **44**, 341–361.

Heitz, T., Bergey, D. R. and Ryan, C. A. (1997). A gene encoding a chloroplast-targeted lipoxygenase in tomato leaves is transiently induced by wounding, systemin, and methyl jasmonate. *Plant Physiology* **114**, 1085–1093.

Herianus, J. D., Singh, L. Z. and Tan, S. C. (2003). Aroma volatiles production during fruit ripening of 'Kensington Pride' mango. *Postharvest Biology and Technology* **27**, 323–336.

Homatidou, V., Karvouni, S., Dourtouglou, V. and Poulos, C. (1992). Determination of total volatile components of *Cucumis melo* L. Variety Cantaloupensis. *Journal of Agricultural and Food Chemistry* **40**, 1385–1388.

Howe, G. A., Lee, G. I., Itoh, A., Li, L. and DeRocher, A. E. (2000). Cytochrome P450-dependent metabolism of oxylipins in tomato. Cloning and expression of allene oxide synthase and fatty acid hydroperoxide lyase. *Plant Physiology* **123**, 711–724.

Huang, F. C. and Schwab, W. Functional characterization of MdCCD4, a carotenoid cleavage dioxygenase from apple. Submitted (DEC-2007) to the EMBL/GenBank/DDBJ databases.

Ibdah, M., Azulay, Y., Portnoy, V., Wasserman, B., Bar, E., Meir, A., Burger, Y., Hirschberg, J., Schaffer, A. A., Katzir, N., Tadmor, Y. and Lewinsohn, E. (2006). Functional characterization of CmCCD1, a carotenoid cleavage dioxygenase from melon. *Phytochemistry* **67**, 1579–1589.

Jacob-Wilk, D., Holland, D., Goldschmidt, E. E., Riov, J. and Eyal, Y. (1999). Chlorophyll break-down by chlorphyllase: Isolation and functional expression of the Chlase I gene from ethylene-treated Citrus fruits and its regulation during development. *Plant Journal* **20**, 653–661.

Jörnvall, H., Persson, B., Krook, M., Atrian, S., Gonzalez-Duarte, R., Jeffery, J. and Ghosh, D. (1995). Short-chain dehydrogenases/reductases (SDR). *Biochemistry* **34**, 6003–6013.

Kanellis, A. K. and Pasentsis, K. Cloning and characterization of *Citrus sinensis* flavedo pyruvate decarboxylase cDNA induced by low oxygen stress. Submitted (APR-2005) to the EMBL/GenBank/DDBJ databases.

Kato, A., Hayashi, M., Takeuchi, Y. and Nishimura, M. (1996). cDNA cloning and expression of a gene for 3-ketoacyl-CoA thiolase in pumpkin cotyledons. *Plant Molecular Biology* **31**, 843–852.

Kato, M., Matsumoto, H., Ikoma, Y., Okuda, H. and Yano, M. (2006). The role of carotenoid cleavage dioxygenases in the regulation of carotenoid profiles during maturation in citrus fruit. *Journal of Experimental Botany* **57**, 2153–2164.

Kendrick, M. D. and Chang, C. (2008). Ethylene signaling: New levels of complexity and regulation. *Current Opinion in Plant Biology* **11**, 1–7.

Lara, I., Miró, R. M., Fuentes, T., Sayez, G., Graell, J. and López, M. L. (2003). Biosynthesis of volatile aroma compounds in pear fruit stored under long-term controlled-atmosphere conditions. *Postharvest Biology and Technology* **29**, 29–39.

Lelièvre, J. M., Latché, A., Jones, B., Bouzayen, M. and Pech, J. C. (1997). Ethylene and fruit ripening. *Physiologia Plantarum* **101**, 727–739.

Leone, A., Verdesca, I., Mita, G., Gerardi, C. and Zacheo, G. Isolation and characterization of a lipoxygenase expressed in strawberry fruit. Submitted (JUL-2003) to the EMBL/GenBank/DDBJ databases.

Lewinsohn, E., Schalechet, F., Wilkinson, J., Matsui, K., Tadmor, Y., Nam, K.-H., Amar, O., Lastochkin, E., Larkov, O., Ravid, U., Hiatt, W. Gepstein, S. *et al.* (2001). Enhanced levels of the aroma and flavor compound *S*-Linalool by metabolic engineering of the terpenoid pathway in tomato fruits. *Plant Physiology* **127**, 1256–1265.

Lewinsohn, E., Sitrit, Y., Bar, E., Azulay, Y., Meir, A., Zamir, D. and Tadmor, Y. (2005a). Carotenoid pigmentation affects the volatile composition of tomato and watermelon fruits, as revealed by comparative genetic analyses. *Journal of Agricultural and Food Chemistry* **53**, 3142–3148.

Lewinsohn, E., Sitrit, Y., Bar, E., Azulay, Y., Ibdah, M., Meir, A., Yosef, E., Zamir, D. and Tadmor, Y. (2005b). Not just colors – Carotenoid degradation as a link between pigmentation and aroma in tomato and watermelon fruit. *Trends in Food Science and Technology* **16**, 407–415.

Li, D., Xu, Y., Xu, G., Gu, L., Li, D. and Shu, H. (2006). Molecular cloning and expression of a gene encoding alcohol acyltransferase (MdAAT2) from apple (cv. Golden Delicious). *Phytochemistry* **67**, 658–667.

Li, M., Zhang, Y., Sui, N., Guo, S. and Meng, Q. Cloning and functional expression of an (E,E)-alpha-farnesene synthase (PFS) cDNA in pear. Submitted (MAR-2004) to the EMBL/GenBank/DDBJ databases.

Lichtenthaler, H. K., Rohmer, M. and Schwender, J. (1997). Two independent biochemical pathways for isopentenyl diphosphate and isoprenoid biosynthesis in higher plants. *Physiologia Plantarum* **101**, 643–652.

Longhurst, T., Lee, E., Hinde, R., Brady, C. and Speirs, J. (1994). Structure of the tomato Adh2 gene and Adh2 pseudogenes, and a study of Adh2 gene expression in fruit. *Plant Molecular Biology* **26**, 1073–1084.

Lucchetta, L., Manríquez, D., El-Sharkawy, I., Flores, F. B., Sanchez-Bel, P., Zouine, M., Ginies, C., Bouzayen, M., Rombaldi, C., Pech, J. C. and Latché, A. (2006). Biochemical and catalytic properties of three recombinant alcohol acyltransferases of melon. Sulfur-containing esters formation, regulatory role of CoA-SH in activity and sequence elements conferring substrate preference. *Journal of Agricultural and Food Chemistry* **55**, 5213–5220.

Lücker, J., El Tamer, M. K., Schwab, W., Verstappen, F. W. A., van der Plas, L. H. W., Bouwmeester, H. J. and Verhoeven, H. A. (2002). Monoterpene biosynthesis in lemon (*Citrus limon*): cDNA isolation and functional analysis of four monoterpene synthases. *European Journal of Biochemistry* **269**, 3160–3171.

Lücker, J., Bowen, P. and Bohlmann, J. (2004). *Vitis vinifera* terpenoid cyclases: Functional identification of two sesquiterpene synthase cDNAs encoding (+)-valencene synthase and (−)-germacrene D synthase and expression of mono- and sesquiterpene synthases in grapevine flowers and berries. *Phytochemistry* **65**, 2649–2659.

Lurie, S., Pre-Aymard, C., Ravid, U., Larkov, O. and Fallik, E. (2002). Effect of 1-methylcyclopropene on volatile emission and aroma in cv. Anna apples. *Journal of Agricultural and Food Chemistry* **50**, 4251–4256.

Maccarone, E., Campisi, S., Fallico, B., Rapisarda, P. and Sgarlata, R. (1998). Flavor components of Italian orange juices. *Journal of Agricultural and Food Chemistry* **46**, 2293–2298.

MacLeond, A. and Gonzalez de Troconis, N. (1982). Volatile flavour components of mango fruit. *Phytochemistry* **21**, 2523–2526.

Manríquez, D., El-Sharkawy, I., Flores, F. B., El-Yahyaoui, F., Regad, F., Bouzayen, M., Latche, A. and Pech, J. C. (2006). Two highly divergent alcohol dehydrogenases of melon exhibit fruit ripening-specific expression and distinct biochemical characteristics. *Plant Molecular Biology* **61**, 675–685.

Martin, D. M. and Bohlmann, J. (2004). Identification of *Vitis vinifera* (−)-alpha-terpineol synthase by in silico screening of full-length cDNA ESTs and functional characterization of recombinant terpene synthase. *Phytochemistry* **65**, 1223–1229.

Mathieu, S., Terrier, N., Procureur, J., Bigey, F. and Gunata, Z. (2005). A Carotenoid Cleavage Dioxygenase from *Vitis vinifera* L.: Functional characterization and expression during grape berry development in relation to C13-norisoprenoid accumulation. *Journal of Experimental Botany* **56**, 2721–2731.

Matsui, K., Shibutani, M., Hase, T. and Kajiwara, T. (1996). Bell pepper fruit fatty acid hydroperoxide lyase is a cytochrome P450 (CYP74B). *FEBS Letters* **394**, 21–24.

Matsui, K., Ujita, C., Fujimoto, S., Wilkinson, J., Hiatt, B., Knauf, V., Kajiwara, T. and Feussner, I. (2000). Fatty acid 9- and 13-hydroperoxide lyases from cucumber. *FEBS Letters* **481**, 183–188.

Matton, D. P., Constabel, P. and Brisson, N. (1990). Alcohol dehydrogenase gene expression in potato following elicitor and stress treatment. *Plant Molecular Biology* **14**, 775–783.

McGarvey, D. J. and Croteau, R. (1995). Terpenoid metabolism. *The Plant Cell* **7**, 1015–1026.

Meigh, D. F. and Hulme, A. C. (1965). Fatty acid metabolism in the apple fruit during the respiration climacteric. *Phytochemistry* **4**, 863–871.

Menager, I., Jost, M. and Aubert, C. (2004). Changes in physicochemical characteristics and volatile constituents of strawberry (Cv. cigaline) during maturation. *Journal of Agricultural and Food Chemistry* **52**, 1248–1254.

Mir, N. A., Pérez, R., Schwallier, P. and Beaudry, R. (1999). Relationship between ethylene response manipulation and volatile production in Jonagold variety apples. *Journal of Agricultural and Food Chemistry* **47**, 2653–2659.

Mita, G., Gallo, A., Greco, V., Zasiura, C., Casey, R., Zacheo, G. and Santino, A. (2001). Molecular cloning and biochemical characterization of a lipoxygenase in almond (*Prunus dulcis*) seed. *European Journal of Biochemistry* **268**, 1500–1507.

Morton, I. D. and McLeod, A. J. (1990). "Food flavors. Part C. The flavor of fruits." Elsevier, Amsterdam.

Moshonas, M. G. and Shaw, P. E. (1994). Quantitative determination of 46 volatile constituents in fresh, unpasteurized orange juices using dynamic headspace gas chromatography. *Journal of Agricultural and Food Chemistry* **42**, 1525–1528.

Moyano, E., Encinas-Villarejo, S., Lopez-Raez, J. A., Redondo-Nevado, J., Blanco-Portales, R., Bellido, M. L., Sanz, C., Caballero, J. L. and Munoz-Blanco, J. (2004). Comparative study between two strawberry pyruvate decarboxylase genes along fruit development and ripening, post-harvest and stress conditions. *Plant Science* **166**, 835–845.

Munoz-Blanco, J., Garcia-Limones, C. and Caballero, J. L. FaCCD1 gene coding a dioxygenase involved in the biosynthesis of flavor compounds in ripe strawberry fruits (*Fragaria* × *anannasa*). Submitted (DEC-2007) to the EMBL/GenBank/DDBJ databases.

Nakane, E., Kawakita, K., Doke, N. and Yoshioka, H. (2003). Elicitation of primary and secondary metabolism during defense in the potato. *Journal of General Plant Pathology* **69**, 378–384.

Ninio, R., Lewinsohn, E., Mizrahi, Y. and Sitrit, Y. (2003). Quality attributes of storage koubo (*Cereus peruvianus* (L.) Miller) fruit. *Postharvest Biology and Technology* **30**, 273–280.

Obando-Ulloa, J. M., Moreno, E., García-Mas, J., Nicolai, B., Lammertyn, J., Monforte, A. J. and Fernandéz-Trujillo, J. P. (2008). Climacteric or non-climacteric behavior in melon fruit 1. Aroma volatiles. *Postharvest Biology and Technology* **49**, 27–37.

Obenland, D., Collin, S., Sievert, J., Fjeld, K., Doctor, J. and Arpaia, M. L. (2008). Commercial packing and storage of navel oranges alters aroma volatiles and reduces flavor quality. *Postharvest Biology and Technology* **47**, 159–167.

Oeller, P. W., Wong, L. M., Taylor, L. P., Pike, D. A. and Theologis, A. (1991). Reversible inhibition of tomato fruit senescence by antisense RNA. *Science* **254**, 437–439.

Or, E., Baybik, J., Sadka, A. and Ogrodovitch, A. (2000). Fermentative metabolism in grape berries: Isolation and characterization of pyruvate decarboxylase cDNA and analysis of its expression throughout berry development. *Plant Science* **156**, 151–158.

Paillard, N. M. M. (1986). Evolution of the capacity of aldehydes production by crushed apple tissues, during an extended storage of fruits. *In* "The shelf-life of foods and beverages" (G. Charalambous, ed.), pp. 369–378. Elsevier Science, Amsterdam.

Paillard, N. M. M. (1990). The flavour of apples, pears and quinces. *In* "Food Flavours, PartC. The Flavour of Fruit" (I. D. Morton and A. J. McLeod, eds.), pp. 1–4. Elsevier, Amsterdam, The Netherlands.

Pech, J. C., Latché, A. and Bouzayen, M. (2004). Ethylene biosynthesis. *In* "Plant Hormones: Biosynthesis, signal transduction, action" (P. J. Davies, ed.), pp. 115–136. Kluwer Academic Publishers, London.

Pech, J. C., Bouzayen, M. and Latché, A. (2008). Climacteric fruit ripening: Ethylene-dependent and independent regulation of ripening pathways in melon fruit. *Plant Science* **175**, 114–120.

Pelayo, C., Ebeler, S. E. and Kader, A. A. (2003). Postharvest life and flavor quality of three strawberry cultivars kept at 5 degrees C in air or air + 20 kPa CO2. *Postharvest Biology and Technology* **27**, 171–183.

Pérez, A. G., Rios, J. J., Sanz, C. and Olias, J. M. (1992). Aroma components and free amino-acids in strawberry variety chandler during ripening. *Journal of Agricultural and Food Chemistry* **40**, 2232–2235.

Pérez, A. G., Sanz, C. and Olías, J. M. (1993). Partial purification and some properties of alcohol acyltransferase from strawberry fruits. *Journal of Agricultural and Food Chemistry* **41**, 1462–1466.

Pérez, A. G., Sanz, C., Olías, R., Rios, J. J. and Olias, J. M. (1996). Evolution of strawberry alcohol acyltransferase activity during fruit development and storage. *Journal of Agricultural and Food Chemistry* **44**, 3286–3290.

Pérez, A. G., Sanz, C., Olías, R. and Olías, J. (1999). Lipoxygenase and hydroperoxide lyase activities in ripening strawberry fruits. *Journal of Agricultural and Food Chemistry* **47**, 249–253.

Peters, J. S. and Frenkel, C. (2004). Relationship between alcohol dehydrogenase activity and low-temperature in two maize genotypes, Silverado F_1 and *Adh1–Adh2–* doubly null. *Plant Physiology and Biochemistry* **42**, 841–846.

Petro-Turza, M. (1987). Flavors of tomato and tomato products. *Food Reviews International* **2**, 309–351.

Phi, N. T. L., Nishiyama, C., Choi, H. S. and Sawamura, M. (2006). Evaluation of characteristic aroma compounds of *Citrus natsudaidai* Hayata (Natsudaidai) cold-pressed peel oil. *Bioscience, Biotechnology, and Biochemistry* **70**, 1832–1838.

Pichersky, E. and Gang, D. R. (2000). Genetics and biochemistry of secondary metabolites in plants: An evolutionary perspective. *Trends in Plant Science* **5**, 439–445.

Pichersky, E. and Gershenzon, J. (2002). The formation and function of plant volatiles: Perfumes for pollinator attraction and defense. *Current Opinion in Plant Biology* **5**, 237–243.

Pichersky, E., Noel, J. P. and Dudareva, N. (2006). Biosynthesis of plant volatiles: Nature's diversity and ingenuity. *Science* **311**, 808–811.

Pino, J. A. and Marbot, R. (2001). Volatile flavor constituents of acerola (*Malpighia emarginata* DC) fruit. *Journal of Agricultural and Food Chemistry* **49**, 5880–5882.

Pino, J. A., Marbot, R. and Vázquez, C. (2002). Characterization of volatiles is Costa Rican guava (*Psidium friedrichsthalianum* (Berg)) fruit. *Journal of Agricultural and Food Chemistry* **50**, 6023–6026.

Porta, H. and Rocha-Sosa, M. (2002). Plant lipoxygenases: Physiological and molecular features. *Plant Physiology* **130**, 15–21.

Preisig-Muller, R. and Kindl, H. (1993). Thiolase mRNA translated *in vitro* yields a peptide with a putative N-terminal presequence. *Plant Molecular Biology* **22**, 59–66.

Prestage, S., Linforth, R. S. T., Taylor, A. J., Lee, E., Speirs, J. and Schuch, W. (1999). Volatile production in tomato fruit with modified alcohol dehydrogenase activity. *Journal of the Science of Food and Agriculture* **79**, 131–136.

Pyysalo, T., Honkanen, E. and Hirvi, T. (1979). Volatiles of wild strawberries, *Fragaria vesca* L., compared to those of cultivated berries, *Fragaria* × *ananassa* cv Senga Sengana. *Journal of Agricultural and Food Chemistry* **27**, 19–22.

Reid, S. J., Watkins, C. B., Janssen, B. and Ross, G. S. (1996). An alcohol dehydrogenase cDNA clone (Accession No. Z48234) isolated from apple fruit. (PGR96–043). *Plant Physiology* **111**, 947.

Rodríguez-Concepción, M. and Boronat, A. (2002). Elucidation of the methylerythritol phosphate pathway for isoprenoid biosyntesis in bacteria and plastids. A metabolic milestone achieved through genomics. *Plant Physiology* **130**, 1079–1089.

Rowan, D. D., Lane, H. P., Allen, J. M., Fielder, S. and Hunt, M. (1996). Biosynthesis of 2-methylbutyl, 2-methyl-2-butenyl, and 2- methylbutanoate esters in Red Delicious and Granny Smith apples using deuterium-labeled substrates. *Journal of Agricultural and Food Chemistry* **44**, 3276–3285.

Rowan, D. D., Allen, J. M., Fielder, S. and Hunt, M. (1999). Biosynthesis of straight-chain ester volatiles in Red Delicious and Granny Smith apples using deuterium-labeled precursors. *Journal of Agricultural and Food Chemistry* **47**, 2553–2562.

Rudell, D. R., Mattinson, D. S., Fellman, J. K. and Mattheis, J. P. (2000). The progression of ethylene production and respiration in the tissues of ripening 'Fuji' apple fruit. *HortScience* **35**, 1300–1303.

Rudell, D. R., Mattinson, D. S., Mattheis, J. P., Wyllie, S. G. and Fellman, J. K. (2002). Investigations of aroma volatile biosynthesis under anoxic conditions and in different tissues of "Redchief Delicious" apple fruit (*Malus domestica* Borkh.). *Journal of Agricultural and Food Chemistry* **50**, 2627–2632.

Sanz, C., Olías, J. M. and Pérez, A. G. (1997). Aroma biochemistry of fruits and vegetables. *In* "Phytochemistry of fruit and vegetables" (F. A. Tomás-Barberán and R. J. Robins, eds.), pp. 125–155. New York, Oxford University Press.

Sato, T. and Theologis, A. (1989). Cloning the mRNA Encoding 1-aminocyclopropane-1-carboxylate Synthase, the Key Enzyme for Ethylene Biosynthesis in Plants. *Proceedings of the National Academy of Sciences USA* **86**, 6621–6625.

Schieberle, P. and Hofmann, T. (1997). Evaluation of the character impact odorants in fresh strawberry juice by quantitative measurements and sensory studies on model mixtures. *Journal of Agricultural and Food Chemistry* **45**, 227–232.

Schreier, P. (1984). "Chromatographic studies of biogenesis of plant volatiles." Alfred Hiithig Verlag GmbH. Heidelberg, Basel, NY.
Schwab, W., Davidovich-Rikanati, R. and Lewinsohm, E. (2008). Biosynthesis of plant-derived compounds. *The Plant Journal* **54**, 712–732.
Schwartz, S. H., Qin, X. and Zeevart, J. A. (2001). Characterization of a novel carotenoid cleavage dioxygenase from plants. *Journal of Biological Chemistry* **276**, 25208–25211.
Shalit, M., Katzir, N., Tadmor, Y., Larkov, O., Burger, Y., Shalekhet, F., Lastochkin, E., Ravid, U., Amar, O., Edelstein, M., Karchi, Z. and Lewinsohn, E. (2001). Acetyl Co-A: Alcohol acetyltransferase activity and aroma formation in ripening melon fruits. *Journal of Agricultural and Food Chemistry* **49**, 794–799.
Shalit, M., Guterman, I., Volpin, H., Bar, E., Tamari, T., Menda, N., Adam, Z., Zamir, D., Vainstein, A., Weiss, D., Pichersky, E. and Lewinsohn, E. (2003). Volatile ester formation in roses. Identification of an acetyl-coenzyme A geraniol/citronellol_ acetyl transferase in developing rose petals. *Plant Physiology* **131**, 1868–1876.
Sharon-Asa, L., Shalit, M., Frydman, A., Bar, E., Holland, D., Or, E., Lavi, U., Lewinsohn, E. and Eyal, Y. (2003). Citrus fruit flavor and aroma biosynthesis: Isolation, functional characterization, and developmental regulation of Cstps1, a key gene in the production of the sesquiterpene aroma compound valencene. *Plant Journal* **36**, 664–674.
Shiota, H. (1993). New esteric components in the volatiles of banana fruit (*Musa sapientum* L.). *Journal of Agricultural and Food Chemistry* **41**, 2056–2062.
Simkin, A. J., Schwartz, S. H., Auldridge, M., Taylor, M. G. and Klee, H. J. (2004a). The tomato carotenoid cleavage dioxygenase 1 genes contribute to the formation of the flavor volatiles beta-ionone, pseudoionone, and geranylacetone. *Plant Journal* **40**, 882–892.
Simkin, A. J., Underwood, B. A., Auldridge, M., Loucas, H. M., Shibuya, K., Schmelz, E. A., Clark, D. G. and Klee, H. J. (2004b). Circadian regulation of the phccd1 carotenoid cleavage dioxygenase controls emission of beta-ionone, a fragrance volatile of petunia flowers. *Journal of Plant Physiology* **136**, 3504–3514.
Singh, Z., Label, H. J. D. and Nair, S. (2004). A review of mango fruit aroma volatile compounds – state of the art research. In VII international mango symposium, Recife, Brazil. Internacional society for horticultural science. *Acta Horticulturae* **645**, 519–527.
Sitrit, Y., Ninio, R., Bar, E., Golan, E., Larkov, O., Ravid, U. and Lewinsohn, E. (2004). *S*-Linalool synthase activity in developing fruit of the columnar cactus koubo (*Cereus peruvianus* (L.) Miller). *Plant science* **167**, 1257–1262.
Song, J. and Bangerth, F. (1994). Production and development of volatile aroma compounds of apple fruits at different times of maturity. *Acta Horticulturae* **368**, 150–159.
Song, J. and Bangerth, F. (2003). Fatty acids as precursors for aroma volatile biosynthesis in pre-climacteric and climacteric apple fruit. *Postharvest Biology and Technology* **30**, 113–121.
Song, J. and Forney, C. F. (2008). Flavour volatile production and regulation in fruit. *Canadian Journal of Plant Sciences* **88**, 537–550.
Souleyre, E. J., Greenwood, D. R., Friel, E. N., Karunairetnam, S. and Newcomb, R. D. (2005). An alcohol acyl transferase from apple (cv. Royal Gala), MpAAT1, produces esters involved in apple fruit flavor. *FEBS Journal* **272**, 3132–3144.

Speirs, J., Lee, E., Holt, K., Yong-Duk, K., Scott, N. S., Loveys, B. and Schuch, W. (1998). Genetic manipulation of alcohol dehydrogenase levels in ripening tomato fruit affects the balance of some flavor aldehydes and alcohols. *Plant Physiology* **117**, 1047–1058.

Stepanova, A. N. and Ecker, J. R. (2000). Ethylene signaling: From mutants to molecules. *Current Opinion in Plant Biology* **3**, 341–435.

Tan, B. C., Schwartz, S. H., Zeevaart, J. A. D. and McCarty, D. R. (1997). Genetic control of abscisic acid biosynthesis in maize. *Proceeding of the National Academy of Sciences* **94**, 12235–12240.

Tesniere, C. and Verries, C. (2000). Molecular cloning and expression of cDNAs encoding alcohol dehydrogenases from *Vitis vinifera* L. during berry development. *Plant Science* **157**, 77–88.

Theologis, A. (1992). One rotten apple spoils the whole bushel: The role of ethylene in fruit ripening. *Cell* **70**, 181–184.

Tijet, N., Waspi, U., Gaskin, D. J., Hunziker, P., Muller, B. L., Vulfson, E. N., Slusarenko, A., Brash, A. R. and Whitehead, I. M. (2000). Purification, molecular cloning, and expression of the gene encoding fatty acid 13-hydroperoxide lyase from guava fruit (*Psidium guajava*). *Lipids* **35**, 709–720.

Tijet, N., Schneider, C., Muller, B. L. and Brash, A. R. (2001). Biogenesis of volatile aldehydes from fatty acid hydroperoxides: Molecular cloning of a hydroperoxide lyase (CYP74C) with specificity for both the 9- and 13-hydroperoxides of linoleic and linolenic acids. *Archives of Biochemistry and Biophysics* **386**, 281–289.

Trebitsh, T., Goldschmidt, E. E. and Riov, J. (1993). Ethylene induces de novo synthesis of chlorophyllase, a chlorophyll degrading enzyme, in Citrus fruit peel. *Proceedings of the National Academy of Sciences USA* **90**, 9441–9445.

Ueda, Y., Fujishita, N. and Chachin, K. (1997). Presence of alcohol acetyltransferase in melons (*Cucumis melo* L.). *Postharvest Biology and Technology* **10**, 121–126.

Underwood, B., Tieman, D., Shibuya, K., Dexter, R., Loucas, H., Simkin, A., Sims, C., Schmelz, E., Klee, H. and Clark, D. (2005). Ethylene-regulated floral volatile synthesis in *Petunia corollas*. *Plant Physiology* **138**, 255–266.

Van Der Straeten, D., Rodrigues Poussada, R. A., Gielen, J. and Van Montagu, M. (1991). Tomato alcohol dehydrogenase. Expression during fruit ripening and under hypoxic conditions. *FEBS Letters* **295**, 39–42.

van Schie, C. C., Haring, M. A. and Schuurink, R. C. (2007). Tomato linalool synthase is induced in trichomes by jasmonic acid. *Plant Molecular Biology* **64**, 251–263.

Velasco, R., Zharkikh, A., Troggio, M., Cartwright, D. A., Cestaro, A., Pruss, D, Pindo, M, FitzGerald, L. M., Vezzulli, S., Reid, J., Malacarne, G. Iliev, D. *et al.* (2007). A high quality draft consensus sequence of the genome of a heterozygous grapevine variety. *PLoS One* **2**(12), e1326. doi:10.1371/journal.pone.0001326.

Wang, C., Xing, J., Chin, C. K., Ho, C. T. and Martin, C. E. (2001). Modification of fatty acids changes the flavor volatiles in tomato leaves. *Phytochemistry* **58**, 227–232.

Waters, D., Holton, T., Ablett, E., Lee, L. and Henry, R. (2005). cDNA microarray analysis of developing grape (*Vitis vinifera* cv. Shiraz) berry skin. *Functional & Integrative Genomics* **5**, 40–58.

Winterhalter, P. and Rouseff, R. (2002). Carotenoid-derived aroma compounds: An introduction. ACS Symposium Series 802. *In* "Carotenoid derived aroma compounds" (P. Winterhalter and R. L. Rouseff, eds.), pp. 1–19. American Chemical Society, Washington, DC.

Wolyn, D. J. and Jelenkovic, G. (1990). Nucleotide sequence of an alcohol dehydrogenase gene in octoploid strawberry (*Fragaria* × *Ananassa* Duch). *Plant Molecular Biology* **14**, 855–857.

Wu, Z. and Burns, J. K. Molecular cloning and expression of fatty acid hydroperoxide lyase gene isolated from orange leaf. Submitted (FEB-2003) to the EMBL/GenBank/DDBJ databases.

Wyllie, S. G. and Fellman, J. K. (2000). Formation of volatile branched chain esters in bananas (*Musa sapientum* L.). *Journal of Agricultural and Food Chemistry* **48**, 3493–3496.

Yahia, E. M. (1994). Apple flavor. *Horticultural Reviews* **16**, 197–234.

Yahyaoui, F. E. L., Wongs-Aree, C., Latche, A., Hackett, R., Grierson, D. and Pech, J. C. (2002). Molecular and biochemical characteristics of a gene encoding an alcohol acyl-transferase involved in the generation of aroma volatile esters during melon ripening. *European Journal of Biochemistry* **269**, 2359–2366.

Zabetakis, I. and Holden, M. A. (1997). Strawberry flavour: Analysis and biosynthesis. *Journal of the Science of Food and Agriculture* **74**, 421–434.

Zabetakis, I., Gramshaw, J. W. and Robinson, D. S. (1999). 2,5-dimethyl-4-hydroxy-2H-furan-3-one and its derivatives: Analysis, synthesis and biosynthesis – A review. *Food Chemistry* **65**, 139–151.

Zhang, B., Chen, K., Bowen, J., Allan, A., Espley, R., Karunairetnam, S. and Ferguson, I. (2006). Differential expression within the LOX gene family in ripening kiwifruit. *Journal of Experimental Botany* **57**, 3825–3836.

Zheng, X. Y. and Wolff, D. W. (2000). Ethylene production, shelf-life and evidence of RFLP polymorphisms linked to ethylene genes in melon (*Cucumis melo* L.). *Theoretical and Applied Genetics* **101**, 613–624.

Jatropha curcas: A Review

NICOLAS CARELS

Fundação Oswaldo Cruz (FIOCRUZ), Instituto Oswaldo Cruz (IOC), Laboratório de Genômica Funcional e Bioinformática, Rio de Janeiro, RJ, Brazil
Universidade Estadual de Santa Cruz (UESC), Núcleo de Biologia Computacional e Gestão de Informações Biotecnológicas, Ilhéus, BA, Brazil

ABSTRACT

Since the ratification of the Kyoto protocol, a significant effort has been made worldwide to boost biofuels with the expectation of a positive contribution to renewable fuel and greenhouse gas reduction. The initial recommendations were generally not acted upon except in some particular cases, such as Brazil and Vietnam. The positive

Advances in Botanical Research, Vol. 50

0065-2296/09 $35.00
DOI: 10.1016/S0065-2296(08)00802-1

contribution of first generation biofuels has been challenged because they rely on limited arable land availability, they need substantial energy inputs, and they compete with wildlife and food crops. However, bioethanol and biodiesel proved their compatibility with existing technologies and prepared the transition to second and third generation biofuels. Countries such as Europe, with adverse climatic conditions and limited availability of arable land and crop options, are already investigating technologies for cellulosic ethanol and microdiesel. Developing countries, on the other hand, often have extended areas of land that are not usable for agriculture with currently available crops. *Jatropha curcas* L. proved to be an opportunistic crop in tropical areas in these unfavorable environments. For this reason, a review on the features and technological achievements obtained with this new crop is welcome. This review covers the (i) agronomy, (ii) oil production, (iii) alkyl ester production, (iv) biofuel features, (v) toxicity, (vi) plant breeding, and (vii) crop expansion of *J. curcas*.

I. INTRODUCTION

Among the biofuels that will progressively contribute to alleviate the dependency from fossil energy, bioethanol and biodiesel are considered first generation since they can be produced by current agriculture techniques. Currently, about 84% of the world's biodiesel production is due to rapeseed oil. The remaining portion is from sunflower oil (13%), palm oil (1%), soybean oil, and others (2%) (Gui *et al.*, 2008). More than 95% of biodiesel is made from edible oil. By converting edible oils into biodiesel, food resources are actually being converted into automotive fuels. It is believed that large-scale production of biodiesel from edible oils may bring global imbalance to food supply and demand. For instance, India imports 70% of its fuel consumption (~111 million tons), and any renewable energy is welcome. Consequently, India adopted a strong position promoting biofuels and promising a market with stable prices for 10% biodiesel blends (Kumar and Sharma, 2008). On the other hand, India is also a net importer of edible oil, and therefore emphasizes non-edible oils from plants that are able to sustain biofuel production on marginal land. *Jatropha curcas* L. and karanja are the two leaders on the Indian list of oil plants. Similarly, in China the area of arable land per capita is lower than the world average. As a result, most edible oils need to be imported, and edible oil demand in 2010 is projected to be 13.5 million tons.

Among the non-edible oleaginous plants that have been described in the literature as sources of alkyl ester (biodiesel)—i.e., neem (*Azadirachta indica* A. Juss.; Meher *et al.*, 2005), karanja (*Pongamia pinnata* L.; Modi *et al.*, 2007; Raheman and Ghadge, 2007; Raheman and Phadatare, 2004; Sharma and Singh, 2008; Srivastava and Verma 2008), mahua (*Madhuca* sp.; Puhan *et al.*, 2005b), undi (also known as Nagchampa or polanga; *Calophyllum inophyllum* L.; Azam *et al.*, 2005; Banapurmath *et al.*, 2008; Sahoo *et al.*, 2007), castor bean

(*Ricinus communis* L.; Goodrum and Geller, 2005; Scholz and da Silva, 2008), simarouba (*Simarouba glauca* DC.; Adjaye *et al.*, 1995; Azam *et al.*, 2005), Ben-oil tree (*Moringa oleifera* Lam.; Rashid *et al.*, 2008), jojoba (*Simmondsia chinensis* (Link) Schneid.; Bouaid *et al.*, 2007; Canoira *et al.*, 2006), rubber tree (*Hevea brasiliensis* (Willd. ex Adr. Juss.) Muell. Arg.; Ramadhas *et al.*, 2005), Chinese tallow tree (*Sapium sebiferum* (L.) Roxb.; Gao *et al.*, 2008), Babassu (*Attalea speciosa* C. Martius.; Oliveira *et al.*, 1999), tucum (*Astrocaryum vulgare* Mart.; de Oliveira Lima *et al.*, 2008), Zanthoxylum (*Zanthoxylum bungeanum* Max.; Zhang and Jiang, 2008), sea mango (*Cerbera odollam* Gartn. Engl. or *Cerbera manghas* L.; Gaillard *et al.*, 2004)—*J. curcas* (Kumar and Sharma, 2008) deserves special attention due to the leading role that it is expected to play in the near future.

A. *JATROPHA*

The genus *Jatropha* is native to the tropical Americas with the oldest remains found by E.W. Berry (1929) in geological formations from Peru corresponding to the early Tertiary age (upper Eocene or lower Oligocene). The classical understanding of continental drift is that Tethys sea had been flooding the land corresponding to Central America since the upper Jurassic (−150 million years), which matches the beginning of the angiosperm eclosion, until around the late Eocene (Gheerbrant and Rage, 2006; Santos *et al.*, 2008) some 30 million years ago, i.e., after angiosperm radiation (−70 million years). If one believes Berry's observation, the genus *Jatropha* would have been installed in tropical South America before its connection to Central America. In addition, Jatropha must have migrated only recently to Central America since it was not found in Africa before its human introduction and Africa separated from South America (Brazil) about 65 million years ago. This suggests that the origin of *Jatropha* is tropical South America—in contrast to Central America, as is generally accepted.

The genus *Jatropha* belongs to the tribe *Jatropheae* in the *Euphorbiaceae* family and contains approximately 170 known species. *Euphorbiaceae* is an ancient and diverse family in the large rosid order Malpighiales and includes in addition to Jatropha such familiar members as rubber, cassava, castor bean, poinsettia, and leafy spurge (Wurdack, 2008).

A preliminary study showed that *Jatropha glandulifera* Roxb. is rather different from the core *Jatropha* species (Ganesh Ram *et al.*, 2008). However, natural hybrids seem to be common in the *Jatropha* genus. For example, *Jatropha tanjorensis* Ellis & Soraja is considered a spontaneous hybrid between *J. curcas* L. and *Jatropha gossypifolia* L. Other naturally occurring hybrids, such as *J. curcas* × *Jatropha canascens* and *Jatropha integerrima*

Jacq. × *Jatropha hastate* Jacq., have been reported in Mexico and Cuba, respectively (see in Sujatha *et al.*, 2008).

J. curcas or "physic nuts," which will be called "Jatropha" below, is a small tree or large shrub that normally reaches a height of 3–5 m, but can reach a height of 8–10 m under favorable conditions (Fig. 1).

The plant root system proceeds through the development of a main taproot and four shallow lateral roots. The leaves are smooth with five lobes spanning 10–15 cm and may fall once a year, depending on their genotype. The plant is monoecious and the terminal inflorescences contain unisexual flowers (Fig. 2) on the same inflorescence (raceme). The inflorescence is a panicle, with the female flowers (about 10–20%) at the apices of the main stem and branches of the inflorescence. Male flowers are more numerous (about 80–90%) and occupy subordinate positions on the inflorescence (Raju *et al.*, 2002; see ref. in Jongschaap *et al.*, 2007). Male flowers open for a period of 8–10 days, whereas female flowers open for 2–4 days only. The continuous flowering results in a sequence of reproductive development stages from yellow mature fruits at the base of the branch, to green fruits in the middle, and flowers at the top.

After pollination, the inflorescences form grapes of 10 green fruits, 2–3 cm long with an ovoid shape. Each fruit typically has three carpels and the potential for two seeds per carpel (Kochhar *et al.*, 2008). However, there are usually not more than three seeds per fruit and often only one, which gives 2.5 seeds per fruit, on average (personal communication from Nagashi Tominaga, Biojan-MG Agro Industrial ltda, Janaúba, MG, Brazil).

Fig. 1. Structure of a native Jatropha tree—in the middle (courtesy from N. Tominaga, Biojan, Brazil).

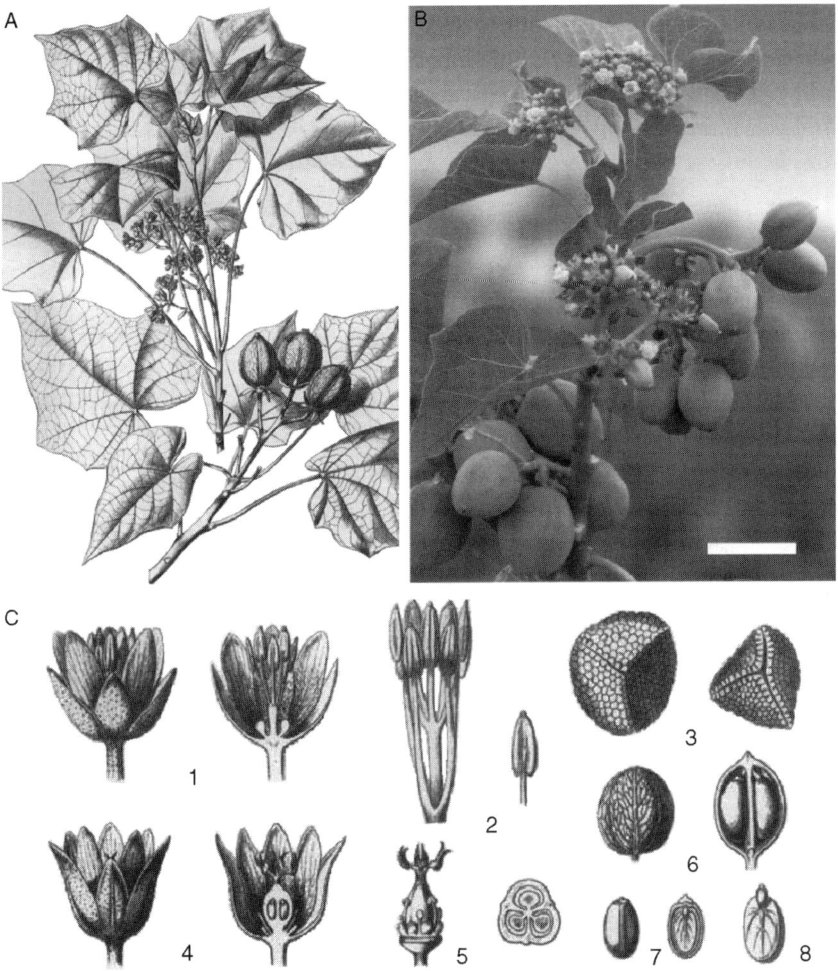

Fig. 2. Inflorescence of *Jatropha curcas* L. (A) as drawn by Johannes Müller Argoviensis (1874). (B) *In situ*. The white bar is 3 cm. (C) Details of male flower (1) anthers (2), pollen grain (3), female flowers (4), Stigma, style, ovary and nectar glands from top to bottom and, on the left, a section through the ovary showing the three carpels (5), fruit (6), seed (7), and embryo (8).

B. AGRONOMICAL DATA

1. Ecology

Jatropha is widely distributed in the wild and cultivated tropical areas of Central America, South America, Africa, India, South Eastern Asia, and Australia. Therefore, it typically grows between 15 and 40 °C with rainfall between 250 and 3000 mm and is more altered by lower temperatures than by altitude or day length (Foidl *et al.*, 1996). However, it is an open-field plant species that requires intense sun such as found in the savanna or desert periphery. It is not adapted to grow under the shadow of forest canopy and does not compete efficiently with fast-growing species of the rain forest.

It is well adapted to arid and semi-arid climates with demonstrated molecular mechanisms for resistance to drought (Zhang *et al.*, 2008). It can also grow on a large range of soils provided they are well drained and aerated (Kumar and Sharma, 2008).

In Egyptian growth conditions, the average rate of water consumption by Jatropha was found to be 6 l/week throughout the growing season, which means that Jatropha can survive and produce significant seed yield of acceptable quality with minimum water requirements compared to other crops (Abou Kheira and Atta, 2008). A water regime of 700 mm/year seems to be around the optimum. Above this threshold fungal diseases may affect the vegetation.

Even though Jatropha can resist to adverse environmental conditions, it is obvious that for high level of oil production it needs at least 45 cm deep soils and suitable nitrate, phosphate, and potassium (NPK) fertilization for fruit growth and maturation. If fertilization is not available, fungal mycorrhization has been shown to help to sustain growth and development (see in Achten *et al.*, 2008).

2. *Propagation*

Jatropha is propagated through cuttings or seeds. Cuttings are typically prepared with one-year-old terminal branches of 25–30 cm. It is good practice to inoculate cuttings with mycorrhizal fungi when establishing them into nursery. This treatment improves the quality of the plant-fungal symbiosis in the field conditions especially in soil with poor fertility (Carvalho *et al.*, 2007). Endo-mycorrhizal fungi were demonstrated to be commonly found in association with Jatropha in natural conditions (Carvalho *et al.*, 2007).

The rooting of cuttings reaches 100% after 45 days when pre-treated for 24 h with 10–100 mg/l of indole-3-butyric acid (IBA). IBA is more effective than 1-naphtaleneacetic acid (NAA) and gives a cutting survival yield >90%. Also the number of leaves is higher and the flowering time earlier when cuttings are treated with IBA (Kochhar *et al.*, 2008). In practice, one chooses the healthier cuttings for field acclimatization. The advantage of cutting propagation is that it offers the possibility to grow elite accessions. However, large-scale plantations are only possible through sowing. Seeds are pre-soaked for 24 h in water, and germinate in 5–10 days at 27–30 °C with humidity saturation. The cuttings and seedlings are grown in nurseries for

2 months and transplanted to the field at the beginning of the wet season. During the dry season, the plant enters dormancy and losses its leaves.

3. Planting density

There are two basic models for planting, Indian and Cuban. Under the Indian model, the tree density is high, i.e., 2500 ha^{-1}, with a planting distance of 2×2 m^2. The total seed production with three harvests per year ranges between 2000 and 4000 kg/ha and is typically around 3200 kg/ha, corresponding to about 1.5 t oil/ha with an annual rainfall of 700–800 mm. For comparison, palm trees produce 3.7 t oil/ha and soybean produces 0.38 t oil/ha. Unfortunately, high planting densities at 2×2 m^2 make tractor pulverization for weeding, or pests and diseases control impossible. By contrast, under the Cuban model, the density of trees is lower, i.e., 357 ha^{-1} with a planting distance of 7×4 m^2. Under this model, the number of fruit grapes per tree is higher and reaches 14 kg seeds/plant or 5000 kg/ha. It seems that a density of 5×2 m^2 (1000 ha^{-1}) would be a good compromise (Fig. 3) for production and mechanization (N. Tominaga, personal communication).

4. Pests and diseases

As expected from any type of monoculture, large Jatropha fields are susceptible to pests and diseases (Banjo *et al.*, 2006; Regupathy and Ayyasamy, 2006; Shanker and Dhyani, 2006). The major problems reported are caused by the scutellarid bug *Scutellera nobilis* Fabr., the capsule-borer *Pempelia*

Fig. 3. Jatropha field. The first pruning of these trees has been carried out at 80 cm. They are >2.5 m after 4 years (courtesy from N. Tominaga, Biojan, Brazil).

morosalis Saalm & Üller, *Pachycoris klugii* Burmeister (*Scutelleridae*), *Leptoglossus zonatus* Dallas (*Coreidae*), the blister miner *Stomphastis thraustica* Meyrick (Acrocercops), the semi-looper *Achaea janata* L., and the flower beetle *Oxycetonia versicolor* Fabr. In Brazil, the pests and diseases that are affecting Jatropha in industrial cultures are typically: (i) Empoasca leafhoppers (*Empoasca* sp.), broad mites (*Polyphagotarsonemus latus* Banks), capsule-borers (*Pachycoris torridus* Scopoli), and thrips, for pests (Saturnino *et al.*, 2005); (ii) antracnose (*Colletotrichum gloeosporioides* Penzig and *Colletotrichum capsici* Syd.), oidium, phythophtora, and fusariose for fungi (Freire and Parente, 2006).

5. Flowering and fructification

In tropical, humid regions, or under irrigated conditions, Jatropha comes into bloom for a large part of the year. Because of continuous flowering, fruit production occurs for 4 months per year and the fruit should be harvested three times during this period, which complicates mechanization. The low number of female flowers reduced branching and inadequate pollination are the major factors that limit Jatropha seed production and thus oil yield. Fruits from cross-pollinated flowers are significantly larger, heavier and more numerous than those produced by autogamous self-pollinated flowers. Therefore, it is not surprising that honeybees play a positive role in the Jatropha pollination (Abdelgadir *et al.*, 2008).

Under optimal conditions, flowering and fruiting start 4–5 months after transplantation and the first crop occurs about 7 months after transplantation, but may last into the second year if the plant is pruned to increase production. Pruning is recommended for building tree architecture (Openshaw, 2000).

Pruning is performed by eliminating the branches below 50 cm and over 80 cm. This process allows the mechanization of the fruit cropping (N. Tominaga, personal communication). Since pruning delays flowering, a compromise must be found on a case-by-case basis. In fact, pruning promotes branching as does nitrate (N) fertilization and water. Since vegetation is inversely correlated with flowering and fruiting, a balance must be made between potassium (P), phosphate (K) and water to take full benefit of the culture management. Since K promotes cell wall thickness, it improves resistance to diseases, compared with N that has the opposite effect. Furthermore, N fertilization only makes sense if water is available. However, NPK fertilization should at least compensate the NPK removal by fruit cropping, which is 14.3–34.3 kg N, 0.7–7.0 kg P, and 14.3–31.6 kg K/t of seeds (Jongschaap *et al.*, 2007).

Plant productivity starts to be stable after the first year, typically when trees are 2–4 years old. The economic production of the Jatropha plants extends from the first year after planting to 40 years. However, the tree life and fruit production span over ~100 years.

6. *Oil extraction and properties*

Fruits are harvested at seed maturity, which occurs ~40 days after flowering and then dried to <8% humidity for 2 days in the sun. Understanding the physical properties of the Jatropha fruit is essential to design equipment for harvesting, drying, cleaning, grading, decorticating, and storing them (Pradhan *et al.*, 2008; Sirisomboon *et al.*, 2007). When dried, the fruits are (i) about 400–425 kg^{-1}, and (ii) composed of 35–40% shell and 60–65% seeds. The seeds (i) are 1580–1600 seeds/kg of fruit, (ii) weigh 600–640 mg each, (iii) account for 4980 cal/g (20.85 MJ/kg) with an oil content of ~35% and an energy content of 9036 cal/g (37.83 MJ/kg), and (iv) are converted to biodiesel by transesterification with a typical rate of 17–18% and an alkyl ester purity of 95–97% (Augustus *et al.*, 2002; Singh *et al.*, 2008).

Laboratory experiments have reported that the rate of oil extraction carried out by a mechanical press is 40–55% for hulled seeds, 70–75% for kernel (Willems *et al.*, 2008a,b) and by manual ram presses is 60–65% (Beerens 2007). Solvent (*n*-hexane) extraction of the kernel alone yields ~60% oil (Veronique *et al.*, 2002). An enzymatic method of extraction is also available (Shah *et al.*, 2004). It gives essentially the same rate of oil extraction as that obtained with hexane, but is about 12 times faster.

In industrial operating conditions, the rate of oil extraction can be much lower and is typically ~35% by mechanic extraction or ~40% by solvent extraction of the hulled seed.

About >10% oil remains in the seed cake after extraction (N. Tominaga, personal communication). Given the high variability in oil rate extraction found in practical applications, oil production may vary greatly from one place to another, however, in Brazil oil production is usually 1–2 t/ha.

The oil extracted from the Jatropha kernel contains approximately 24.6% crude protein, 47.2% crude fat, and 5.5% moisture (Akintayo, 2004). The seed oil has a good oxidation stability compared to soybean oil, low viscosity compared to castor oil and a low pour point (the temperature where it starts to become solid) compared to palm oil. In addition, the biodiesel is stable upon storage (Augustus *et al.*, 2002). The profile of fatty acid composition corresponding to these features is C14:0 (1.4%), C16:0 (15.6%), C18:0 (9.7%), C18:1 (40.8%), C18:2 (32.1%), and C20:0 (0.4%) (Foidl *et al.*, 1996; Kumar and Sharma, 2008; Nahar *et al.*, 2005).

The fuel properties of Jatropha biodiesel are close to those of fossil diesel and match the American and European standards (Tiwari *et al.*, 2007). Because of its lower pour point, Jatropha oil can be used to complement palm oil and give a mixed product compatible with consumption in Asian countries (Shah and Gupta, 2007).

7. Seed cake

Jatropha generates approximately 1 t of seed cake per hectare. Taking India as an example, it is expected that Jatropha will be grown on more than 20 million hectares in the few coming years and will generate ~20 million tons of seed cake per year. This is a significant biomass potential; however, at the moment, seed cake is devolved to the crop field for mulching (Sharma *et al.*, 2008). Alternatively, it can be fermented for biogas production. It releases more energy than cattle dung. Seed cake can also be converted to briquettes for domestic or industrial combustion. One kilogram of briquettes combusts completely in 35 min at 525–780 °C (Singh *et al.*, 2008; Vyas and Singh, 2007). Mahanta *et al.* (2008) also investigated seed cake as a substrate for the industrial production of enzymes such as proteases and lipases.

II. JATROPHA AS A FUEL

A. OIL

The plant oils are made up of 98% triacylglycerides or triglycerides (TAG) and a small amount of mono- and diacylglycerides (DAG) or diglycerides (Barnwal and Sharma, 2005). The direct use of plant oils and/or blends with fossil fuels was generally considered unsatisfactory and impractical for both direct and indirect diesel engines due to poor fuel atomization, piston ring sticking, fuel injector choking, fuel pump failure, and the dilution of lubricating oil (Agarwal *et al.*, 2003). Features such as the (i) high viscosity (Ramadhas *et al.*, 2005), (ii) engine corrosion due to free fatty acid (FFA) content, (iii) gum formation due to polymerization during storage, (iv) carbon deposits, and (v) thickening of lubricating oil are obvious problems (Ma and Hanna, 1999). Plant oil gums can be removed by acid treatment with the consequence that the viscosity is decreased and the cetane number is increased. Degumming is simple and less expensive than transesterification (Haldar *et al.*, 2008), where this involves an acid treatment whose duration (typically 1 week) is critical.

Even after degumming, the emission of exhaust gas and fume by native oils is significantly higher than for fossil diesel and biodiesel. Nitrogen oxide

(NO_x) is a potent greenhouse gas whose emission by native oil combustion can be decreased by advancing the injection timing that is normally needed for fossil diesel and by increasing the injector opening. These modifications increase the brake thermal efficiency, which is accompanied by a significant reduction in hydrocarbons (HC) and smoke emissions (Reddy and Ramesh, 2006). The re-injection of a part of the exhaust gas in the air admission pipe is effective in reducing NO_x (Pradeep and Sharma, 2007). The positive effect reported is due to the decrease in viscosity induced by the higher biofuel temperature at the admission pipe (Agarwal and Agarwal, 2007).

Despite transformation costs, alkyl esters of oil (biodiesel) are considered better technological and environmental solution. In fact, the combustion characteristics of biodiesel are better because of its reduced viscosity compared to native oils. This is due to the fact that the large TAG molecule is reduced in size by a factor of three by transesterification.

B. BIODIESEL

1. Homogeneous catalysis

The main components of plant oils are fatty acids and their derivatives; mono-, di-, and tri-acylglycerides. The TAGs are esters formed by fatty acid condensation with tri-alcohol glycerol (propanetriol). Depending on the number of fatty acids fixed on the glycerol molecule, one can have mono-, di-, or tri-acylglycerides. Of course, these fatty acids can be the same or different. Biodiesel can be obtained by esterification and transesterification.

Esterification is the process by which a fatty acid reacts with a mono-alcohol to form an ester. Esterification reactions are catalyzed by acids. Esterification is commonly used as a step in the process of biodiesel fabrication to eliminate FFAs from low quality oil with high acid content.

Transesterification, or alcoholysis, is the displacement of alcohol from an ester by another alcohol in a process similar to hydrolysis. This process has been widely used to reduce the viscosity of TAGs. The transesterification reaction is represented by the general equation $RCOOR' + R''OH \rightarrow RCOOR'' + R'OH$. This stepwise reaction occurs through the successive formation of di- and mono-glycerides as intermediate products. Theoretically, transesterification needs three alcohol molecules for one of TAG, however, an excess of alcohol is necessary because the three intermediate reactions are reversible (Freedman *et al.*, 1986; Marchetti *et al.*, 2007). After the reaction period, the glycerol-rich phase is separated from the ester layer either by decantation or centrifugation.

In conventional transesterification and esterification processes for biodiesel production, strong alkalis or acids are used as chemical catalysts.

These processes are energy intensive and the physico-chemical synthesis often results in poor reaction selectivity. Therefore an extra purification step to remove glycerol, water, and other contaminants is required (Jian-Xun *et al.*, 2007).

In the traditional way of performing the transesterification, the catalyst can be a base such as NaOH, KOH, and $NaOCH_3$ or an acid such as H_2SO_4. The alkaline catalysis is, however, much faster than the acid catalysis. Low cost and favorable kinetics turned NaOH into the most-used catalyst in industry (Schwab *et al.*, 1987). However, soaps and emulsions can be formed during the reaction, which complicates the purification process. Oil conversion with an alkalin catalyst usually reaches 95–99% after 1 h Freedman *et al.* (1984). Transesterification is typically obtained with (i) a molar ratio of alcohol to oil of 6:1 (~10 wt%), (ii) a reaction temperature near the boiling point of alcohol (60–70 °C at atmospheric pressure), and (iii) NaOH or KOH catalysts at concentrations <1.0 wt% of the oil. In these conditions, the reaction should yield >95% alkyl ester and ~10 wt% glycerol.

The drawback of acid catalysis is that it is about 400 times slower than alkaline catalysis by conventional technologies (Al-Zuhair *et al.*, 2007; Schuchardt *et al.*, 1998) and typically needs ~48 h at 60 °C with a high alcohol to oil ratio (30:1) to achieve a 98% conversion (Canakci and Van Gerpen, 1999). However, by eliminating mass transfer resistance through cavitation, Kelkar *et al.* (2008) showed that acid catalysis of biodiesel yields >90% in <3 h at room temperature. Low frequency ultrasounds are also effective at speeding up the transesterification reaction (Jian-Xun *et al.*, 2007; Shah *et al.*, 2005).

According to Freedman *et al.* (1984) and Liu (1994), the oil feedstock for alkaline transesterification should not contain more than 1% FFA and 0.05% water. If the FFA level exceeds this threshold, saponification hinders the separation of alkyl esters from glycerin and reduces the alkyl ester yield. The elimination of FFA relies on the esterification reaction that can be carried out by acid catalysis (Berrios *et al.*, 2007).

The specific advantages of the acid esterification and alkalin transesterification can be combined in a sequential process (Berchmans and Hirata 2008; Ghadge and Raheman 2005; Gui *et al.*, 2008; Ramadhas *et al.*, 2005; Sharma and Singh 2008; Tiwari *et al.*, 2007). According to that process the oil is first pre-treated with an alcohol in an acid-catalyzed environment to convert the FFA into alkyl esters. The acid-free oil feedstock can then be treated with an alkali, as described above (Canakci and Van Gerpen, 2001).

Jatropha oil is rich in FFA (15%), therefore, biodiesel is produced through a two-step transesterification process (Berchmans and Hirata 2008; Tiwari *et al.*, 2007). In the first step of this process, the conditions optimum for reducing the FFA levels to less than 1% are 1.43% v/v H_2SO_4, a 0.28 (v/v) ratio of methanol to oil for 90 min at 60 °C. In the second step, the alkalin transesterification is catalyzed with a 0.16 (v/v) ratio of methanol to oil for 24 min at 60 °C. This process gives an average alkyl ester yield >99% (Tiwari *et al.*, 2007).

Of course, a process using a heterogeneous acid-catalyst that did not dissolve would be appealing because the separation and recovery of the solid catalyst after reaction completion would be easier. Numerous solid catalysts have already been described, but research is still in the screening stage.

2. Heterogeneous catalysis using lipase

In oilseeds, lipases cleave TAGs into fatty acids and glycerol during the early stages of germination. The fatty acids released by lipase are then hydrolyzed by β-oxidation in peroxizomes (Pracharoenwattana and Smith, 2008) to supply the energy necessary for embryo germination and seedling growth until the photosynthesis take off (Baud and Lepiniec, 2008). In most cases, lipases are produced only after seed imbibition (Hassanien and Mukherjee, 1986). The use of plant esterases and lipases for the transesterification catalysis of oil up to 98% alkyl esters has been reported by Gu *et al.* (1986) and Ozaki *et al.* (1995).

The enzymatic production of biodiesel offers the advantage that no complex operations are needed for (i) the salt remediation and (ii) the catalyst and glycerol recovery in comparison with homogeneous (non-enzymatic) methods. The cost of lipase is the main hurdle to the industrialization of lipase-catalyzed biodiesel.

Oil fermentation by whole-cell microorganisms is one way to reduce the cost of lipase since it avoids the complex processes of enzyme isolation, purification and immobilization (Ban *et al.*, 2001, 2002). It has been demonstrated that whole-cell *Rhizopus oryzae* Went & Prins can efficiently catalyze the methanolysis of several plant oils (Ban *et al.*, 2001, 2002; Zeng *et al.*, 2006) including Jatropha oil (Tamalampudi *et al.*, 2008). The *Rhizopus* fermentation had a maximum methyl esters yield of 80% after 60 h with Jatropha oil and had more than 90% lipase activity after five cycles (Tamalampudi *et al.*, 2008). However, stability of whole cells during repeated uses is poor and higher *Rhizopus* cell stability is obtained if *tert*-butanol is used in place of methanol or ethanol (Li *et al.*, 2007a).

Batch production of oil transesterification has also been obtained with immobilized lipases. In addition to the costs of producing the lipase, this technique also has the drawback that it is very time consuming. For instance, when lipase is immobilized on hydrotalcite, the methyl ester yield reached a 95% conversion rate after 105 h. This large increase in reaction time is due to the reduction in initial activity because of the process of lipase immobilization (Yagiz *et al.*, 2007). However, such diminution of activity is enzyme-specific since Shah and Gupta (2007) obtained a 98% (w/w) ethyl ester yield from the Jatropha oil in 8 h with *Pseudomonas cepacia* lipase immobilized on celite. These results were even better than those obtained (84%) with commercial preparations of immobilized lipase such as Lipozyme TL IM (*Thermomyces lanuginosus*) and Novozyme 435 (*Candida antarctica*) (Hernández-Martín and Otero, 2008).

The achievement of Shah and Gupta (2007) is important because it shows that a sustainable technology based on ethanol (95% grade) for the enzymatic preparation of biodiesel is possible. Ethanol (95% grade) can be derived from sugarcane or other sugar crops by common distillation techniques. Thus, crude bioethanol can be used without expensive purification steps. In addition, drying the celite support together with *P. cepacia* lipase provides a catalyst that allows oil transesterification by adding ethanol in one step without concern for the initial FFA concentration. Finally, a cross-linked aggregation of *P. cepacia* lipase increases its stability and efficiency (Shah *et al.*, 2006).

Investigations carried out on lipase distribution in the oil/alcohol mixture showed that the short operational lifetime of the enzyme is due to its low solubility in alcohols such as methanol, ethanol, and the by-product glycerol (Belafi-bako *et al.*, 2002; Samukawa *et al.*, 2000; Shimada *et al.*, 2002). To manage the short life of lipase in the oil/alcohol system, Watanabe *et al.* (2007) proposed a two-step process in which excess amounts of methanol are added in the first step, and glycerol in the second step. Under this procedure, Novozyme lipase could be reused for over 100 cycles of 24 h with a satisfactory oil transesterification rate.

A more elegant solution to the problem of lipase stability is the substitution of the alcohol by methyl or ethyl acetate. The maximum ethyl ester yield using ethyl acetate is 91.3% with crude Jatropha oil (Modi *et al.*, 2007). Reusability of the lipase over 12 repeated cycles with ethyl acetate was proved, while the ethyl ester yield reached zero by the sixth cycle when ethanol was used as the acyl acceptor (Modi *et al.*, 2007). This means a reduction in enzyme cost by a factor three. The triacetin that is produced during the interesterification under these conditions has no negative effect on lipase activity (Du *et al.*, 2004). In addition, triacetin is a valuable molecule with widespread applications.

Ultimately, *in situ* transesterification has been achieved by alkalin ethanolysis directly on the seed pulp (Haas *et al.*, 2004) in less than 2 h at 35 °C (Haas *et al.*, 2007) with a greatly reduced ethanol proportion (Haas and Scott, 2007). In an extension of this work, Su *et al.* (2007) carried out the *in situ* transesterification by lipase catalysis in methyl acetate or ethyl acetate. Such a process of fatty acid ester production eliminates the costs associated with solvent extraction and oil cleanup by performing these steps in just one operation. Because methyl acetate and ethyl acetate are the acyl acceptors, they act as solvents and transesterification agents at the same time. In addition, they eliminate the risk of enzyme deactivation since glycerol is not produced in that medium. The methyl/ethyl esters are directly recovered from the extraction process by simply removing the catalyst, filtering and evaporating the solvent. The yield of alkyl ester by *in situ* transesterification is higher than that achieved by the conventional two-step alkaline methanolysis.

The best conditions were obtained with a 7.5:1 ratio of methyl or ethyl acetates to seeds, and 4.26% water. The ester yield increased rapidly before 16 h, and then very slowly until 36 h to reach a yield of ~90% alkyl esters.

In situ transesterification is very convenient in Brazil since bioethanol from sugar cane is abundant. For this reason, Petrobras (Brazil's largest energy company) considers *in situ* transesterification the most viable technique for the commercial transesterification of Jatropha oil (U. Soares, personal communication).

Jatropha seed extract contained five esterases with very high temperature and pH-stabilities compared to other plant lipases and esterases. Lipase activity is only detectable after seed germination. It reaches a maximum at 4 days after germination and decreases thereafter. Lipase can hydrolyze >80% TAGs into fatty acids at very low water activities (0.2). Lipase hydrolyzes short chain and long chain TAGs at about the same rate (Staubmann *et al.*, 1999). In fact, methyl esters of fatty acids are observed in Jatropha fruit right after weight increase, but they disappear when TAGs start to accumulate (Annarao *et al.*, 2008). If lipase synthesis could be induced during maturity when the fruit change from green to yellow (stage VI of Annarao *et al.*, 2008), the oil would probably be more readily convertible to fuel.

C. COMBUSTION

Compared to fossil diesel, biodiesel generally causes a decrease in unburned HC, carbon monoxide (CO) and particulate (PM) emissions along with an increase in NO_x emissions (Antolin *et al.*, 2002; Beer *et al.*, 2002; Cardone *et al.*, 2002; Cheng *et al.*, 2008; Dorado *et al.*, 2003a; Durbin and Norbeck 2002; Kalam *et al.*, 2003; Kalligeros *et al.*, 2003; Kegl 2008; Lin *et al.*, 2006a,b; Wang *et al.*, 2000). However, the abatement of soot emission due to oxygen (~10%) naturally present in the biodiesel might be counterbalanced by the rate of alkyl ester unsaturation (double bonds), which favors soot formation (Klein-Douwel *et al.*, 2009). The US Environmental Protection Agency (EPA) (USEPA 2002) concluded that the average emissions by B20 (a blend of 20% biodiesel and 80% fossil diesel) are 10.1% lower for PM, 21% lower for HC, 11% lower for CO, and 2% higher for NO_x compared to fossil diesel.

The dependence of biodiesel emissions on feedstock, engine technology, and engine operating conditions has also been reported by other authors (Dorado *et al.*, 2003a,b; Knothe and Steidley, 2005; Lin and Lin, 2006; Yamane *et al.*, 2001). Both the injection system and the feedstock play an important role because they influence the fuel spray and consequently the combustion characteristics (Agarwal, 2007; Boehman *et al.*, 2004; Durbin *et al.*, 2000; Kegl, 2006; Krisnangkura *et al.*, 2006; Lee *et al.*, 2005; McCormick *et al.*, 2001; Nabi *et al.*, 2006). For example, Banapurmath *et al.* (2008) reported higher HC and smoke emissions with three biodiesels from karanja, Jatropha, and sesame, compared to fossil diesel. In contrast, the HC emissions are lower with methyl esters from polanca (*C. inophyllum*).

NO_x emission is a direct function of engine load. This is expected because the temperature of the combustion chamber increases with higher load and NO_x formation is positively correlated with temperature. In addition, since the NO_x exhaust from an engine running on biodiesel is higher than one running on fossil diesel, it must be due to the higher combustion temperature of biodiesel (Raheman and Ghadge, 2007; Srivastava and Verma, 2008). By retarding the injection timing with an enhanced injection rate Reddy and Ramesh (2006) succeed in improving the engine performance and emissions levels significantly. The emissions were even lower than with fossil diesel, however this came at the cost of a 3% loss in engine power.

The blending of methanol or ethanol with fossil diesel or biodiesel, in particular Jatropha biodiesel, has been widely investigated as a way to reduce smoke and NO_x (Ajav *et al.*, 1998; Kumar *et al.*, 2003, 2006; Senthil *et al.*, 2003).

III. BREEDING JATROPHA

Jatropha has never been domesticated and its yield is difficult to predict with accuracy. The conditions that best suit its growth are not well defined and the potential environmental impacts of large-scale cultivation are not yet understood. Without understanding the basics of the agronomy of Jatropha, a premature push to cultivate it could lead to very unproductive agriculture.

Genetic variation in Jatropha is sufficient to allow the improvement of crop features such as seed morphology, oil content (Kaushik *et al.*, 2007; Raina and Gaikwad 1987; Sunil *et al.*, 2008), synchronization of fruit maturation, nanism, toxicity, digestibility, resistance to pests and diseases.

Jatropha is a cross-pollinated crop and for this reason, breeding will involve discreet populations rather than discreet breeding lines. Breeding programs and germplasm conservation are principally ongoing in India, China, Thailand, Philippines, Mexico, Guatemala and are just starting in Brazil at EMBRAPA, EPAMIG/CTNM, EBDA and UFS (Table I). In fact, it was only in 2007 that Brazil introduced Jatropha to its list of official crops.

Jatropha accessions available in India show only modest levels of genetic variation, while wide variation has been found between the Indian and Mexican genotypes (Basha and Sujatha, 2007). This was expected since variability is higher at the center of origin, i.e., intertropical Americas. However, the precise center of origin within the intertropical Americas has not yet been established (Sujatha *et al.*, 2008). Jatropha samples were taken to Brazil by Portuguese colonists and introduced in several distant tropical countries such as India and Africa.

The best available practice at the moment for selecting breeding material is to use the best-performing trees in the location of interest. Trees with an annual yield above 2 kg of dry seeds and a seed oil content higher than 35% can be considered good accessions.

PCR profiling for analyzing diversity among a collection of accessions is a common practice nowadays. Ranade *et al.* (2008) evaluated the genomic diversity of Jatropha accessions by single primer amplification reaction (SPAR) (Heath *et al.*, 1993; Welsh and McClelland, 1990; Williams *et al.*, 1990). This method is interesting because it does not depend on the availability of genomic sequences for the plants being analyzed. Eleven primer pairs were proposed for probing polymorphic DNA sequences (RAPD) and minisatellites (DMAD) in Jatropha (Ranade *et al.*, 2008). Jatropha individuals exhibit high phenotypic interaction with the environment, which makes genomic DNA probes with reproducible polymorphisms essential.

Probes for polymorphic genomic sequences are needed for fast identification of specific population abilities. This is typically assessed by correlating the populations, recognized by a lower distance between their constitutive individuals than to the average heterogeneity for a given marker set, to specific phenotypic traits (Sunil *et al.*, 2008). In fact, accessions from distinct populations form clusters according to their genotypic polymorphisms that match their geographic provenance and, therefore, their membership in a given population. Genotype clustering can be detected by UPGMA analysis and ultimately revealed by a dendogram (Ranade *et al.*, 2008). The assessment of specific performances is necessary to ground future breeding programs. Basic data like the genetic background of the accessions and agroclimatic adaptation are not yet available.

TABLE I

Localization and research activities of the most important public institutions involved in the improvement of Jatropha *(germplasm collection is indicated by "+" when available)*

Country	Research institution	Acronym	Activity	Germplasm collection	Locality
India	Biotech park		Diversity	+	Lucknow
	Centre for Research and Application in Plant Tissue Culture	CRAPTC (CCS HAU)	Tissue culture	+	Hissar
	Central Research Institute for Dryland Agriculture	CRIDA	Diversity	+	Hyderabad
	Central Salt & Marine Chemicals Research Institute	CSMCRI	RAPD/AFLP/SSR		Gujarat
	Forest Research Institute	FRI	Breeding	+	Dehradun
	National Bureau of Plant Genetic Research	NBPGR	Germplasm		New Delhi
	National Botanical Research Institute	NBRI	Germplasm	+	Lucknow
	National Research Center for Agro Forestry	NRCAF	Genetics		Jhansi
	Panjab Agricultural University	PAU	Breeding	+	Panjab
	The Energy and Research Institute	TERI	Oil characterization		New Delhi
	Thapar Institute	TI	Breeding	+	Patilala
China	Xishuangbanna Botanical Garden	XSBG	Breeding		Yunnan
Thailand	Lao Institute for Renewable Energy	LIRE	Breeding	+	Ventiane
Philippines	University of the Philippines Los Baños	UPLBFI	Breeding	+	Los Baños
Brazil	Empresa Brasileira de Pesquisa Agropecuária	EMBRAPA Agroenergia	Breeding	+	Brasilia

	Empresa de pesquisa agropecuária de Minas Gerais	EPAMIG	Breeding	+	Montes Claros
	Empresa Bahiana de Desenvolvimento Agrícola	EBDA	Breeding	+	Salvador
	Universidade Estadual de Santa Cruz	UESC	EST		Ilhéus
	Universidade Federal de Sergipe	UFS	Isozymes	+	Sergipe
Mexico	Centro de Desarrollo de Productos Bióticos del Instituto Politécnico Nacional	CEPROBI-IPN	Breeding	+	Morelos
Guatemala	Alianza em Energia y Ambiente	AEA	Breeding	+	Guatemala
The Netherlands	Plant Research International B.V.	PRIBV	Molecular genetics		Wageningen

Qualitative and quantitative traits, such as the amount of FFAs, unsaponifiables, acid number and carbon residues, show a wide range of variation. This indicates that the oil quality is dependent on the interaction of environmental and genetic factors (Kaushik *et al.*, 2006, 2007). Recently, fatty acid accumulation during seed maturation has been assessed. Seven stages were found. During stage I–V, the fresh weight of developing seeds increases from 32.4 to 1061 mg in nearly 27 days (Annarao *et al.*, 2008). The analysis of the data shows that oil content raises in two major steps with a very clear transition occurring around stage IV when the oil content raises from 3 to 18% and the TAGs from 30 to >90%. Fresh weight reaches its maximum (1061 mg) at stage V and then decreases until ~640 mg during stage VI and VII. These last two stages match fruit ripening, which is accompanied by a color change from green to yellowish. Fruit maturation is also accompanied by a decrease in the saturated fatty acid concentration and an increase in unsaturated fatty acids. Methyl esters of fatty acids and sterols also exist until stage IV and then disappear. Sterols have an essential role at the cellular level in hormonal signaling and are obviously involved in the process of fruit organogenesis (Annarao *et al.*, 2008).

Non-synchronous fruit maturation in Jatropha is an important drawback to this crop. Fruit should be continuously harvested, which is only possible in developing countries such as Brazil or India where Jatropha can be used as a tool for social incentive and job creation in unfavored economic areas. Therefore, this negative agronomic trait can be transformed into social value. For now, the Brazilian government guarantees a minimum price for Jatropha seeds. It has been calculated that this incentive should sustain 40 000 farmer families in the state of Bahia. This position is necessary to (i) reach critical seed mass in the short term, (ii) feed biodiesel refineries that are being built by Petrobras in the country, and (iii) meet the government recommendation of B5 by 2012. This will mean a production of ~2.4 billon liters of biodiesel from various oil feedstocks. At the moment, the major oil feedstock for B2 is soybean seeds.

There is a critical need for scientific breeding of Jatropha using advanced DNA mapping technologies (Sudheer Pamidiamarri *et al.*, 2008a,b). Recently, a new full-length cDNA of stearoyl-acyl carrier protein desaturase was obtained by RTPCR and RACE techniques from developing Jatropha seeds and the gene was functionally expressed in *E. coli* (Tong *et al.*, 2006). This is an important enzyme for fatty acid biosynthesis in higher plants, and it also plays an important role in determining the ratio of saturated fatty acids to unsaturated fatty acids in plants (Lindqvist *et al.*, 1996). Expressed sequence tags (ESTs) from immature and mature seeds are also being used to derive markers from quantitative trait loci (QTLs) to assist Jatropha breeding (Carels *et al.*, CNPq project 471214/2006-0). In fact, key genes involved in complex traits such as oil synthesis can be mapped using a combination of EST and AFLP techniques (Suárez *et al.*, 2000).

TAG synthesis occurs through complex pathways involving multiple metabolic steps and cell compartments. The endoplasmic reticulum has a complex architecture that is formed by sheets and tubular structures (tER) by interaction with the actin cytoskeleton framework. Actin is also involved in the transport of specific transcripts unto tERs. Non-soluble TAGs are synthesized by ER-localized diacylglyceride transferases and accumulated by forming lipid bodies at tER extremities. When a critical size is attained, the lipid body is released. At this point, it is made of a lipid core wrapped in a phospholipid monolayer that may include specific membrane proteins (Herman, 2008).

TAGs occur through the condensation of two fatty acid molecules with one molecule of phosphoric acid and one molecule of glycerol. The process of esterification involved produces various types of phosphatidic acids. Examples of phosphatidic acids are lecitines, cefalines, and serines. Then, phosphatidic acids are dephosphorilated in the endoplasmic reticulum to form DAG and finally a fatty acid chain is added to the DAG by an acyltransferase resulting in a TAG that is sequestered in lipid bodies for storage. The endoplasmic reticulum is the principal place of phosphatidic acid synthesis, however plastids also produce these acids (Cahoon *et al.*, 2007). The fatty acids incorporated into TAGs can suffer desaturation, elongation or other additional modifications in parallel with the esterification of acyl units in phosphatidilcoline or coenzyme A. Alternative mechanisms for fatty acid transfer to the DGATs are also possible (Cahoon *et al.*, 2007; Cahoon and Schmid 2008).

Acetyl-coenzyme A (acetyl-CoA) carboxylase (ACCase, EC 6.4.1.2) plays an important role in fatty acid synthesis because the enzyme catalyzes the irreversible carboxylation of acetyl-CoA to produce malonyl-CoA through its two catalytic activities, biotin carboxylase and carboxyltransferase (Davis *et al.*, 2000). ACCase is a large, multi-domain enzyme in most eukaryotes. The enzyme catalyzes the fatty acid synthesis reaction by which acetyl-CoA is carboxylated to form malonyl-CoA. ACCase activity can be controlled at the transcriptional level, as well as by small molecule modulators and covalent modifications. Transgenic plants with 10–20 times higher ACCase activity in mature seeds have a total increase in oil content of $\sim$5% (Ohlrogge *et al.*, 1999). In addition, the reduction of ACCase activity in seeds of *Brassica napus* L. by antisense expression reduces their lipid content (Sellwood *et al.*, 2000).

Phosphoenolpyruvate carboxylase (PEPC, EC 4.1.1.31) is an enzyme that catalyzes the addition of CO_2 to phosphoenolpyruvate (PEP) to form oxaloacetate. PEP can be converted into pyruvate by pyruvate kinase. Pyruvate is then transformed into acetyl-CoA by pyruvate dehydrogenase and malonyl-CoA by the ACCase. The relative activity of PEPC and ACCase determines the PEP flow direction and the ratio of proteins to lipids produced in oilseeds crops. Chen *et al.* (1999a,b) found that antisense expression of *B. napus* PEPC reduced the activity of PEPC and that the PEPC transgenic plants were 6.4–18% richer in oil.

Genomic information is providing the starting point for understanding the instructions for molecular machines and the systems needed to control and operate them. Through a labyrinth of pathways, networks, chemistry, and mechanics, this machinery makes the cell and the organism come alive (Santos Mendoza *et al.*, 2008). Understanding the operation, function, and coordination of genomic information is a necessary step for the emerging field of systems biology. The gene networks triggering seed maturation in the model plant Arabidopsis are being characterized. Until now, four loci have been identified: FUS3, ABI3, LEC1, and LEC2. The induction of LEC2 correlates with the onset of oil accumulation in the seed. LEC2 controls the gene WRI1, which itself controls several genes encoding enzymes for the late glycolysis and the plastidial machinery of fatty acid biosynthesis (Baud and Lepiniec, 2008).

A draft sequence of the genome of castor bean is now available (Chan *et al.*, 2008). Like Jatropha, this species belongs to the family of the *Euphorbiaceae*, which will facilitate future investigations on Jatropha genome. The genome of castor has been assembled with 4× coverage using a whole genome shotgun strategy and the gene annotation has being assisted by ~50 000 ESTs (Sujatha *et al.*, 2008).

The rapid growth of *Euphorbiaceae* genomics projects along with the relationship of that family to Populus (both are members of Malpighiales) means that comparative genomics with non-model plant within at least the *Euphorbiaceae* is now possible (Wurdack, 2008). In castor, variability has been assessed worldwide using AFLPs, SSRs and SNPs, and a genetic map was also constructed (Allan *et al.*, 2008). Considering cassava (*Manihot esculenta* Crantz), for instance, genetic maps are being constructed. More than 3000 RFLPs, 800 SSRs, 9 isozymes (Fregene *et al.*, 2007), 8577 unique gene clusters (Lokko *et al.*, 2007), 646 microsatellites (Lokko *et al.*, 2007), and 80 SNPs (Lopez *et al.*, 2005) were developed and used in genetic mapping. Molecular markers linked to pest and disease resistance were identified. More than 100 primary and secondary QTLs for plant architecture are used to assist cassava breeding given their correlations with productivity (Okogbenin and Fregene, 2003). Among these QTLs, plant height, branching height, branching levels, branching index, stem portion with leaves, and leaf area index could be interesting for Jatropha breeding. Bacterial artificial chromosome libraries (BAC) have been developed for positional cloning and investigation of resistance (Fregene *et al.*, 2007). Rubber (*H. brasiliensis*), another species of *Euphorbiaceae* with considerable economic interest, also enters the biotechnology era with genetic maps (Lespinasse *et al.*, 2000) and the sequencing of 20 000 ESTs (Chow, 2008).

With DNA sequencing techniques, it has become obvious that the genes associated with fatty acid biosynthesis are expressed at a medium to high rate in the developing seed (van de Loo *et al.*, 1995). This property is particularly interesting for the identification of the cDNA involved in fatty acid biosynthesis through EST profiling. The 5′ side sequencing of ~1000 random cDNAs from developing seeds accumulating fatty acids at a high rate identified the key enzymes involved in fatty acid biosynthesis (Cahoon and Kinney, 2005). In addition, EST sequencing, associated with real-time PCR and macro- or micro-arrays during seed development, allows the investigation of the gene regulation network both in the lipid pathway (Chen *et al.*, 2007; Holter *et al.*, 2000) and among genotypes (Nelson *et al.*, 2004). Even when the metabolic pathway is known, as it is for the fatty acid pathway, the analysis of EST datasets is often a faster process for the identification of corresponding enzymes. For instance, the involvement of cytochrome P450 in vernolic acid synthesis from linoleic acid has been demonstrated using ESTs (Bafor *et al.*, 1993). Since enzymes like cytochrome P450 are associated with membranes in plants, their purification is difficult. In addition, the protein family in which P450 belongs is large since Arabidopsis alone has >250 genes for that enzyme (Nelson *et al.*, 2004). This makes PCR investigation difficult. The transcriptome investigation using ESTs has also allowed for the identification of functional acyl-CoA desaturases in plants, such as Limnanthes (Cahoon *et al.*, 2000).

Transcriptomics, proteomics and other "omics" open the way to metabolome engineering (Frazier *et al.*, 2003). Omics are now used in assisting plant breeding since they allow the correlation of sequence and phenotype. However, our understanding of the network of interactions that is taking place is still lacking. The identification of QTL alleles that improve agricultural production allows breeders to introgress the traits into modern genetic backgrounds and explore the molecular mechanisms that regulate these effects (Zamir, 2008). For instance, map-based cloning of a diacylglycerol acyltransferase (DGAT) that catalyzes the final step in the glycerol biosynthetic pathway allowed the selection of a new protein variant influencing oil content and composition in maize seeds (Zheng *et al.*, 2008).

Jatropha (i) is a diploid species with $2n = 22$ chromosomes, (ii) has a small genome whose size (416 Mb) is approximately the same as that of rice

($\sim$400 Mb) and close to that of castor, which has been estimated at 323 Mb (Arumuganathan and Earle, 1991), and (iii) has an average base composition of GC = 38% (Carvalho *et al.*, 2008), which is typical for core dicots (Carels 2005). Methodologies have been developed for Jatropha tissue culture propagation (Datta *et al.*, 2007; Kalimuthu *et al.*, 2007; see Sujatha *et al.*, 2008 for a review of 14 references on the subject) and genetic transformation with *Agrobacterium tumefaciens* (Li *et al.*, 2007b). Genetic transformation could be used to generate RNAi transformants to produce accessions with zero curcin and phorbol (see Section IV). It is clear that the basal resistance to pests could be improved by transferring the toxin gene from *Bacillus thuringiensis*. In addition, there is room for linking genetic transformation and metabolomics since antinutritional compounds are a recalcitrant problem associated with the Jatropha seed cake. Since phorbol is a diterpene, it has been speculated that GGDP synthase could be an appropriate target for RNAi (Gressel, 2008). In fact, nontoxic varieties from different regions of Mexico were shown to be toxic in the agro-climatic conditions of Brazil (N. Tominaga, personal communication). In general, it is believed that environmental conditions have a large impact on genetic modulation (Achten *et al.*, 2008; Kaushik *et al.*, 2006, 2007; Raina and Gaikwad, 1987).

IV. SECONDARY METABOLITES

Jatropha oil is not edible, which raises concerns about toxic components such as curcin (Lin *et al.*, 2003) and phorbol esters (0.03 and 3.4%) (Aderibigbe *et al.*, 1997; Adolf *et al.*, 1984; Aregheore *et al.*, 2003; Haas and Mittelbach 2000; Hirota *et al.*, 1988; Kinghorn and Evans, 1975; Makkar *et al.*, 1997, 1998a,b, 2007; Martínez-Herrera *et al.*, 2006). Phorbol esters are a serious problem since they withstand roasting temperatures as high as 160 °C for 30 min. In addition, even if the chemical treatments that reduce their concentration in the meal are promising, phorbol esters raise potential problems for coproduct valorization in Jatropha (Aregheore *et al.*, 2003).

Jatropha poisoning causes vomiting and symptoms resembling those induced by organophosphate insecticides (Joubert *et al.*, 1984). The Jatropha kernel is reported to be highly toxic to animals (Adam and Magzoub, 1975; Goonasekera *et al.*, 1995). El-Badwi *et al.* (1995) reported high mortality in chicks fed with a diet containing 0.5% Jatropha seeds. Calves fed with 0.25 g seed/kg of body weight also showed a rapid onset of toxic manifestations and death occurring within hours of administration (Ahmed and Adam, 1979). The same diet was shown to be lethal for six- to eight-month-old goats between days 7 and 21. It caused bloody diarrhea, dyspnea, dehydration, and hind limb paring before death. Internal lesions included enterohepatonephrotoxicity, pulmonary hemorrhage, emphysema and cyanosis, tracheal froths, ascites and hydropericardium. These lesions were accompanied by (i) an increase in the activity of serum aspartate aminotransferase, (ii) an increase in urea concentration, (iii) a decrease in total protein and albumin, and (iv) anemia and leucopenia (Abdel Gadir *et al.*, 2003). Widespread hemorrhaging and congestion could be due to an altered permeability of the capillaries. The alteration in the activity of serum aspartate aminotransferase, urea concentration, and total proteins reflects hepatorenal damage. Leukocytosis and macrocytic normochromic anemia were also observed (Abdel Gadir *et al.*, 2003). The acute oral LD 50 of the oil was 6 ml/kg body weight in rats. In rabbits and rats, the toxic fraction produced severe skin irritation followed by necrosis. In mice, the same fraction had dermal toxicity with hemolytic activity (Gandhi *et al.*, 1995; Kinghorn and Evans, 1975). In fact, phorbol esters in Jatropha oil have been reported to promote skin tumors (Hirota *et al.*, 1988; Horiuchi *et al.*, 1987). These skin irritants could affect people that process oil extraction. Since the Jatropha economy relies heavily on manual work, this issue should not be neglected.

The seed kernel also contains antinutritional factors such as saponin, phytate, trypsine inhibitors, lectines (Cano *et al.*, 1989), glucosinolates, amylase inhibitors, cyanogenic glucosides (Rakshit *et al.*, 2008), β-D-glycosides of β-D-sitosterol (Chhabra *et al.*, 1990; Mampane *et al.*, 1987), flavonoids (Subramanian *et al.*, 1971), vitexine, and isovitexine (Sankara *et al.*, 1971). Phytates decrease mineral bioavailability and protein digestibility by forming complexes and by interacting with enzymes such as trypsin and pepsin (Reddy and Pierson, 1994).

Rakshit *et al.* (2008) found that treating kernel pulp with alkalinity and heat reduces the phorbol ester content by up to 89%. This treatment also inactivates trypsin inhibitors, but not other antinutritional compounds such as saponins, lectins, and phytate (Aderibigbe *et al.*, 1997). Even though the mortality of rats fed with this preparation was reduced, they still showed lower appetite, diarrhea, anemia, and organ atrophy.

Though various processing techniques have been attempted, no treatment has been successful in the complete elimination of antinutritional and toxic factors from nontoxic and toxic varieties (Martínez-Herrera *et al.*, 2006). Saponins and lectins can also be reduced by ethanol and $NaHCO_3$ extraction, respectively. However, these treatments do not reach complete cake deodorization, which is another factor responsible for decreased appetite (Haas and Mittelbach, 2000). Despite all these treatments, the digestibility remains lower than it should be by ~14% (Martínez-Herrera *et al.*, 2006).

This is a potential problem for the valorization of Jatropha accessions that could have interesting agronomical features, and plant breeding will be necessary. For instance, these toxins prevent the seed cake from being used as animal feed, with the consequence that the cost of biodiesel from Jatropha will be higher than it could be.

In addition to its toxicity, Jatropha is rich in active compounds with interesting medicinal properties (Tables II and III) that should be investigated in more details given the massive amount of seed cake that will be soon available from biodiesel production. These medicinal activities are principally found in the *curcas, multifida, gossypifolia, macrorhiza*, and *cinerea* species.

V. MAJOR WORLDWIDE INITIATIVES OF JATROPHA IMPLEMENTATION

Due to its great biodiversity and diversified climate and soil conditions, Brazil has access to many different oil feedstock plants, mainly including soybean, sunflower, coconut, castor bean, cottonseed, oil palm, Jatropha, babassu, and others. The National Program of Production and Use of Biodiesel (PNPB) was launched in 2004 with the objective of guaranteeing the economic

TABLE II
Some active compounds in the genus Jatropha

Secondary metabolites	Name	References
Pyrazine	Tetramethylpirazine	Ojewole and Odebiyi (1984)
Peroxide		Sutthivaiyakit *et al.* (2003)
Diterpene	Phorbol	Adolf *et al.* (1984)
	Jatraphone	Kupchan *et al.* (1976)
	Jatrophone	Taylor *et al.* (1983)
	Jatropholone	Purushothaman *et al.* (1979)
	Jatrophatrione	Rahman *et al.* (1990)
	Others	Brum *et al.* (1998), Das and Venkataiah (1999), Naengchomnong *et al.* (1986), and Ravindranath *et al.* (2003)
Polyphenol	Lignans	Das and Anjani (1999) and Das and Das (1995)
Cyclic peptide	Curcacycline	Auvin *et al.* (1997) and van den Berg *et al.* (1995, 1996)
	Labaditin	Kosasi *et al.* (1989)
	Mahafacyclin	Baraguey *et al.* (2000)
	Pohfianin	Auvin *et al.* (1999)
Phytate		Aderibigbe *et al.* (1997)

TABLE III
Biological activities found in the genus Jatropha

Biologic activity	References
Antiviral	
HIV inhibitior	Matsuse *et al.* (1999) and Wender *et al.* (2008)
Antibiotic and fungicide	Aiyelaagbe (2001), Aiyelaagbe *et al.* (2001), Canales *et al.* (2005), de Lima *et al.* (2006), Fagbenro *et al.* (1998), Faria *et al.* (2006), Hamza *et al.* (2006), Kumar *et al.* (2006), Madhumathi *et al.* (2000), Marquez *et al.* (2005), Mothana and Lindequist (2005), Nwosu and Okafor (1995), Odebiyi (1985), Ravindranath *et al.* (2004), Rug and Ruppel (2000), Sánchez-Medina *et al.* (2001), Songjang and Wimolwattanasarn (2004), and Tequida-Meneses *et al.* (2002)
Insecticide	Chatterjee *et al.* (1980), Sauerwein *et al.* (1993), Karmegam *et al.* (1997), Li *et al.* (2005), Adebowale and Adedire (2006), Rao and Kumar (2006), and Georges *et al.* (2008)
Molluscide	Adewunmi and Adesogan (1986), Amin *et al.* (1972), Liu *et al.* (1997), and Yadav and Singh (2006)
Parasite control	
Schistosomiasis	Al-Zanbagi *et al.* (2000)
Anti-malaria	Ankrah *et al.* (2003), Gbeassor *et al.* (1989), and Koehler *et al.* (2002)
Anti-leishmania	Akendengue *et al.* (1999)
Onco-regulator	
Antitumoral	Lin *et al.* (2003), Muangman *et al.* (2005), Torrance *et al.* (1976, 1977), and Valente *et al.* (2004),
Antileukemia	Abreu *et al.* (2003), Paquette *et al.* (2002), and Taylor *et al.* (1983)
Tissue-regulator	
Anticoagulating agent	Oduola *et al.* (2005) and Osoniyi and Onajobi (2003)
Wound-healing promotor	Villegas *et al.* (1997)
Anti-inflammatory	Mujumdar and Misar (2004)
Immuno-modulator	Kosasi *et al.* (1989)
Aphrodisiac	Benavides *et al.* (2006)
Antinutrition	
Trypsin inhibitor	Makkar *et al.* (1998a,b)

viability of biodiesel production together with social and regional development. The current diesel consumption in Brazil is approximately 40 billion liters/year, and the potential market for biodiesel is currently 800 million liters and should achieve two billion liters by 2013. The auction prices vary between 0.3 and 0.8 US$/l, according to the area of production (Barros *et al.*, 2006). Since 2008, the use of B2 is mandatory and B5 will be mandatory by 2013. More than 4000 gas stations were commercializing B2 by 2006 (Pousa *et al.*,

2007). The characteristics of pure biodiesel (B100) and blends are defined by two resolutions (ANP 2004 and 2006). US$ 5 billion were invested in bioenergy over 14 years, resulting in the creation of 700 000 new jobs and US$ 43 billion in savings over the last 25 years in gasoline imports (Goldemberg *et al.*, 2008). Petrobras is now processing a mixture of plant oils and crude under the name of H-Bio with a 425 000 t capacity. The country has the potential for 90 million hectares for oleaginous crop production that extends over Mato Grosso (South–West), Goiás, Tocantins, Minas Gerais (Center), Bahia, Piauí, and Maranhão (North–East) under tropical climates.

In China, the arable land area per capita is lower than the world average. As a result, most edible oils need to be imported and edible oil demand in 2010 is projected to be 13.5 million tons. Since Jatropha is not demanding in terms of soil quality, it is a good option for China, which planted ~71 000 hectares and announced the planting of an additional 13 million hectares across its southern states by 2010 (Weyerhaeuser *et al.*, 2007) and the production of 12 million tons of B15 by 2020.

In other Asian countries with high dependency (>80%) from foreign fuel such as Thailand, Cambodia, Vietnam, Myanmar, Laos, Indonesia, and India, eroded land areas from deforested soil are also available for Jatropha cultivation. In Thailand, for instance, the government hopes that 10% of all diesel used in the country will soon come from biodiesel. This will require the production of 6.7 million liters/day biodiesel by growing Jatropha over 6 million hectares. 30 academic institutes and private industry groups are involved in developing Jatropha for this purpose (http://www.akha.org/content/environment/jatrophacurcasagriculturetoindustrythailand.pdf).

Since more than 7 years, the Indonesian government is supporting various national and international agencies as well as research institutes for the investigation of Jatropha with a target biodiesel production covering 10% of the national fuel consumption. The reasoning is that growing 2500 plants/ha will provide 1.5 t of biodiesel. With over 23 million hectares of potential land for Jatropha cultivation, this amounts to more than 18 million liters of biodiesel. However, at moment about 1.5 million hectares are effectively planted (Legowo, 2007).

Since India imports 70% of its fuel consumption (~111 Mio t), any renewable energy is welcome. Therefore, India guarantees a market and prices for 10% biodiesel blends (Kumar and Sharma, 2008). Because India is a net importer of edible oil, it emphasizes non-edible oils from plants such as Jatropha, karanja, neem, mahua, simarouba, and so on. Jatropha and karanja are the two leaders of the Indian plant list for biodiesel production.

Of its 306 million hectares of land, 173 million are already under cultivation, but the rest are classified as either eroded farmland or non-arable wasteland (Fairless, 2007). Nearly 40% (80–100 million hectares) of the land is degraded because of improper land use and population pressure over several years. These wasted areas are being considered for recuperation with Jatropha (Kumar and Sharma, 2008). In the first phase, the Indian government managed the planting of ~500 000 hectares with Jatropha (Francis *et al.*, 2005). Should this first phase produce according to plan, India's government would embark on planting 12 million hectares and privatizing the production of Jatropha biodiesel (Fairless, 2007). This should ensure the production of 13 million tons of biodiesel per year and meet the B10 targets (Sarin *et al.*, 2007).

Pilot experiments for Jatropha implementation are also being carried out throughout Africa and were reviewed by Henning (2005).

VI. CONCLUSIONS

The general feeling is that first generation biofuels are not far from reaching saturation because of the limited availability of arable land. Brazil still has some additional room for sugarcane and Jatropha, while India is promoting Jatropha on its extensive waste lands. Despite these limitations, the development of first generation biofuels has already been a success because it demonstrated that motor technology running on ethanol or biodiesel is feasible.

If one agrees that Jatropha biomass production would sequester 5.5 t CO_2/ha-year (see in Achten *et al.*, 2008), a substantial proportion of the carbon should end up in the soil because of seed cake mulching. This organic manure is expected to have a positive effect on the soil structure and eventually slow down erosion.

Since Jatropha can grow on wasteland, where poor populations are generally found, it is also expected to have a positive social effect by (i) attracting government investment, (ii) stabilizing the population in rural areas, and (iii) providing the population with incomes and energy.

Under low-intensity Jatropha cultivation, the total primary energy input would account for 17% of the total energy released by this crop, i.e., the ratio of energy output versus input would be ~5.8. This ratio is <5 for European oil crops and >8 for sugar cane (Bourne, 2007). Under high-intensity cultivation, the picture is much less favorable and the total primary energy input would account for 38% of the total energy output, i.e., a ratio of 2.6 closer to that of temperate crops. Achten *et al.* (2008) showed that irrigation, fertilization, oil extraction, and transesterification account for a lot of energy consumption. Despite the uncertainty of the data, the life-cycle energy balance of

Jatropha biodiesel seems to be generally positive. How positive the balance is, will mainly depend on the by-product valorization (Achten *et al.*, 2008). However, the concern about Jatropha's profitability that was estimated to US$ 480 ha^{-1} in 2000 by Openshaw (2000) will depend ultimately on technological breakthroughs in biofuel processing and by-product valorization.

Glycerol is now the main by-product of the oleochemical industry and accounts for 10% of total biodiesel production (Valliyappan *et al.*, 2008; Yazdani and Gonzalez, 2007). With the increasing production of biodiesel, market prices of glycerol went down, allowing new alternatives for its use. Although there is extensive utilization of high purity glycerol (>99.0%) in the food, cosmetic, and pharmaceutical industries, it is technically difficult and costly to obtain the pure glycerol from biodiesel (Sun and Chen, 2008). However, glycerol can be used in other applications. It is a potential feedstock for syngas in a H_2/CO ratio of 2:1 and could be used for synthetic diesel production by the Fischer–Tropsch reaction (Steynberg and Nel, 2004). Alternatively, glycerol can also be used for the production of 1,3-propanediol, polyglycerols and polyurethanes (Claude, 1999). Propanediol and propanol are produced by fermentation (Yazdani and Gonzalez, 2007) and can be used in blends with gasoline or ethanol (Fernando *et al.*, 2006).

The lack of seed cake valorization for animal feeding because of its toxicity is a drawback of Jatropha that cancels a significant part of its potential value. If cake could be used for animal feeding it would help to produce meat and milk locally.

The environmental impacts discussed are lower than the fossil alternative as long as no natural ecosystems are removed in favor of Jatropha as it occurs along with sugarcane and palm oil (Darussalam, 2007; Laurance, 2007; Malhi *et al.*, 2008; Stone 2007; Venter *et al.*, 2008) and as long as the byproducts of the bio-diesel production are efficiently used (Achten *et al.*, 2008).

ACKNOWLEDGMENTS

We thank the Coordenação de Aperfeiçoamento de Pessoal de Nível Superior (CAPES) and Fundação Oswaldo Cruz (FIOCRUZ) for providing a research fellowship from the Centro de Desenvolvimento Tecnológico em Saúde (CDTS) to N. Carels. This work received financial support from Conselho Nacional de Desenvolvimento Científico e Tecnológico (CNPq), Brazil (no. 471214/2006-0).

REFERENCES

Abdel Gadir, W. S., Onsa, T. O., Ali, W. E. M., El Badwi, S. M. A. and Adam, S. E. I. (2003). Comparative toxicity of Croton macrostachys, *Jatropha curcas* and *Piper abyssinica* seeds in Nubian goats. *Small Ruminant Research* **48**, 61–67.

Abdelgadir, H. A., Johnson, S. D. and Van Staden, J. (2008). Approaches to improve seed production of *Jatropha curcas* L. *SAJB Annual Meeting Abstracts* **74**, 359, doi:10.1016/j.sajb.2008.01.023.

Abou Kheira, A. A. and Atta, N. M. M. (in press). Response of *Jatropha curcas* L. to water deficit: Yield, water use efficiency and oilseed characteristics. *Biomass and Bioenergy* doi:10.1016/j.biombioe.2008.05.015.

Abreu, I. C., Marinho, A. S. S., Paes, A. M. A., Freire, S. M. F., Olea, R. S. G., Borges, M. O. R. and Borges, A. C. R. (2003). Hypotensive and vasorelaxant effects of ethanolic extract from *Jatropha gossypiifolia* L. in rats. *Fitoterapia* **74**, 650–657.

Achten, W. M. J., Verchot, L., Franken, Y. J., Mathijs, E., Singh, V. P., Aerts, R. and Muys, B. (2008). Jatropha bio-diesel production and use. *Biomass and Bioenergy* **35**, 1063–1084. doi:10.1016/j.biombioe.2008.03.003.

Adam, S. E. I. and Magzoub, M. (1975). A toxicity of *Jatropha curcas* for goats. *Toxicology* **4**, 347–354.

Adebowale, K. O. and Adedire, C. O. (2006). Chemical composition and insecticidal properties of the underutilized *Jatropha curcas* seed oil. *African Journal of Biotechnology* **5**, 901–906.

Aderibigbe, A. O., Johnson, C. O. L. E., Makkar, H. P. S., Becker, K. and Foidl, N. (1997). Chemical composition and effect of heat on organic matter and nitrogen-degradability and some antinutritional components of *Jatropha* meal. *Animal Feed Science and Technology* **67**, 223–243.

Adewunmi, C. O. and Adesogan, E. K. (1986). Toxicology of some Nigerian plants used in Schistosomiasis control I. The effect of molluscicides on molluscan hearts. *Fitoterapia* **57**, 353–358.

Adjaye, J. D., Katikaneni, S. P. R. and Bakhsi, N. N. (1995). Catalytic conversion of canola oil to fuels and chemicals over various cracking catalysts. *Canadian Journal of Chemical Engineering* **73**, 484–497.

Adolf, W., Opferkuch, H. J. and Hecker, E. (1984). Irritant phorbol derivatives from four *Jatropha* species. *Phytochemistry* **23**, 129–132.

Agarwal, A. K. (2007). Biofuels (alcohols and biodiesel) applications as fuels for internal combustion engines. *Progress in Energy and Combustion Science* **33**, 233–271.

Agarwal, D. and Agarwal, A. K. (2007). Performance and emissions characteristics of *Jatropha* oil (preheated and blends) in a direct injection compression ignition engine. *Applied Thermal Engineering* **27**, 2314–2323.

Agarwal, A. K., Bijwe, J. and Das, L. M. (2003). Effect of bio-diesel utilization on wear of vital parts in compression ignition engine. *Journal of Engineering for Gas Turbines and Power* **125**, 604–611.

Ahmed, O. M. M. and Adam, S. E. I. (1979). Effects of *Jatropha curcas* on calves. *Veterinary Pathology* **16**, 476–481.

Aiyelaagbe, O. O. (2001). Antibacterial activity of *Jatropha multifida* roots. *Fitoterapia* **72**, 544–546.

Aiyelaagbe, O. O., Adesogan, K., Ekundayo, O. and Gloer, J. B. (2007). Antibacterial diterpenoids from *Jatropha podagrica* Hook. *Phytochemistry* **68**, 2420–2425.

Ajav, E. A., Singh, B. and Bhattacharya, T. K. (1998). Performance of a stationary diesel engine using vapourized ethanol as supplementary fuel. *Biomass and Bioenergy* **15**, 493–502.

Akendengue, B., Ngou-Milama, E., Laurens, A. and Hocquemiller, R. (1999). Recent advances in the fight against leishmaniasis with natural products. *Parasite* **6**, 3–8.

Akintayo, E. T. (2004). Characteristics and composition of *Parkia biglobbossa* and *Jatropha curcas* oils and cakes. *Bioresource Technology* **92**, 307–310.

Allan, G. J., Foster, J. T., Williams, A., Rabinowicz, P. D., Chan, A. P., Ravel, J. and Keim, P. (2008). Worldwide genotyping of castor bean (*Ricinus communis*) using AFLPs, SSRs And SNPs. *Genetic Resources and Crop Evolution* **55**, 365–378.

Al-Zanbagi, N. A., Banaja, A. E. A. and Barrett, J. (2000). Molluscicidal activity of some Saudi Arabian Euphorbiales against the snail *Biomphalaria pfeifferi*. *Journal of Ethnopharmacology* **70**, 119–125.

Al-Zuhair, S., Ling, F. W. and Jun, L. S. (2007). Proposed kinetic mechanism of the production of biodiesel from palm oil using lipase. *Process Biochemistry* **42**, 951–960.

Amin, M. A., Dafalla, A. A. and Abdel Moniem, O. (1972). Preliminary report on the molluscicidal properties of habat elmeluk *Jaropha* sp. *Transaction of the Royal Society of Tropical Medicine and Hygiene* **66**, 805.

Ankrah, N. A., Nyarko, A. K., Addo, P. G., Ofosuhene, M., Dzokoto, C., Marley, E., Addae, M. M. and Ekuban, F. A. (2003). Evaluation of efficacy and safety of a herbal medicine used for the treatment of malaria. *Phytotherapy Research* **17**, 697–701.

Annarao, S., Sidhu, O. P., Roy, R., Tuli, R. and Khetrapal, C. L. (2008). Lipid profiling of developing *Jatropha curcas* L. seeds using ^{1}H NMR spectroscopy. *Bioresource Technology* **99**, 9032–9035.

Antolin, G., Tinaut, F. V., Briceno, Y., Castano, V., Perez, C. and Ramírez, A. I. (2002). Optimisation of biodiesel production by sunflower oil transesterification. *Bioresource Technology* **83**, 111–114.

Aregheore, E. M., Becker, K. and Makkar, H. P. S. (2003). Detoxification of a toxic variety of *Jatropha curcas* using heat and chemical treatments, and preliminary nutritional evaluation with rats. *South Pacific Journal of Natural Science* **21**, 50–56.

Arumuganathan, K. and Earle, E. D. (1991). Nuclear DNA content of some important plant species. *Plant Molecular Biology Reporter* **9**, 211–215.

Augustus, G. D. P. S., Jayabalan, M. and Seiler, G. J. (2002). Evaluation and bioinduction of energy components of *Jatropha curcas*. *Biomass and Bioenergy* **23**, 161–164.

Auvin, C., Baraguey, C., Blond, A., Lezenven, F., Pousset, J. L. and Bodo, B. (1997). Curcacycline B, a cyclic nonapeptide from *Jatropha curcas* enhancing rotamase activity of cyclophilin. *Tetrahedron Letters* **38**, 2845–2848.

Auvin, C., Baraguey, C., Blond, A., Xavier, H. S., Pousset, J. L. and Bodo, B. (1999). Pohfianins A, B and C, Cycfic Peptides from the Latex of *Jatropha pohliana* ssp. *Molissima*. *Tetrahedron* **55**, 11495–11510.

Azam, M. M., Waris, A. and Nahar, N. M. (2005). Prospects and potential of fatty acid methyl esters of some non-traditional seed oils for use as biodiesel in India. *Biomass and Bioenergy* **29**, 293–302.

Bafor, M., Smith, M. A., Jonsson, L., Stobart, K. and Stymne, S. (1993). Biosynthesis of vernoleate (*cis*-12-epoxyoctadeca-cis-9-enoate) in microsomal preparations from developing endosperm of *Euphorbia lagascae*. *Archives of Biochemistry and Biophysics* **303**, 145–151.

Ban, K., Kaieda, M., Matsumoto, T., Kondo, A. and Fukuda, H. (2001). Whole cell biocatalyst for biodiesel fuel production utilizing *Rhizopus oryzae* cells immobilized within biomass support particles. *Biochemical Engineering Journal* **8**, 39–43.

Ban, K., Hama, S., Nishizuka, K., Kaieda, M., Matsumoto, T., Kondo, A., Noda, H. and Fukuda, H. (2002). Repeated use of whole-cell biocatalysts immobilized within biomass support particles for biodiesel fuel production. *Journal of Molecular Catalysis B: Enzymatic* **17**, 157–165.

Banapurmath, N. R., Tewari, P. G. and Hosmath, R. S. (2008). Performance and emission characteristics of a DI compression ignition engine operated on Honge, *Jatropha* and sesame oil methyl esters. *Renewable Energy* **33**, 1982–1988.

Banjo, A. D., Lawal, O. A. and Aina, S. A. (2006). The entomofauna of two medicinal *Euphorbiacae* in southwestern Nigeria. *Journal of Applied Sciences Research* **2**, 858–863.

Baraguey, C., Blond, A., Correia, I., Pousset, J. L., Bodo, B. and Auvin-Guette, C. (2000). Mahafacyclin A, a cyclic heptapeptide from *Jatropha mahafalensis* exhibiting β-bulge conformation. *Tetrahedron Letters* **41**, 325–329.

Barnwal, B. K. and Sharma, M. P. (2005). Prospects of biodiesel production from vegetable oils in India. *Renewable and Sustainable Energy Reviews* **9**, 363–78.

Barros, G. S. A. C., Silva, A. P., Ponchio, L. A., Alves, L. A., Osaki, M. and Cenamo, M. (2006). Custos de produção de biodiesel no Brasil. *Revista de Política Agrícola* **3**, 36–50.

Basha, S. D. and Sujatha, M. (2007). Inter and intra-population variability of *Jatropha curcas* (L.) characterized by RAPD and ISSR markers and development of population-specific SCAR markers. *Euphytica* **156**, 375–386.

Baud, S. and Lepiniec, L. (2008). Compared analysis of the regulatory systems controlling lipogenesis in hepatocytes of mice and in maturing oilseeds of Arabidopsis. *Comptes Rendus en Biologies* **331**, 737–745.

Beer, T., Grant, T., Williams, D. and Watson, H. (2002). Fuel-cycle greenhouse gas emissions from alternative fuels in Australian heavy vehicles. *Atmospheric Environment* **36**, 753–763.

Beerens, P. (2007). Screw-pressing of *Jatropha* seeds for fueling purposes in less developed countries. Master of Science dissertation Eindhoven University of Technology.

Belafi-bako, K., Kovacs, F., Gubicza, L. and Hancsok, J. (2002). Enzymatic biodiesel production from sunflower oil by *Candida antarctica* lipase in a solvent-free system. *Biocatalisys and Biotransformation* **20**, 437–439.

Benavides, A., Montoro, P., Bassarello, C., Piacente, S. and Pizza, C. (2006). Catechin derivatives in *Jatropha macrantha* stems: Characterisation and LC/ESI/MS/MS quali-quantitative analysis. *Journal of Pharmaceutical and Biomedical Analysis* **40**, 639–647.

Berchmans, H. J. and Hirata, S. (2008). Biodiesel production from crude *Jatropha curcas* L. seed oil with a high content of free fatty acids. *Bioresource Technology* **99**, 1716–1721.

Berrios, M., Siles, J., Martín, M. A. and Martín, A. (2007). A kinetic study of the esterification of free fatty acids (FFA) in sunflower oil. *Fuel* **86**, 2383–2388.

Berry, E. W. (1929). An eogene tropical forest in the Peruvian desert. *Proceedings of the National Academy of Sciences USA* **15**, 345–346.

Boehman, A. L., Morris, D. and Szybist, J. (2004). The impact of the bulk modulus of diesel fuels on fuel injection timing. *Energy and Fuels* **18**, 1877–1882.

Bouaid, A., Bajo, L., Martinez, M. and Aracil, J. (2007). Optimization of biodiesel production from Jojoba oil. *Transactions Institution Chemical Engineers part B* **85**, 378–382.

Bourne, J. K., Jr. (2007). Biofuels: Green dreams. *National Geographic Magazine* (October), 41–59.

Brum, R., Honda, N. K., Mazarin, S. M., Hess, S. C., Cavalheiro, A. J. and Delle Monache, F. (1998). Jatrowedione, a lathyrane diterpene from *Jatropha weddeliana*. *Phytochemistry* **48**, 1225–1227.

Cahoon, E. B. and Kinney, A. J. (2005). The production of vegetable oils with novel properties: Using genomic tools to probe and manipulate plant fatty acid metabolism. *European Journal of Lipid Science and Technology* **107**, 239–243.

Cahoon, E. B. and Schmid, K. M. (2008). Metabolic engineering of the content and fatty acid composition of vegetable oils. *Advances in Plant Biochemistry and Molecular Biology* **1**, 161–200.

Cahoon, E. B., Marrillia, E. F., Stecca, K. L., Hall, S. E., Taylor, D. C. and Kinney, A. J. (2000). Production of fatty acid components of meadowfoam oil in somatic soybean embryos. *Plant Physiology* **124**, 243–251.

Cahoon, E. B., Shockey, J. M., Dietrich, C. R., Gidda, S. K., Mullen, R. T. and Dyer, J. M. (2007). Engineering oilseeds for sustainable production of industrial and nutritional feedstocks: Solving bottlenecks in fatty acid flux. *Current Opinion in Plant Biology* **10**, 236–244.

Canakci, M. and Van Gerpen, J. (1999). Biodiesel production via acid catalysis. *Transactions American Society of Agricultural Engineers* **42**, 1203–10.

Canakci, M. and Van Gerpen, J. (2001). Biodiesel production from oils and fats with high free fatty acids. *Transactions American Society of Agricultural Engineers* **44**, 1429–1436.

Canales, M., Hernández, T., Caballero, J., de Vivar, A. R., Avila, G., Duran, V. and Lira, R. (2005). Informant consensus factor and antibacterial activity of the medicinal plants used by the people of San Rafael Coxcatlán, Puebla, México. *Journal of Ethnopharmacology* **97**, 429–439.

Cano, L. M., Plumbley, R. A. and Hylands, P. J. (1989). Purification and partial characterization of a hemagglutinin from seeds of *Jatropha curcas*. *Journal of Food Biochemistry* **13**, 1–20.

Canoira, L., Alcántara, R., García-Martínez, M. J. and Carrasco, J. (2006). Biodiesel from Jojoba oil-wax: Transesterification with methanol and properties as a fuel. *Biomass and Bioenergy* **30**, 76–81.

Cardone, M., Prati, M. V., Rocco, V., Seggiani, M., Senature, A. and Vitolo, S. (2002). *Brassica carinata* as an alternative oil crop for the production of biodiesel in Italy: Engine performance and regulated and unregulated exhaust emission. *Environmental Science and Technology* **36**, 4656–4662.

Carels, N. (2005). The genome organization of angiosperms. *In* "Recent Research Developments in Plant Science" (S. G. Pandalai, ed.), pp. 129–194. Research Signpost, Trivandrum, Kerala, India.

Carvalho, A. M. X., Moreira, B. C., Silva, M. C. S., Silva, N. D. and Kasuya, M. C. M. (2007). Fungos Micorrízicos Arbusculares em Plantios de Pinhão Manso (*Jatropha curcas* L.). XXXI Congresso Brasileiro de Ciência do Solo. Conquistas e Desafios da Ciência do Solo Brasileira Gramado-RS. http://www6.ufrgs.br/cbcs/trabalhos/trabalhos/trab_9897-1924.pdf.

Carvalho, C. R., Clarindo, W. R., Praça, M. M., Araújo, F. S. and Carels, N. (2008). Genome size, base composition and karyotype of *Jatropha curcas* L., an important biofuel plant. *Plant Science* **174**, 613–617.

Chan, A. P., Melake, A., Zhao, Q., Orvis, J., Wortman, J., Fraser-Liggett, C., Ravel, J. and Rabinowicz, P. D. (2008). The genome of an oilseed plant – castor bean (*Ricinus communis*). Plant and Animal Genomes XVI Conference, San Diego, CA. http://www.intl-pag.org/16/abstracts/PAG16_W21_157.html.

Chatterjee, A., Das, B., Adityachaudhury, N. and Debkirtaniya, S. (1980). Note on the insecticidal properties of the seeds of *Jatropha gossypifolia. Indian Journal of Agricultural Sciences* **50**, 637–638.

Chen, J. Q., Huang, Y. Z., Lang, C. X., Hu, Z. H. and Liu, Z. H. (1999a). Molecular cloning and sequencing of the PEP gene from *Brassia napus* and the construction of the antisense PEP gene. *Journal of Zhejiang Univiversity* **25**, 365–367.

Chen, J. Q., Lang, C. X., Hu, Z. H., Liu, Z. H. and Huang, Z. (1999b). Antisense PEP gene regulates t ratio of protein and lipid content in *Brassica napus* seeds. *Chinese Journal of Agricultural Biotechnology* **7**, 316–320.

Chen, G. Q., Turner, C., He, X., Nguyen, T., McKeon, T. A. and Laudencia-Chingcuanco, D. (2007). Expression profiles of genes involved in fatty acid and triacylglycerol synthesis in castor bean (*Ricinus communis* L.). *Lipids* **42**, 263–274.

Cheng, C. H., Cheung, C. S., Chan, T. L., Lee, C., Yao, D. and Tsang, K. S. (2008). Comparison of emissions of a direct injection diesel engine operating on biodiesel with emulsified and fumigated methanol. *Fuel* **87**, 1870–1879.

Chhabra, S. C., Mahunnah, R. L. A. and Mshiu, E. N. (1990). Plants used in Traditional Medicine in Eastern Tanzania. III. Angiosperms (*Euphorbiaceae–Menispermaceae*). *Journal of Ethnopharmacology* **28**, 255–283.

Chow, K.-S. (2008). Transcriptomics analysis of *Hevea brasiliensis* rubber tree latex. Plant and Animal Genomes XVI Conference, San Diego, CA. http://www.intl-pag.org/16/abstracts/PAG16_W21_156.html.

Claude, S. (1999). Research of new outlets for glycerol – recent developments in France. *Lipid* **101**, 101–104.

Darussalam, B. (2007). Last-Gasp effort to save Borneo's tropical rainforests. *Science* **317**, 192.

Das, B. and Anjani, G. (1999). Gossypidien, a lignan from stems of *Jatropha gossypifollia. Phytochemistry* **51**, 115–117.

Das, B. and Das, R. (1995). Gossypifan, a lignan from *Jatropha gossypifoli. Phytochemistry* **40**, 931–932.

Das, B. and Venkataiah, B. (1999). A rare diterpene from *Jatropha gossypifolia. Biochemical Systematics and Ecology* **27**, 759–760.

Datta, M. M., Mukherjee, P., Ghosh, B. and Jha, T. B. (2007). *In vitro* clonal propagation of biodiesel plant (*Jatropha curcas* L.). *Current Science* **93**, 1438–1442.

Davis, M. S., Solbiati, J. and Cronan, J. E. (2000). Overproduction of Acetyl-CoA carboxylase activity increases the rate of fatty acid biosynthesis in *Escherichia coli. Journal of Biological Chemistry* **275**, 28593–28598.

de Lima, M. R. F., Luna, J., dos Santos, A. F., de Andrade, M. C. C., Sant'Ana, A. E. G., Genet, J.-P., Marquez, B., Neuville, L. and Moreau, N. (2006). Anti-bacterial activity of some Brazilian medicinal plants. *Journal of Ethnopharmacology* **105**, 137–147.

de Oliveira Lima, J. R., da Silva, R. B., de Moura, E. M. and de Moura, C. V. R. (2008). Biodiesel of tucum oil, synthesized by methanolic and ethanolic routes. *Fuel* **87**, 1718–1723.

Dorado, M. P., Ballesteros, E., Arnal, J. M., Gomez, J. and Lopez, F. J. (2003a). Exhaust emissions from a diesel engine fueled with transesterified waste olive oil. *Fuel* **82**, 1311–1315.

Dorado, M. P., Ballesteros, E., Arnal, J. M., Gomez, J. and Gimenez, F. J. L. (2003b). Testing waste olive oil methyl ester as a fuel in a diesel engine. *Energy and Fuels* **17**, 1560–1565.

Du, W., Xu, Y., Liu, D. and Zeng, J. (2004). Comparative study on lipasecatalyzed transformation of soybean oil for biodiesel production with different acyl acceptors. *Journal of Molecular Catalysis B: Enzymatic* **30**, 125–129.

Durbin, T. D. and Norbeck, J. M. (2002). The effects of biodiesel blends and ARCO EC-diesel on emissions from light heavyduty diesel vehicles. *Environmental Science and Technology* **36**, 1686–1691.

Durbin, T. D., Collins, J. R., Norbeck, J. M. and Smith, M. R. (2000). Effects of biodiesel, biodiesel blends, and a synthetic diesel on emissions from light heavy-duty diesel vehicles. *Environmental Science and Technology* **34**, 349–355.

El-Badwi, D. M. A., Adam, S. E. I. and Hapke, H. J. (1995). Comparative toxicity of *Ricinus Communis* and *Jatropha curcas* in Brown Hisex Chicks. *Deutsche Tierärztliche Wochenschrift* **102**, 75–77.

Fagbenro-Beyioku, A. F., Oyibo, W. A. and Anuforom, B. C. (1998). Disinfectant/antiparasitic activities of *Jatropha curcas*. *East African Medical Journal* **75**, 508–511.

Fairless, D. (2007). The little shrub that could – maybe. *Nature* **449**, 652–655.

Faria, M. H. G., Carvalho, T. G., Rabenhorst, S. H. B., Sidrim, J. J. C. and Moraes-Filho, M. O. (2006). Cytotoxic and antifungal properties of medicinal plants from Ceara, Brazil. *Brazilian Journal of Biology* **66**, 1133–1135.

Fernando, S., Hall, C. and Jha, S. (2006). NO_x reduction from biodiesel fuels. *Energy and Fuels* **20**, 376–382.

Foidl, N., Foidl, G., Sanchez, M., Mittelbach, M. and Hackel, S. (1996). *Jatropha curcas* L. as a source for the production of biofuel in Nicaragua. *Bioresource Technology* **58**, 77–82.

Francis, G., Edinger, R. and Becker, K. A. (2005). Concept for simultaneous wasteland reclamation, fuel production, and socioeconomic development in degraded areas in India: Need, potential and perspectives of *Jatropha* plantations. *Natural Resources Forum* **29**, 12–24.

Frazier, M. E., Johnson, G. M., Thomassen, D. G., Oliver, C. E. and Patrinos, A. (2003). Realizing the potential of the genome revolution: The Genomes to Life Program. *Science* **300**, 290–293.

Freedman, B., Pryde, E. H. and Mounts, T. L. (1984). Variables affecting the yields of fatty esters form transesterified vegetable oils. *Journal of the American Oil Chemistry Society* **61**, 1638–1643.

Freedman, B., Butterfield, R. O. and Pryde, E. H. (1986). Transesterification kinetics of soybean oil. *Journal of the American Oil Chemistry Society* **63**, 1375–1380.

Fregene, M., Moreno, M. I., Ospina, C., Marin, J., Hurtado, P., Santos, L. G., Barrera, E., Gutierrez, J. P., Okogbenin, E., Egesi, C., Mba, C. Buitrago, C. *et al.* (2007). Cassava Genomics research at CIAT. Plant and Animal Genomes XV Conference San Diego, CA. http://www.intl-pag.org/15/abstracts/PAG15_W18_117.html.

Freire, F. C. O. and Parente, G. B. (2006). As doenças das Jatrofas (*Jatropha curcas* L. e *J. podagrica* Hook.) no estado do Ceará. Comunicado Técnico Embrapa 120. http://www.cnpat.embrapa.br/home/down/index.php?pub/cot_120.pdf.

Gaillard, Y., Krishnamoorthy, A. and Bevalot, F. (2004). *Cerbera odollam*: A 'suicide tree' and cause of death in the state of Kerala, India. *Journal of Ethnopharmacology* **95**, 123–126.

Gandhi, V. M., Cherian, K. M. and Mulky, M. J. (1995). Toxicological Studies on Ratanjyot Oil. *Food and Chemical Toxicology* **33**, 39–42.

Ganesh Ram, S., Parthiban, K. T., Senthil, K. R., Thiruvengadam, V. and Paramathma, M. (2008). Genetic diversity among *Jatropha* species as revealed by RAPD markers. *Genetic Resources and Crop Evolution* **55**, 803–809. doi:10.1007/s10722-007-9285-7.

Gao, Y.-Y., Chen, W.-W., Lei, H., Liu, Y., Lin, X. and Ruan, R. (in press). Optimization of transesterification conditions for the production of fatty acid methyl ester (FAME) from Chinese tallow kernel oil with surfactant-coated lipase. *Biomass and Bioenergy* doi:10.1016/j.biombioe.2008.05.013.

Gbeassor, M., Kossou, Y., Amegbo, K., De Souza, C., Koumaglo, K. and Denke, A. (1989). Antimalarial effects of eight African medicinal plants. *Journal of Ethnopharmacology* **25**, 115–118.

Georges, K., Jayaprakasam, B., Dalavoy, S. S. and Nair, M. G. (2008). Pest-managing activities of plant extracts and anthraquinones from *Cassia nigricans* from Burkina Faso. *Bioresource Technology* **99**, 2037–2045.

Ghadge, S. V. and Raheman, H. (2005). Biodiesel production from mahua (*Madhumica indica*) oil having high free fatty acids. *Biomass Bioenergy* **28**, 601–605.

Gheerbrant, E. and Rage, J.-C. (2006). Paleobiogeography of Africa: How distinct from Gondwana and Laurasia? *Palaeogeography, Palaeoclimatology, Palaeoecology* **241**, 224–246.

Goldemberg, J., Coelho, S. T. and Guardabassi, P. (2008). The sustainability of ethanol production from sugarcane. *Energy Policy* **36**, 2086–2097.

Goodrum, J. W. and Geller, D. P. (2005). Influence of fatty acid methyl esters from hydroxylated vegetable oils on diesel fuel lubricity. *Bioresource Technology* **96**, 851–855.

Goonasekera, M. M., Gunawardana, V. K., Jayasena, K., Mohammed, S. G. and Balasubramaniam, S. (1995). Pregnancy terminating effect of *Jatropha curcas* in rats. *Journal of Ethnopharmacology* **47**, 117–123.

Gressel, J. (2008). Transgenics are imperative for biofuel crops. *Plant Science* **174**, 246–263.

Gu, Q. M., Chen, C. S. and Sih, C. J. (1986). A facile enzymatic resolution process for the preparation of (+) S 2 (6methoxy 2 naphthyl)propionic acid (Naproxen). *Tetrahedron Letters* **27**, 1763–1766.

Gui, M. M., Lee, K. T. and Bhatia, S. (2008). Feasibility of edible oil vs. non-edible oil vs. waste edible oil as biodiesel feedstock. *Energy* **33**, 1646–1653. doi:10.1016/j.energy.2008.06.002.

Haas, W. and Mittelbach, M. (2000). Detoxification experiments with the seed oil from *Jatropha curcas* L. *Industrial Crops and Products* **12**, 111–118.

Haas, M. J. and Scott, K. M. (2007). Moisture removal substantially improves the efficiency of *in situ* biodiesel production from soybeans. *Journal of the American Oil Chemistry Society* **84**, 197–204.

Haas, M. J., Scott, K. M., Marmer, W. N. and Foglia, T. A. (2004). *In situ* alkaline transesterification: An effective method for the production of fatty acid esters from vegetable oils. *Journal of the American Oil Chemistry Society* **81**, 83–89.

Haas, M. J., Scott, K. M., Foglia, T. A. and Marmer, W. N. (2007). The general applicability of *in situ* transesterification for the production of fatty acid esters from a variety of feedstocks. *Journal of the American Oil Chemistry Society* **84**, 963–970.

Haldar, S. K., Ghosh, B. B. and Nag, A. (in press). Studies on the comparison of performance and emission characteristics of a diesel engine using three degummed non-edible vegetable oils. *Biomass and Bioenergy* doi:10.1016/j.biombioe.2008.01.021.

Hamza, O. J. M., van den Bout-van den Beukel, C. J. P., Matee, M. I. N., Moshi, M. J., Mikx, F. H. M., Selemani, H. O., Mbwambo, Z. H., Van der Ven, A. J. A. M. and Verweij, P. E. (2006). Antifungal activity of some Tanzanian plants used traditionally for the treatment of fungal infections. *Journal of Ethnopharmacology* **108**, 124–132.

Hassanien, F. R. and Mukherjee, K. D. (1986). Isolation of lipase from germinating oilseeds for biotechnological processes. *Journal of the American Oil Chemistry Society* **63**, 893–897.

Heath, D. D., Iwana, G. K. and Delvin, R. H. (1993). PCR primed with VNTR core sequences yield species specific patterns and hypervariable probes. *Nucleic Acids Research* **21**, 5782–5785.

Henning, R. K. (2005). Assessment of the impact of the dissemination of "the Jatropha System" on the ecology of the rural area and the social and economic situation of the rural population (target group) in selected countries in Africa. http://www.underutilized-species.org/Documents/PUBLICATIONS/jatropha_curcas_africa.pdf.

Herman, E. M. (2008). Endoplasmic reticulum bodies: Solving the insoluble. *Current Opinion in Plant Biology* **11**, 672–679. doi:10.1016/j.pbi.2008.08.004.

Hernández-Martín, E. and Otero, C. (2008). Different enzyme requirements for the synthesis of biodiesel: Novozym 435 and Lipozyme TL IM. *Bioresource Technology* **99**, 277–286.

Hirota, M., Suttajit, M., Suguri, H., Endo, Y., Shudo, K., Wongchai, V., Hecker, E. and Fujiki, H. (1988). A new tumor promoter from the seed oil of *Jatropha curcas* L., an intramolecular diester of 12-deoxy-16-hydroxyphorbol. *Cancer Research* **48**, 5800–5804.

Holter, N. S., Maritan, A., Cieplak, M., Fedoroff, N. V. and Banavar, J. R. (2000). Dynamic modeling of gene expression data. *Proceedings of the National Academy of Sciences USA* **98**, 1693–1698.

Horiuchi, T., Fujiki, H., Hirota, M., Suttajit, M., Suganuma, M., Yoshioka, A., Wongchai, V., Hecker, E. and Sugimura, T. (1987). Presence of tumor promoters in the seed oil of *Jatropha curcas* L. from Thailand. *Japanese Journal of Cancer Research* **78**, 223–226.

Jian-Xun, W., Qing-De, H., Feng-Hong, H., Jiang-Wei, W. and Qin-Jie, H. (2007). Lipase-catalyzed production of biodiesel from high acid value waste oil using ultrasonic assistant. *Chinese Journal of Biotechnology* **23**, 1121–1128.

Jongschaap, R. E. E., Corré, W. J., Bindraban, P. S. and Brandenburg, W. A. (2007). Claims and facts on *Jatropha curcas* L. Plant Research International B.V., Wageningen, The Netherlands. http://www.fact-fuels.org/media_en/Claims_and_Facts_on_Jatropha_-WUR?session = isgsklbna58j7grrfst888n5r7.

Joubert, P. H., Brown, J. M. M., Hay, I. T. and Sebata, P. D. B. (1984). Acute poisoning with *Jatropha curcas* (purging nut tree) in children. *South African Medical Journal* **65**, 729–730.

Kalam, M. A., Husnawan, M. and Masjuki, H. H. (2003). Exhaust emission and combustion evaluation of coconut oil-powered indirect injection diesel engine. *Renewable Energy* **28**, 2405–2415.

Kalimuthu, K., Paulsamy, S., Senthilkumar, R. and Sathya, M. (2007). *In vitro* propagation of the biodiesel plant *Jatropha curcas* L. *Plant Tissue Culture and Biotechnology* **17**, 137–147.

Kalligeros, S., Zannikos, F., Stroumas, S., Lois, E., Anastopoulos, G., Teas, C. H. and Sakellarppoulos, F. (2003). An investigation of using biodiesel/matine diesel blends on the performance of a stationary diesel engine. *Biomass and Bioenergy* **24**, 141–149.

Karmegam, N., Sakthivadivel, M., Anuradha, V. and Daniel, T. (1997). Indigenous-plant extracts as larvicidal agents against *Culex quinquefasciatus* Say. *Bioresource Technology* **59**, 137–140.

Kaushik, N., Roy, S. and Biswas, G. C. (2006). Screening of Indian germplasm of *Jatropha curcas* for selection of high oil yielding plants. *Indian Journal of Agroforestry* **8**, 54–57.

Kaushik, N., Kumar, K., Kumar, S., Kaushik, N. and Roy, S. (2007). Genetic variability and divergence studies in seed traits and oil content of Jatropha (*Jatropha curcas* L.) accessions. *Biomass and Bioenergy* **31**, 497–502.

Kegl, B. (2006). Numerical analysis of injection characteristics using biodiesel fuel. *Fuel* **85**, 2377–2387.

Kegl, B. (2008). Effects of biodiesel on emissions of a bus diesel engine. *Bioresource Technology* **99**, 863–873.

Kelkar, M. A., Gogate, P. R. and Pandit, A. B. (2008). Intensification of esterification of acids for synthesis of biodiesel using acoustic and hydrodynamic cavitation. *Ultrasonics Sonochemistry* **15**, 188–194.

Kinghorn, A. D. and Evans, F. J. (1975). A biological screen of selected species of genus *Euphorbia* for skin irritant effect. *Planta Medica* **28**, 325–335.

Klein-Douwel, R. J. H., Donkerbroek, A. J., van Vliet, A. P., Boot, M. D., Somers, L. M. T., Baert, R. S. G., Dam, N. J. and ter Meulen, J. J. (2009). Soot and chemiluminescence in diesel combustion of bio-derived, oxygenated and reference fuels. *Proceedings of the Combustion Institute* **32**, 1–17. doi:10.1016/j.proci.2008.06.140.

Knothe, G. and Steidley, K. R. (2005). Kinematic viscosity of biodiesel fuel components and related compounds, influence of compound structure and comparison to petrodiesel fuel components. *Fuel* **84**, 1059–1065.

Kochhar, S., Singh, S. P. and Kochhar, V. K. (2008). Effect of auxins and associated biochemical changes during clonal propagation of the biofuel plant – *Jatropha curcas*. *Biomass and Bioenergy* **32**, 1136–1143. doi:10.1016/j.biombioe.2008.02.014.

Koehler, I., Jenett-Siems, K., Siems, K., Hernandez, M. A., Ibarra, R. A., Berendsohn, W. G., Bienzle, U. and Eich, E. (2002). In vitro antiplasmodial investigation of medicinal plants from El Salvador. *Zeitschrift für Naturforschung* **57C**, 277–81.

Kosasi, S., van der Sluis, W. G., Boelens, R., 't Hart, L. A. and Labadie, R. P. (1989). Labaditin, a novel cyclic decapeptide from the latex of *Jatropha multzfida* L. (*Euphorbiaceae*). Isolation and sequence determination by means of two-dimensional NMR. *FEBS Letters* **256**, 91–96.

Krisnangkura, K., Yimsuwan, T. and Pairintra, R. (2006). An empirical approach in predicting biodiesel viscosity at various temperatures. *Fuel* **85**, 107–113.

Kumar, A. and Sharma, S. (2008). An evaluation of multipurpose oil seed crop for industrial uses (*Jatropha curcas* L.): A review. *Industrial Crops and Products* **28**, 1–10.

Kumar, A., Bhattacharya, S. C. and Pham, H. L. (2003). Greenhouse gas mitigation potential of biomass energy technologies in Vietnam using the long range energy alternative planning system model. *Energy* **28**, 627–654.

Kumar, M. S., Kerihuel, A., Bellttre, J. and Tazerout, M. (2006). Ethanol animal fat emulsions as a diesel engine fuel – Part 2: Engine test analysis. *Fuel* **85**, 2646–2652.

Kupchan, S. M., Sigel, C. W., Matz, M. J., Gilmore, C. J. and Bryan, R. F. (1976). Structure and stereochemistry of jatrophone, a novel macrocyclic diterpenoid tumor inhibitor. *Journal of American Chemical Society* **98**, 2295–2300.

Laurance, W. F. (2007). Switch to corn promotes amazon deforestation. *Science* **318**, 1721.

Lee, C. S., Park, S. W. and Kwon, S. I. (2005). An experimental study on the atomization and combustion characteristics of biodiesel-blended fuels. *Energy and Fuels* **19**, 2201–2208.

Legowo, E. H. (2007). Development of alternative energy in Indonesia. 5th Asian Petroleum Technology Symposium Jakarta.

Lespinasse, D., Rodier-Goud, M., Grivet, L., Leconte, A., Legnate, H. and Seguin, M. (2000). A saturated genetic linkage map of rubber tree (*Hevea* spp.) based on RFLP, AFLP, microsatellite, and isozyme markers. *Theoretical and Applied Genetics* **100**, 127–138.

Li, J., Yan, F., He, W.-Z., Xiao, M., Chen, Y.-Y. and Chen, F. (2005). Toxicity and mechanism of Jatropherol I to silkworm, *Bombys mori* L. *Nongyaoxue Xuebao* **7**, 29–34.

Li, X., Lu, G., Guo, Y., Guo, Y., Wang, Y., Zhang, Z., Liu, X. and Wang, Y. (2007). A novel solid superbase of Eu_2O_3/Al_2O_3 and its catalytic performance for the transesterification of soybean oil to biodiesel. *Catalysis Communications* **8**, 1969–1972.

Li, M., Li, H., Jiang, H., Pan, X. and Wu, G. (2008). Establishment of an Agrobacterium-mediated cotyledon disc transformation method for *Jatropha curcas*. *Plant Cell Tissue and Organ Culture* **92**, 173–181. doi:10.1007/s11240-007-9320-6.

Lin, C. Y. and Lin, H. A. (2006). Diesel engine performance and emission characteristics of biodiesel characteristics of biodiesel produced by the peroxidation process. *Fuel* **85**, 298–305.

Lin, J., Yan, F., Tang, L. and Chen, F. (2003). Antitumor effects of curcin from seeds of *Jatropha curcas*. *Acta Pharmacologica Sinica* **24**, 241–246.

Lin, Y. C., Lee, W. J. and Hou, H. C. (2006a). PAH emissions and energy efficiency of palm-biodiesel blends fueled on diesel generator. *Atmospheric Environment* **40**, 3930–3940.

Lin, Y. C., Lee, W. J., Wu, T. S. and Wang, C. T. (2006b). Comparison of PAH and regulated harmful matter emissions from biodiesel blends and paraffinic fuel blends on engine accumulated mileage test. *Fuel* **85**, 2516–2523.

Lindqvist, Y., Huang, W., Schneider, G. and Shanklin, J. (1996). Crystal structure of delta 9 stearoyl-acyl carrier protein desaturase from castor seed and its relationship to other di-iron proteins. *EMBO Journal* **15**, 4081–4092.

Liu, K. (1994). Preparation of fatty acid methyl esters for gas-chromatographic analysis of lipids in biological materials. *Journal of the American Oil Chemistry Society* **71**, 1179–87.

Liu, S. Y., Sporer, F., Wink, M., Jourdane, J., Henning, R., Li, Y. L. and Ruppel, A. (1997). Anthraquinones in *Rheum palmatum* and *Rumex dentatus* (*Polygonaceae*), and phorbol esters in *Jatropha curcas* (*Euphorbiaceae*) with molluscicidal activity against the schistosome vector snails Oncomelania, *Biomphalaria* and *Bulinus*. *Tropical Medicine and International Health* **2**, 179–188.

Lokko, Y., Anderson, J. V., Rudd, S., Raji, A., Horvath, D., Mikel, M. A., Kim, R., Liu, L., Hernandez, A., Dixon, A. G. and Ingelbrecht, I. L. (2007). Characterization of an 18,166 EST dataset for cassava (*Manihot esculenta* Crantz) enriched for drought-responsive genes. *Plant Cell Report* **26**, 1605–1618.

Lopez, C., Piégu, B., Cooke, R., Delseny, M., Tohme, J. and Verdier, V. (2005). Using cDNA and genomic sequences as tools to develop SNP strategies in cassava (*Manihot esculenta* Crantz). *Theoretical and Applied Genetics* **110**, 425–431.

Ma, F. and Hanna, M. A. (1999). Biodiesel production: A review. *Bioresource Technology* **70**, 1–15.

Madhumathi, S., Mohan, M. S. S. and Radha, R. (2000). Anti-microbial properties of crude fruit extracts of *Jatropha gossypifolia*. *Journal of Medicinal and Aromatic Plant Science* **22**, 717–20.

Mahanta, N., Gupta, A. and Khare, S. K. (2008). Production of protease and lipase by solvent tolerant *Pseudomonas aeruginosa* PseA in solid-state fermentation using *Jatropha curcas* seed cake as substrate. *Bioresource Technology* **99**, 1729–1735.

Makkar, H. P. S., Becker, K., Sporer, F. and Wink, M. (1997). Studies on nutritive potential and toxic constituents of different provenances of *Jatropha curcas*. *Journal of Agricultural and Food Chemistry* **45**, 3152–3157.

Makkar, H. P. S., Aderibigbe, A. O. and Becker, K. (1998a). Comparative evaluation of non-toxic and toxic varieties of *Jatropha curcas* for chemical composition, digestibility, protein degradability and toxic factors. *Food Chemistry* **62**, 207–215.

Makkar, H. P. S., Becker, K. and Schmook, B. (1998b). Edible provenances of *Jatropha curcas* from Quintana Roo state of Mexico and effect of roasting on antinutrient and toxic factors in seeds. *Plant Foods for Human Nutrition* **52**, 31–36.

Makkar, H. P. S., Francis, G. and Becker, K. (2007). Bioactivity of phytochemicals in some lesser-known plants and their effects and potential applications in livestock and aquaculture production systems. *Animal* **1**, 1371–1391.

Malhi, Y., Roberts, J. T., Betts, R. A., Killeen, T. J., Li, W. and Nobre, C. A. (2008). Climate change, deforestation, and the fate of the Amazon. *Science* **319**, 169–172.

Mampane, K. J., Joubert, P. H. and Hay, I. T. (1987). *Jatropha curcas*: Use as a traditional tswana medicine and its role as a cause of acute poisoning. *Phytotherapy Research* **1**, 50–51.

Marchetti, J., Miguel, V. and Errazu, A. (2007). Possible methods for biodiesel production. *Renewable and Sustainable Energy Reviews* **11**, 1300–1311.

Marquez, B., Neuville, L., Moreau, N. J., Genet, J.-P., dos Santos, A. F., de Andrade, M. C. C. and Sant'Ana, A. E. G. (2005). Multidrug resistance reversal agent from *Jatropha elliptica*. *Phytochemistry* **66**, 1804–1811.

Martínez-Herrera, J., Siddhuraju, P., Francis, G., Dávila-Ortíz, G. and Becker, K. (2006). Chemical composition, toxic/antimetabolic constituents, and effects of different treatments on their levels, in four provenances of *Jatropha curcas* L. from Mexico. *Food Chemistry* **96**, 80–89.

Matsuse, I. T., Lim, Y. A., Hattori, M., Correa, M. and Gupta, M. P. (1999). A search for anti-viral properties in Panamanian medicinal plants. The effects on HIV and its essential enzymes. *Journal of Ethnopharmacology* **64**, 15–22.

McCormick, R. L., Graboski, M. S., Alleman, T. L. and Herring, A. M. (2001). Impact of biodiesel source material and chemical structure on emissions of criteria pollutants from a heavy-duty engine. *Environmental Science and Technology* **35**, 1742–1747.

Meher, L. C., Dharmagadda, V. S. S. and Naik, S. N. (2005). Optimization of alkali-catalyzed transesterification of *Pongamia pinnata* oil for production of biodiesel. *Bioresource Technology* **96**, 1425–1429.

Modi, M. K., Reddy, J. R. C., Rao, B. V. S. K. and Prasad, R. B. N. (2007). Lipase-mediated conversion of vegetable oils into biodiesel using ethyl acetate as acyl acceptor. *Bioresource Technology* **98**, 1260–1264.

Mothana, R. A. A. and Lindequist, U. (2005). Antimicrobial activity of some medicinal plants of the island Soqotra. *Journal of Ethnopharmacology* **96**, 177–181.

Muangman, S., Thippornwong, M. and Tohtong, R. (2005). Anti-metastatic effects of curcusone B, a diterpene from *Jatropha curcas*. *In Vivo (Attiki)* **19**, 265–268.

Mujumdar, A. M. and Misar, A. V. (2004). Anti-inflammatory activity of *Jatropha curcas* roots in mice and rats. *Journal of Ethnopharmacology* **90**, 11–15.

Nabi, Md.N. S., Akhter, Md.S. and Shahadat, M. M. Z. (2006). Improvement of engine emissions with conventional diesel fuel and diesel — biodiesel blends. *Bioresource Technology* **97**, 372–378.

Naengchomnong, W., Thebtaranonth, Y., Wiriyachitra, P., Okamoto, K. T. and Clardy, J. (1986). Isolation and structure determination of 4 novel diterpenes from *Jatropha curcas*. *Tetrahedron Letters* **27**, 2439–2442.

Nahar, N. M., Waris, A. and Azam, M. M. (2005). Prospects and potential of fatty acid methyl esters of some non-traditional seed oils for use as biodiesel in India. *Biomass and Bioenergy* **29**, 293–302.

Nelson, D. R., Schuler, M. A., Paquette, S. M., Werck-Reichhart, D. and Bak, S. (2004). Comparative genomics of rice and Arabidopsis. Analysis of 727 cytochrome P450 genes and pseudogenes from a monocot and a dicot. *Plant Physiology* **135**, 756–772.

Nwosu, M. O. and Okafor, J. I. (1995). Preliminary studies of the antifungal activities of some medicinal plants against *Basidiobolus* and some other pathogenic fungi. *Mycoses* **38**, 191–195.

Odebiyi, O. O. (1985). Antimicrobial and antifungal properties of the extractives of *Jatropha podagrica*. *Fitoterapia* **56**, 297–299.

Oduola, T., Avwioro, O. G. and Ayanniyi, T. B. (2005). Suitability of the leaf extract of *Jatropha gossypifolia* as an anticoagulant for biochemical and haematological analyses. *African Journal of Biotechnology* **4**, 679–681.

Ohlrogge, J. B., Roesler, K. R. and Shorrosh, B. S. (1999). Methods of increasing oil content of seeds. United States Patent 5925805. http://www.freepatentsonline.com/5925805.html.

Ojewole, J. A. O. and Odebiyi, O. O. (1984). Some studies on the pharmacology of Tetramethylpyrazine an alkaloid from the stem bark of *Jatropha podagrica*. *Fitoterapia* **55**, 213–216.

Okogbenin, E. and Fregene, M. (2003). Genetic mapping of QTLs affecting productivity and plant architecture in a full-sib cross from non-inbred parents in Cassava (*Manihot esculenta* Crantz). *Theoretical and Applied Genetics* **107**, 1452–1462.

Oliveira, A. L. A., Gioielli, L. A. and Oliveira, M. N. (1999). Hidrólise parcial enzimática da gordura de babaçu. *Ciência e Tecnologia de Alimentos* **19**, 270–276.

Openshaw, K. (2000). A review of *Jatropha curcas*: An oil plant of unfulfilled promise. *Biomass and Bioenergy* **19**, 1–15.

Osoniyi, O. and Onajobi, F. (2003). Coagulant and anticoagulant activities in *Jatropha curcas* latex. *Journal of Ethnopharmacology* **89**, 101–105.

Ozaki, E., Uragaki, T., Sakashita, K. and Sakimae, A. (1995). Enzymatic preparation of alkanedicarboxylic acid monoesters. *Chemistry Letters* **24**, 539–540.

Paquette, L. A., Yang, J. and Long, Y. O. (2002). Concerning the antileukemic agent jatrophatrione: The first total synthesis of a [5.9.5] tricyclic diterpene. *Journal of American Chemical Society* **124**, 6542–6543.

Pousa, G. P. A. G., Santos, A. L. F. and Suarez, P. A. Z. (2007). History and policy of biodiesel in Brazil. *Energy Policy* **35**, 5393–5398.

Pracharoenwattana, I. and Smith, S. M. (2008). When is a peroxisome not a peroxisome? *Trends in Plant Science* **13**, 522–525. doi:10.1016/j.tplants.2008.07.003.

Pradeep, V. and Sharma, R. P. (2007). Use of HOT EGR for NO_x control in a compression ignition engine fuelled with bio-diesel from *Jatropha* oil. *Renewable Energy* **32**, 1136–1154.

Pradhan, R. C., Naik, S. N., Bhatnagar, N. and Vijay, V. K. (2008). Moisture-dependent physical properties of jatropha fruit. *Industrial Crops and Products* doi:10.1016/j.indcrop.2008.07.002.

Puhan, S., Vedaraman, N., Ram, B. V. B., Sankarnarayanan, G. and Jeychandran, K. (2005b). Mahua oil (*Madhuca indica* seed oil) methyl ester as biodiesel-preparation and emission characteristics. *Biomass and Bioenergy* **28**, 83–87.

Purushothaman, K. K., Chandrasekharan, S., Cameron, A. F., Connoly, J. D., Labbé, C., Maltz, A. and Rycroft, D. S. (1979). Jatropolones A and B, new diterpenoids from the roots of *Jatropha gossipiifolia* (*Euphorbiaceae*). Crystal structure analysis of Jatropholone B. *Tetrahedron Letters* **2**, 979–980.

Raheman, H. and Ghadge, S. V. (2007). Performance of compression ignition engine with mahua (*Madhuca indica*) biodiesel. *Fuel* **86**, 2568–2573.

Raheman, H. and Phadatare, A. G. (2004). Diesel engine emissions and performance from blends of karanja methyl ester and diesel. *Biomass and Bioenergy* **27**, 393–397.

Rahman, M., Ismail, M. H. and Yin, L. T. (1990). Jatropholone and jatrophatrione two diterpenes from *Jatropha gossypifolia*. *Pertanika* **13**, 405–408.

Raina, A. K. and Gaikwad, B. R. (1987). Chemobotany of *Jatropha* species in India and further characterisation of curcas oil. *Journal of Oil Technology of India* **19**, 81–85.

Raju, A. J. S. and Ezradanam, V. (2002). Pollination ecology and fruiting behaviour in a monoecious species, *Jatropha curcas* L. (*Euphorbiaceae*). *Current Science* **83**, 1395–1398.

Rakshit, K. D., Darukeshwara, J., Raj, K. R., Narasimhamurthy, K., Saibaba, P. and Bhagya, S. (2008). Toxicity studies of detoxified *Jatropha* meal (*Jatropha curcas*) in rats. *Food and Chemical Toxicology* **46**, 3621–3625. doi: 10.1016/j.fct.2008.09.010.

Ramadhas, A. S., Jayaraj, S. and Muraleedharan, C. (2005). Biodiesel production from high FFA rubber seed oil. *Fuel* **84**, 335–40.

Ranade, S. A., Srivastava, A. P., Rana, T. S., Srivastava, J. and Tuli, R. (2008). Easy assessment of diversity in *Jatropha curcas* L. plants using two single-primer amplification reaction (SPAR) methods. *Biomass and Bioenergy* **32**, 533–540.

Rao, P. N. and Kumar, K. P. (2006). White fly management in cotton agro-ecocystem with botanical pesticides: An ecofriendly approach. *Journal of Ecophysiology and Occupational Health* **6**, 57–60.

Rashid, U., Anwar, F., Moser, B. R. and Knothe, G. (2008). *Moringa oleifera* oil: A possible source of biodiesel. *Bioresource Technology* **99**, 8175–8179.

Ravindranath, N., Ramesh, C. and Das, B. (2003). A rare dinorditerpene from *Jatropha curcas*. *Biochemical Systematics and Ecology* **31**, 431–432.

Ravindranath, N., Reddy, M. R., Mahender, G., Ramu, R., Kumar, K. R. and Das, B. (2004). Deoxypreussomerins from *Jatropha curcas*: Are they also plant metabolites? *Phytochemistry* **65**, 2387–2390.

Reddy, N. R. and Pierson, M. D. (1994). Reduction in antinutritional and toxic components in plant foods by fermentation. *Food Research International* **27**, 281–290.

Reddy, J. N. and Ramesh, A. (2006). Parametric studies for improving the performance of a *Jatropha* oil-fuelled compression ignition engine. *Renewable Energy* **31**, 1994–2016.

Regupathy, A. and Ayyasamy, R. (2006). Need for generating baseline data for monitoring insecticide resistance in leaf Webber cum fruit borer, *Pempelia morosalis* (Saalm Uller), the key pest of biofuel crop, *Jatropha curcas*. *Resistant Pest Management Newsletter* **16**, 1–5.

Rug, M. and Ruppel, A. (2000). Toxic activities of the plant *Jatropha curcas* against intermediate snail hosts and larvae of schistosomes. *Tropical Medicine and International Health* **5**, 423–430.

Sahoo, P. K., Das, L. M., Babu, M. K. G. and Naik, S. N. (2007). Biodiesel development from high acid value polanga seed oil and performance evaluation in a CI engine. *Fuel* **86**, 448–54.

Samukawa, T., Kaieda, M., Matsumoto, T., Ban, K., Kondo, A., Shimada, Y., Noda, H. and Fukuda, H. (2000). Pretreatment of immobilized *Candida antarctica* lipase for biodiesel production from plant oil. *Journal of Bioscience and Bioengineering* **90**, 180–183.

Sánchez-Medina, A., García-Sosa, K., May-Pat, F. and Peña-Rodríguez, L. M. (2001). Evaluation of biological activity of crude extracts from plants used in Yucatecan Traditional Medicine Part I. Antioxidant, antimicrobial and b-glucosidase inhibition activities. *Phytomedicine* **8**, 144–151.

Sankara, S. S., Maharajan, S. and Sulochana, N. (1971). Flavonoids of some euphorbiaceous plants. *Phytochemistry* **10**, 2548–2549.

Santos, C., Jaramillo, C., Bayona, G., Rueda, M. and Torres, V. (2008). Late Eocene marine incursion in north-western South America. *Palaeogeography, Palaeoclimatology, Palaeoecology* **264**, 140–146.

Santos Mendoza, M., Dubreucq, B., Baud, S., Parcy, F., Caboche, M. and Lepiniec, L. (2008). Deciphering gene regulatory networks that control seed development and maturation in Arabidopsis. *Plant Journal* **54**, 608–620.

Sarin, R., Sharma, M., Sinharay, S. and Malhotra, R. K. (2007). *Jatropha*–Palm biodiesel blends: An optimum mix for Asia. *Fuel* **86**, 1365–1371.

Saturnino, H. M., Pacheco, D. D., Kakida, J., Tominaga, N. and Gonçalves, N. P. (2005). Cultura do pinhão-manso (*Jatropha curcas* L.). *Informe Agropecuário* **26**, 44–78.

Sauerwein, M., Sporer, F. and Wink, M. (1993). Insect-toxicity of phorbol esters from *Jatropha curcas* seed oil. *Planta Medica* **59**, 686.

Scholz, V. and da Silva, J. N. (2008). Prospects and risks of the use of castor oil as a fuel. *Biomass and Bioenergy* **32**, 95–100.

Schuchardt, U., Sercheli, R. R. and Vargas, M. (1998). Transesterification of vegetable oils: A review. *Journal of the Brazilian Chemical Society* **9**, 199–210.

Schwab, A. W., Baghy, M. O. and Freedman, B. (1987). Preparation and properties of diesel fuels from vegetable oils. *Fuel* **66**, 1372.

Sellwood, C., Slabas, A. R. and Rawsthorne, S. (2000). Effects of manipulating expression of acetyl-CoA carboxylase I in *Brassica napus* L. embryos. *Biochemical Society Transactions* **28**, 598–600.

Senthil, K. M., Ramesh, A. and Nagalingam, B. (2003). An experimental comparison of methods to use methanol and *Jatropha* oil in a compression ignition engine. *Biomass and Bioenergy* **25**, 309–318.

Shah, S. and Gupta, M. N. (2007). Lipase catalyzed preparation of biodiesel from *Jatropha* oil in a solvent free system. *Process Biochemistry* **42**, 409–414.

Shah, S., Sharma, A. and Gupta, M. N. (2004). Extraction of oil from *Jatropha curcas* L. seed kernels by enzyme assisted three phase partitioning. *Industrial Crops and Products* **20**, 275–279.

Shah, S., Sharma, A. and Gupta, M. N. (2005). Extraction of oil from *Jatropha curcas* L. seed kernels by combination of ultrasonication and aqueous enzymatic oil extraction. *Bioresource Technology* **96**, 121–123.

Shah, S., Sharma, A. and Gupta, M. N. (2006). Preparation of cross-linked enzyme aggregates by using bovine serum albumin as a proteic feeder. *Analytical Biochemistry* **351**, 207–213.

Shanker, C. and Dhyani, S. K. (2006). Insect pests of *Jatropha curcas* L. and the potential for their management. *Current Science* **91**, 162–163.

Sharma, Y. C. and Singh, B. (2008). Development of biodiesel from karanja, a tree found in rural India. *Fuel* **87**, 1740–1742. doi:10.1016/j.fuel.2007.08.001.

Sharma, D. K., Pandey, A. K. and Lata (2008). Use of *Jatropha curcas* hull biomass for bioactive compost production. *Biomass and Bioenergy* **33**, 159–162. doi:10.1016/j.biombioe.2008.05.002.

Shimada, Y., Watanabe, Y., Sugihara, A. and Tominaga, Y. (2002). Enzymatic alcoholysis for biodiesel fuel production and application of the reaction to oil processing. *Journal of Molecular Catalysis B: Enzymatic* **17**, 133–142.

Singh, R. N., Vyas, D. K., Srivastava, N. S. L. and Narra, M. (2008). SPRERI experience on holistic approach to utilize all parts of *Jatropha curcas* fruit for energy. *Renewable Energy* **33**, 1868–1873.

Sirisomboon, P., Kitchaiya, P., Pholpho, T. and Mahuttanyavanitch, W. (2007). Physical and mechanical properties of *Jatropha curcas* L. fruits, nuts and kernels. *Biosystems Engeneering* **97**, 201–207.

Songjang, K. and Wimolwattanasarn, P. (2004). Antibacterial activity from the latex of *Jatropha curcas*. *Chiang Mai Journal of Science* **31**, 309–312.

Srivastava, P. K. and Verma, M. (2008). Methyl ester of karanja oil as an alternative renewable source energy. *Fuel* **87**, 1673–1677.

Staubmann, R., Ncube, I., Gübitz, G. M., Steiner, W. and Read, J. S. (1999). Esterase and lipase activity in *Jatropha curcas* L. seeds. *Journal of Biotechnology* **75**, 117–126.

Steynberg, A. P. and Nel, H. G. (2004). Clean coal conversion options using Fischer–Tropsch technology. *Fuel* **83**, 765–770.

Stone, R. (2007). Can palm oil plantations come clean? *Science* **317**, 1491.

Su, E.-Z., Xu, W.-Q., Gao, K.-L., Zheng, Y. and Wei, D.-Z. (2007). Lipase-catalyzed in situ reactive extraction of oilseeds with short-chained alkyl acetates for fatty acid esters production. *Journal of Molecular Catalysis B: Enzymatic* **48**, 28–32.

Suárez, M. C., Bernal, A., Gutiérrez, J., Tohme, J. and Fregene, M. (2000). Developing expressed sequence tags (ESTs) from polymorphic transcript-derived fragments (TDFs) in cassava (*Manihot esculenta* Crantz). *Genome* **43**, 62–67.

Subramanian, S. S., Nagarajan, S. and Sulochana, N. (1971). *Euphorbiaceae*. Flavonoids of some Euphorbiaceous plants. *Phytochemistry* **10**, 2548–2549.

Sudheer Pamidiamarri, D. V. N., Pandya, N., Reddy, M. P. and Radhakrishnan, T. (2008a). Comparative study of interspecific genetic divergence and phylogenic analysis of genus *Jatropha* by RAPD and AFLP genetic divergence and phylogenic analysis of genus *Jatropha*. *Molecular Biology Reporter* doi:10.1007/s11033-008-9261-0. http://www.springerlink.com/content/m642ml6440657117/.

Sudheer Pamidimarri, D. V. N., Singh, S., Mastan, S. G., Patel, J. and Reddy, M. P. (2008b). Molecular characterization and identification of markers for toxic and non-toxic varieties of *Jatropha curcas* L. using RAPD, AFLP and SSR markers. *Molecular Biology Reports* doi:10.1007/s11033-008-9320-6. http://www.springerlink.com/content/32l63012u767wh32/.

Sujatha, M., Reddy, T. P. and Mahasi, M. J. (2008). Role of biotechnological interventions in the improvement of castor (*Ricinus communis* L.) and *Jatropha curcas* L. *Biotechnology Advances* **26**, 424–435.

Sun, F. and Chen, H. (2008). Organosolv pretreatment by crude glycerol from oleochemicals industry for enzymatic hydrolysis of wheat straw. *Bioresource Technology* **99**, 5474–5479.

Sunil, N., Varaprasad, K. S., Sivaraj, N., Kumar, T. S., Abraham, B. and Prasad, R. B. N. (2008). Assessing *Jatropha curcas* L. germplasm *in situ* – A case study. *Biomass and Bioenergy* **32**, 198–202.

Sutthivaiyakit, S., Mongkolvisut, W., Ponsitipiboon, P., Prabpai, S., Kongsaeree, P., Ruchirawat, S. and Mahidol, C. (2003). A novel 8,9-seco-rhamnofolane and a new rhamnofolane endoperoxide from *Jatropha integerrima* roots. *Tetrahedron Letters* **44**, 3637–3640.

Tamalampudi, S., Talukder, M. R., Hama, S., Numata, T., Kondo, A. and Fukuda, H. (2008). Enzymatic production of biodiesel from *Jatropha* oil: A comparative study of immobilized-whole cell and commercial lipases as a biocatalyst. *Biochemical Engineering Journal* **39**, 185–189.

Taylor, M. D., Smith, A. B., III, Furst, G. T., Gunasekara, S. P., Bevelle, C. A., Cordell, G. A., Farnsworth, N. R., Kupchan, S. M., Uchida, H., Braufman, A. R., Daily, R. G., Jr. and Sneden, A. T. (1983). New antileukemic jatrophone derivatives from *Jatropha gossypiifolia*: Structural and stereochemical assignment through nuclear magnetic resonance spectroscopy. *Journal of American Chemical Society* **105**, 3177–3183.

Tequida-Meneses, M., Cortez-Rocha, M., Rosas-Burgos, E. C., Lopez-Sandoval, S. and Corrales-Maldonado, C. (2002). Effect of alcoholic extracts of wild plants on the inhibition of growth of *Aspergillus flavus, Aspergillus niger, Penicillium chrysogenum, Penicillium expansum, Fusarium moniliforme* and *Fusarium poae* moulds. *Revista Iberoamericana de Micología* **19**, 84–88.

Tiwari, A. K., Kumar, A. and Raheman, H. (2007). Biodiesel production from jatropha oil (*Jatropha curcas*) with high free fatty acids: An optimized process. *Biomass Bioenergy* **31**, 569–575.

Tong, L., Peng, S.-M., Deng, W.-Y., Ma, D.-W., Xu, Y., Xiao, M. and Chen, F. (2006). Characterization of a new stearoyl-acyl carrier protein desaturase gene from *Jatropha curcas*. *Biotechnology Letters* **28**, 657–662.

Torrance, S. J., Wiedhopf, R. M. and Cole, J. R. (1976). Antitumor agents from *Jatropha macrorhiza* (*Euphorbiaceae*). II. Isolation and characterization of jatrophatrione. *Journal of Organic Chemistry* **41**, 1855–1857.

Torrance, S. J., Wiedhopf, R. M. and Cole, J. R. (1977). Anti-Tumor agents from *Jatropha macrorhiza Euphorbiaceae*. Part 3. Acetyl aleuritolic-acid. *Journal of Pharmaceutical Science* **66**, 1348–1349.

USEPA (2002). A comprehensive analysis of biodiesel impacts on exhaust emissions. EPA420-P-02-001. http://www.biodiesel.org/resources/reportsdatabase/reports/gen/20021001_gen-323.pdf.

Valente, C., Pedro, M., Duarte, A., São José Nascimento, M., Abreu, P. M. and Ferreira, M.-J. U. (2004). Bioactive diterpenoids, a new jatrophane and two ent-abietanes, and other constituents from euphorbia pubescens. *Journal of Natural Products* **67**, 902–904.

Valliyappan, T., Bakhshi, N. N. and Dalai, A. K. (2008). Pyrolysis of glycerol for the production of hydrogen or syn gas. *Bioresource Technology* **99**, 4476–4483.

van de Loo, F. J., Broun, P., Turner, S. and Somerville, C. (1995). An oleate 12-hydroxylase from Ricinus communis L. is a fatty acyl desaturase homolog. *Proceedings of the National Academy of Sciences USA* **92**, 6743–6747.

van den Berg, A. J. J., Horsten, S. F. A. J., Kettenes-van den Bosch, J. J., Kroes, B. H., Beukelman, C. J., Leeflang, B. R. and Labadie, R. P. (1995). Curcacycline A – a novel cyclic octapeptide isolated from the latex of *Jatropha curcas* L. *FEBS Letters* **358**, 215–218.

van den Berg, A. J. J., Horsten, S. F. A. J., Kettenes-van den Bosch, J. J., Beukelman, C. J., Kroes, B. H., Leeflang, B. R. and Labadie, R. P. (1996). Podacycline A and B, two cyclic peptides in the latex of *Jatropha podagrica*. *Phytochemistry* **42**, 129–133.

Venter, O., Meijaard, E. and Wilson, K. (2008). Strategies and alliances needed to protect forest from palm-oil industry. *Nature* **451**, 16.

Veronique, J. B., Chornick, T. and James, K. D. (2002). Comparison of methods to measure the oil contents in oilseeds. *Journal of Oleo Science* **51**, 589–597.

Villegas, L. F., Fernfindez, I. D., Maldonado, H., Torres, R., Zavaleta, A., Vaisberg, A. J. and Hammond, G. B. (1997). Evaluation of the wound-healing activity of selected traditional medicinal plants from Peril. *Journal of Ethnopharmacology* **55**, 193–200.

Vyas, D. K. and Singh, R. N. (2007). Feasibility study of *Jatropha* seed husk as an open core gasifier feedstock. *Renewable Energy* **32**, 512–517.

Wang, W. G., Lyons, D. W., Clark, N. N., Gautam, M. and Norton, P. M. (2000). Emissions from nine heavy trucks fuelled by diesel and biodiesel blend without engine modification. *Environmental Science and Technology* **34**, 933–939.

Watanabe, Y., Pinsirodom, P., Nagao, T., Yamauchi, A., Kobayashi, T., Nishida, Y., Takagi, Y. and Shimada, Y. (2007). Conversion of acid oil by-produced in vegetable oil refining to biodiesel fuel by immobilized *Candida antarctica* lipase. *Journal of Molecular Catalysis B: Enzymatic* **44**, 99–105.

Welsh, J. and McClelland, M. (1990). Fingerprinting genomes using PCR with arbitrary primers. *Nucleic Acids Research* **18**, 7213–7218.

Wender, P. A., Kee, J.-M. and Warrington, J. M. (2008). Practical synthesis of Prostratin, DPP, and their analogs, adjuvant leads against latent HIV. *Science* **320**, 649–652.

Weyerhaeuser, H., Tennigkeit, T., Yufang, S. and Kahrl, F. (2007). Biofuels in China: An analysis of the opportunities and challenges of *Jatropha curcas* in South-west China. *ICRAF* Working Paper Number 53, 1–20. http://www.icuc-iwmi.org/files/Publications/WP53_jatropa.pdf.

Willems, P., Kuipers, N. J. M. and De Haan, A. B. (2008a). Gas assisted mechanical expression of oilseeds: Influence of process parameters on oil yield. *Journal of Supercritical Fluids* **45**, 298–305.

Willems, P., Kuipers, N. J. M. and De Haan, A. B. (2008b). Hydraulic pressing of oilseeds: Experimental determination and modeling of yield and pressing rates. *Journal of Food Engineering* **89**, 8–16.

Williams, J. G. K., Kubelik, A. R., Livak, K. J., Rafalski, J. A. and Tingey, S. V. (1990). DNA polymorphisms amplified by arbitrary primers are useful as genetic markers. *Nucleic Acids Research* **18**, 6531–6535.

Wurdack, K. J. (2008). Molecular evolution and phylogenetics of *Euphorbiaceae*: Beyond the model organisms. Plant and Animal Genomes XVI Conference San Diego, CA. http://www.intl-pag.org/16/abstracts/PAG16_W21_155.html.

Yadav, R. P. and Singh, A. (2006). Toxic effects of *Jatropha gosspifolia* and its binary and tertiary combinations with other plant molluscicides in natural ponds. *Iberus* **24**, 47–54.

Yagiz, F., Kazan, D. and Akin, A. N. (2007). Biodiesel production from waste oils by using lipase immobilized on hydrotalcite and zeolites. *Chemical Engineering Journal* **134**, 262–267.

Yamane, K., Ueta, A. and Shimamoto, Y. (2001). Influence of physical and chemical properties of biodiesel fuels on injection, combustion and exhaust emission characteristics in a direct injection compression ignition engine. *International Journal of Engine Research* **2**, 249–261.

Yazdani, S. S. and Gonzalez, R. (2007). Anaerobic fermentation of glycerol: A path to economic viability for the biofuels industry. *Current Opinion in Biotechnology* **18**, 213–219.

Zamir, D. (2008). Plant breeders go back to nature. *Nature Genetics* **40**, 269–270.

Zeng, J., Du, W., Liu, X. and Dai, L. (2006). Study on the effect of cultivation parameters and pretreatment on *Rhizopus oryzae* cell-catalyzed transesterification of vegetable oils for biodiesel production. *Journal of Molecular Catalysis B: Enzymatic* **43**, 15–18.

Zhang, J. and Jiang, L. (2008). Acid-catalyzed esterification of *Zanthoxylum bungeanum* seed oil with high free fatty acids for biodiesel production. *Bioresour Technology* **99**, 8995–8998. doi:10.1016/j.biortech.2008.05.004.

Zhang, F.-L., Niu, B., Wang, Y.-C., Chen, F., Wang, S.-H., Xu, Y., Jiang, L.-D., Gao, S., Wu, J., Tang, L. and Jia, Y.-J. (2008). A novel betaine aldehyde dehydrogenase gene from *Jatropha curcas*, encoding an enzyme implicated in adaptation to environmental stress. *Plant Science* **174**, 510–518.

Zheng, P., Allen, W. B., Roesler, K., Williams, M. E., Zhang, S., Li, J., Glassman, K., Ranch, J., Nubel, D., Solawetz, W., Bhattramakki, D. Llaca, V. *et al.* (2008). A phenylalanine in DGAT is a key determinant of oil content and composition in maize. *Nature Genetic* **40**, 367–372.

You are What You Eat: Interactions Between Root Parasitic Plants and Their Hosts

LOUIS J. IRVING*[,1] AND DUNCAN D. CAMERON[†]

**Graduate School of Agricultural Science, Tohoku University, 1-1, Amamiyamachi Tsutsumidori, Aoba-ku, Sendai 981-8555, Japan*
[†]Department of Animal and Plant Sciences, University of Sheffield, Alfred Denny Building, Western Bank, Sheffield S10 2TN, United Kingdom

[1]Corresponding author: Email: ljirving@gmail.com

Advances in Botanical Research, Vol. 50

0065-2296/09 $35.00
DOI: 10.1016/S0065-2296(08)00803-3

ABSTRACT

In this review we will discuss three main areas of research in plant–parasitic plant interactions. In the first section, we will deal with the physiology and biochemistry of the plant parasitic interaction, primarily focusing on the effects of resource abstraction. A functional split shall be introduced between parasites that are either predominantly xylem- or predominantly phloem feeding, and the implications of these two parasitic strategies discussed. In the second section we shall discuss the events leading up to the mature system, and the defenses employed by host plants against parasites. This section will focus on parasites' life cycles, and on the hosts' defense mechanisms against parasitism. Host plants and parasites are locked in a perpetual evolutionary arms race, with host plants evolving defenses, and parasites evolving counter-defense mechanisms. Significant events, such as parasite germination, and its subsequent attachment (or not) are discussed in reference to the functional groups introduced earlier. Finally, we discuss the evolution of parasitic plants and their ecological effects, from the negative effects of parasitic plants on agro-ecosystems, to their positive effects on plant biodiversity, via the suppression of dominant species, in natural and seminatural ecosystems.

I. INTRODUCTION

The parasitic plants are a taxonomically and biogeographically diverse group. There are approximately 3–4000 parasitic angiosperm species, with parasitism having independently evolved approximately 11 times (Barkman *et al.*, 2007). Parasitic angiosperms have a global distribution, from the arctic to the tropics (Watling and Press, 2001). Only a single parasitic gymnosperm species, *Parasitaxus usta*, has been identified, deriving water and nutrients from its host's xylem, but carbon by mycoheterotrophy (Feild and Brodribb, 2005; Sinclair *et al.*, 2002). Parasitic plants have significant deleterious effects on host plant and agricultural productivity (Adetimirin *et al.*, 2000), especially in developing countries where chemical treatments are not widely available, yet they may also have positive effects on biodiversity in natural and seminatural environments (Ameloot *et al.*, 2005). Likewise, host plant identity has significant effects on parasite performance, with different hosts proving to be either vulnerable to, or exhibiting a variety of defenses against, plant parasitism (Cameron and Seel, 2007; Cameron *et al.*, 2006).

Parasitic plants can be largely split into two distinct groups; facultative parasites and obligate parasites. Facultative parasites have the ability to complete their life cycle independently of a host plant, albeit suffering reduced growth and fecundity, while obligate parasites are unable to complete their life cycle without a host plant. A further division can be made between the achlorophyllous *holoparasites*, which derive all their growth requirements from their host(s), and the chlorophyllous *hemiparasites*, which derive some of their resources from nonhost sources. For example, *Orobanche* species are obligate holoparasites, deriving both carbon and (reduced) nitrogen from their host(s) (Hibberd *et al.*, 1999). Conversely, *Striga hermonthica* represents an obligate hemiparasite which requires a host to support it for the 4–6 weeks before shoot emergence from the soil, yet can photosynthesize after emergence. *Striga*'s ability to survive independently of a host postemergence is unknown. *Rhinanthus minor* exemplifies a facultative hemiparasite, able to germinate, photosynthesize and reproduce independently, or live parasitically, deriving nutrients and water from the host (Table I; Cameron *et al.*, 2008a). Facultative holoparasites do not occur in nature for obvious reasons. Naturally, as with all biological classifications, the hemiparasite–holoparasite distinction is somewhat artificial, with some species, for example *Striga,* exhibiting both holo- and hemiparasitic stages within their life cycle.

We will split this review into three sections; the first section will deal with the physiology and biochemistry of the mature plant–parasite system, with a particular emphasis on the growth and fecundity benefits of the parasitic lifestyle, and the costs to the host. This section will have a particular focus on differences in the cost/benefit ratio for both the parasite and the host between systems where the parasite is predominantly xylem or predominantly phloem feeding. The second section will look at the events leading up to the mature system; germination, haustoria formation, differentiation, and attachment, with special emphases on the interactions at each stage of the host and parasite life cycles, and a particular focus on the host defenses which have evolved as part of that relationship. The final section will focus on the evolution of parasitic plants, and the effects of plant parasitism on ecological and agro-ecological systems. To some readers our review will feel somewhat "backward," with, for example, the range of host species that each group can parasitize not being discussed until the final section. However, we hope the reader will agree with our ordering upon reading the review, since our overall goal is to explain the broad *functional* and *physiological* patterns evident in the literature, with a conceptual basis that *ecological function follows physiological constraints.* For example, a parasitic species' host range is a function of whether the parasite is xylem or phloem feeding, of host defense mechanisms, and the specific events within the species' evolutionary history.

II. THE MATURE PLANT–PARASITE ASSOCIATION

A. HAUSTORIA STRUCTURE AND FUNCTION

Parasitic angiosperms attach to their host by a structure known as a haustorium that provides a physical as well as physiological bridge between host and parasite. In hemiparasites, haustoria are typically secondary haustoria,

TABLE I

Common parasitic species used in this review and their N/C relations

Parasitic habit		Name	% C derived from host	% C lost by host	% N derived from host	% N lost by host	Other substances derived
Xylem feeding	Facultative	*Rhinanthus minor*	10%[a]	18–20%[b]	~87%[b]	~18%[c]	Nutrients, water hormones
		Olax phyllanthi	~40%[d]	27–35%[d,e]		Up to 55%[d]	Nutrients, water
	Obligate	*Striga* spp.	100% preemergence, 25–35% thereafter[f]	25–35%[f]			Germination stimulants, HIFs water, nutrients
Phloem feeding	Obligate	*Orobanche* spp.	100%[g]	~35%[h]		10–30%[h]	Nutrients, water

[a]Jiang *et al.* (2004a).
[b]Jiang *et al.* (2004b).
[c]Jiang *et al.* (2003).
[d]Tennakoon *et al.* (1997).
[e]Tennakoon and Pate (1996).
[f]Press *et al.* (1987).
[g]Assumed.
[h]Hibberd *et al.* (1999).

developing proximally to the root tip, as opposed to primary haustoria, haustoria that form terminally at the root tip. Haustorial ontogeny of the xylem-feeding hemiparasite *R. minor* was described by Cameron and Seel (2007) showing that after contact with the host root, the parasite radical forms a proto-haustorium that encompasses the host root as it differentiates. The penetration peg (endophytic tissue) forces its way through the host cortex, peeling back the endodermis and is driven into the core of the stele, thereby accessing the host's vasculature. This process appears to be facilitated by extra-cellular enzyme activity as Kraushaar and Seel (unpublished data), using immuno-histochemical techniques, have localized the products of pectinase ahead of invading endophytic tissues in *R. minor*. *Parasitaxius usta*, the parasitic gymnosperm, connects to its host not using a haustoria, but using woody root-like sinkers which penetrate the host (*Falcatifolium taxoides*) epidermis, creating a haustoria-like bridge between the host and parasite, although without apparent vascular continuity. Parasite xylem has been noted close to, but not contiguous with, host xylem (Feild and Brodribb, 2005). The mechanism by which the penetration of the sinkers is facilitated by the parasite is unknown.

In phloem-feeding parasites haustoria are highly specialized structures. The first haustoria is terminal, leading to the cessation of elongational root growth (Shen *et al.*, 2006; Tomilov *et al.*, 2005), with further, secondary, haustoria developing along the root axis. The haustorium facilitates the abstraction of mineral nutrients, amino acids, soluble-C and water, either through direct connection to the host xylem utilizing a "penetration peg" and oscula (Dörr, 1997), or indirectly from the phloem (and/or xylem), via interfacial parenchyma cells (Fig. 1; Christensen *et al.*, 2003; Tennakoon and Cameron, 2006). Some species exhibit both mechanisms, for example the root holoparasite *Hydnora triceps*, which spends almost its entire life cycle below ground, with the stem only emerging for flowering and seed dispersal (Tennakoon *et al.*, 2007). *H. triceps* exhibits apparent xylem continuity, although xylem–xylem continuity could not be verified. Phloem conduits were also notable in the haustoria, as were apparent interfacial transfer cells, suggesting transfer of C from the host phloem to the parasite, although likewise a phloem tissue graft could not be visualized, making both phloem and xylem transfer impossible to confirm. However, the transfer of solutes from the host to the parasite in such systems is usually heavily biased toward the phloem, with xylem transfer being a minor flux. After establishing the primary haustoria, the parasite develops a root tubercle from which the shoot develops (Fig. 2).

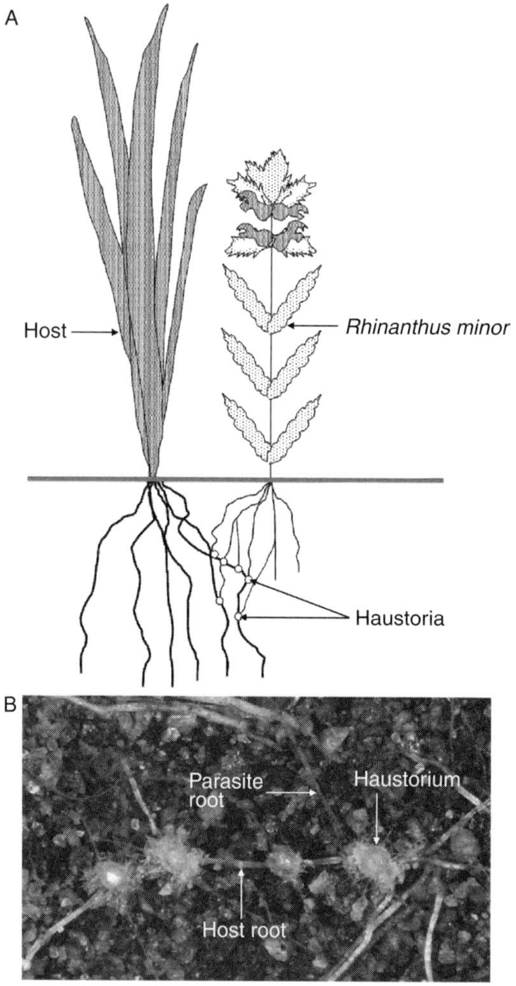

Fig. 1. (A) Schematic representation of the host–parasite system, indicating host and parasite stems, roots, and haustoria. And (B) photograph of *Rhinanthus* haustoria on *Festuca ovina* roots.

B. GENERAL FACTORS INFLUENCING THE EFFICIENCY OF RESOURCE ABSTRACTION

Several factors influence the success with which parasites can abstract resources from their host; ranging from the number and efficiency of haustorial connections (Cameron *et al.*, 2005), and the morphology, size and architecture of the host root system (Keith *et al.*, 2004), to the presence or absence of host defense mechanisms (Fig. 3). Haustorial numbers can vary

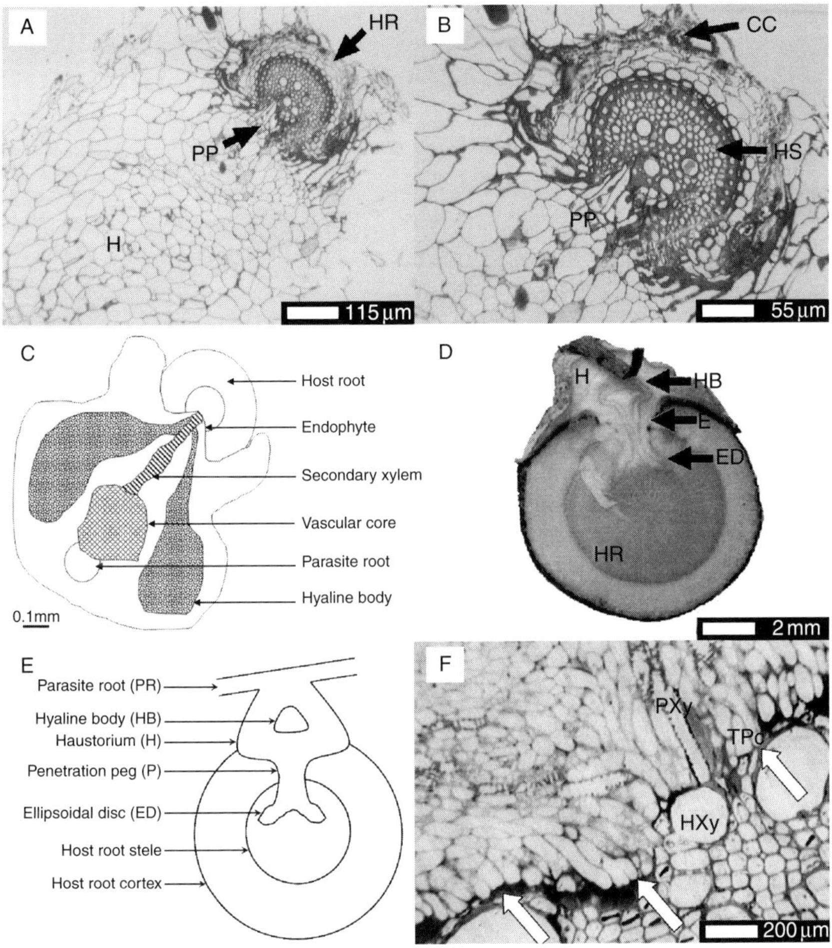

Fig. 2. (A and B) Transverse sections through the haustoria of the parasitic angiosperm *Rhinanthus minor* attacking the roots of the susceptible grass *Poa pratensis*. Sections are 2 μm thick and were stained with toluidine blue for 30 s; (C) a schematic diagram of the model haustoria formed by *Rhinanthus minor* (modified from Rümer *et al*., 2007); (D) transverse hand-sections of the host [HR] at the mid-point of attachment of a haustorium of *Santalum album* [H]. The portion external to the host root is denoted as the hyaline body [HB] and E denotes the area making initial contact with the host root and penetrating the host tissue (endophyte). The endophytic tissue [E] has penetrated the cortex of the host root and flattened out laterally to form a thin ellipsoidal disc [ED] (modified fromTennakoon and Cameron, 2006); (E) schematic diagram of the model haustoria formed by *Santalum album* (modified fromTennakoon and Cameron, 2006); (F) detailed anatomy of the haustorial interface of *S. album* with *Tithonia diversifolia* showing the graft union between the endophytic tissue of haustorium and the xylem tissue of host root. The host–parasite interface is demarcated by arrows. Sections are 2 μm thick and were stained with toluidine blue for 30 s. TPc—irregularly spirally thickened tubular shaped contact parenchyma cells of the haustorium, Pc—parenchyma cells within

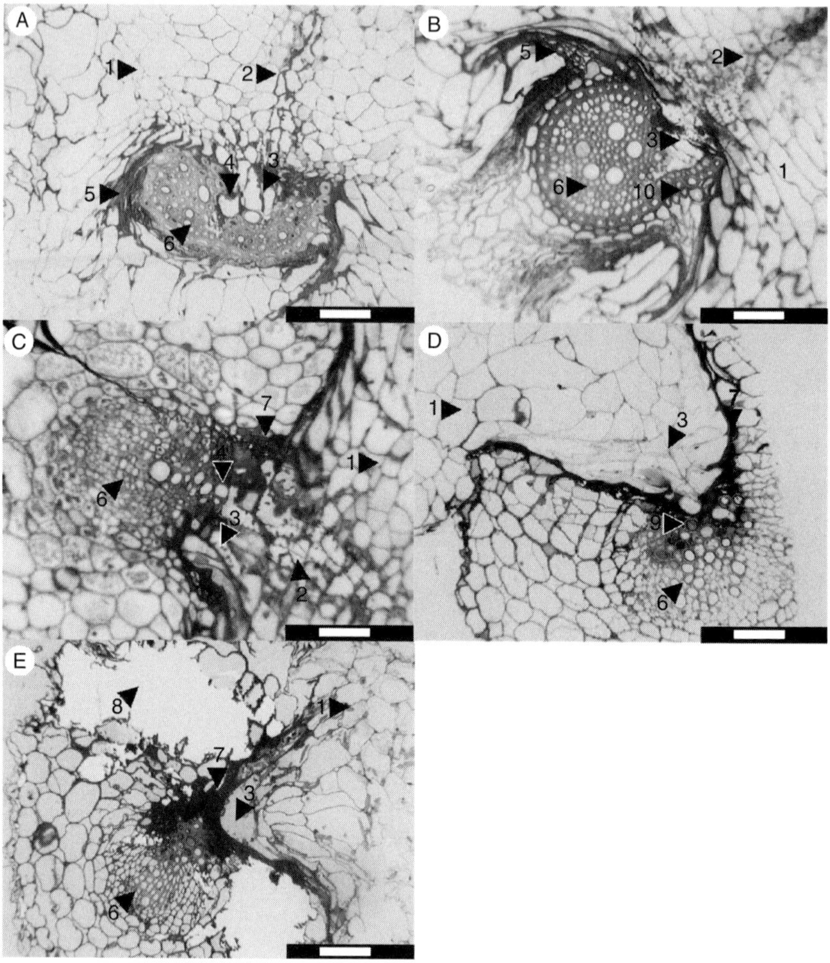

Fig. 3. Transverse sections through the haustoria of the parasitic angiosperm *Rhinanthus minor* attacking the roots of *Cynosurus cristatus* (A), *Phleum bertolonii* (B)—both grasses, *Vicia cracca* (C)—a legume, *Leucanthemum vulgare* (D), and *Plantago lanceolata* (E)—both forbs. Sections are 2 μm thick and were stained with-toluidine blue for 30 s. The following numbers represent: 1 = parasite haustorium; 2 = secondary parasite xylem; 3 = penetration peg (endophyte); 4 = parasite penetrating host xylem; 5 = crushed host cortical cells; 6 = host stele; 7 = darkly staining material; 8 = fragmented host cells; 9 = occluded cells, and 10 = peeled endodermis. Resistance responses in the form of cell death and lignification (identified using FTIR microspectoscopy) are clearly visible in both forbs-*Rhinanthus* associations (reproduced with permission from Cameron *et al*., 2006).

the body of the haustorium, PXy—xylem elements of the parasite present inside the haustorium, HXy—xylem tissue of the host root. Note the terminating file of parasite xylem elements [PXy] present in the endophyte and the absence of direct lumen-to-lumen xylem element continuity between xylem of the host and parasite (modified from Tennakoon and Cameron, 2006).

widely, from 1400 haustoria per plant in *Olax phyllanthi* parasitizing *Acacia littorea,* to 10 haustoria per plant for *R. minor* parasitizing *Poa pratensis* (Cameron *et al.*, 2005). Kawachi *et al.* (2008) noted that *Orobanche*, a phloem-feeding holoparasite, attached to red clover (*Trifolium pratense*) can abstract 73.6% of the nitrate—N recently taken up by the parasitized root, while abstraction from non-parasitized roots on the same plant was negligible, emphasizing the importance of the relationship between the proportion of host roots parasitized per plant and both host and parasite growth. In *R. minor*, Cameron *et al.* (2005) noted significant differences in the amounts of N transferred to the parasite between differing host species. The physiological reasons for these differences remain obscure, yet we may posit various potential explanations, such as differences in host root vasculature, architecture, or morphology, differences in oscula penetration, haustorial cross-sectional area, or differences in host plant xylem composition, which alter the competitive balance between the host and the parasite. However, without further careful investigation of this area, these remain, at best, speculative.

Keith *et al.* (2004) examined the effect of host root architecture and parasite position on the interaction between *Festuca* hosts and their *R. minor* parasites. Two species of *Festuca* were grown in root rhizotrons; *F. rubra*, which has a rhizatomous, spreading root system, and *F. ovina*, which exhibits a clumpy root distribution. Parasites were introduced either 5 or 10 cm from the host plant, and the physical characteristics of the system recorded. Significant competition, apparently for light, occurred close to the host plant stem, which led to a 77% pre-attachment mortality for *R. minor* seedlings. Conversely, in a second experiment, parasites further from the host stem (which had been trimmed to reduce competitive effects) took longer to attach to their hosts, although by 49 days after the start similar numbers of plants had been infected in each treatment. In terms of biomass accumulation, parasitism had significant negative effects on both host species, while host–parasite propinquity had significant ancillary effects on *F. rubra*, the rhizatomous species, only. As expected, parasites located closer to the *F. rubra* host depressed host biomass increase to a greater extent than those further away. Furthermore, parasites planted closer to their host species were significantly taller from 38 to 63 days after parasite infection than those more distally located, with this difference being greater in parasites attached to *F. rubra*. These differences were reflected in the biomass accumulated by the parasites, which was greatest for *R. minor* plants growing close to *F. rubra* stems, and lowest for *R. minor* plants growing more distally from *F. ovina* hosts. Likewise, the numbers of flowers produced by the parasites were greater for plants attached more proximally to the host, suggesting increased fecundity. Correlations were also noted between the number of haustoria produced, with more haustoria produced by parasites located close to their hosts, and parasite biomass, parasite leaf biomass, and host root weight, suggesting a complex set of interactions that determine parasite growth rates. Although some of these correlations may simply be indicative of allometric relationships between plant organs, and yet others are determined more by the duration of attachment rather than the physical position of the parasite, some relationships have more functional significance, such as the correlation between host root mass and haustorial number. Such relationships may be purely probabilistic, with *R. minor* plants growing close to their hosts more likely to intercept a root and form a haustorium than plants growing further from the host plant, although other possible explanations exist too, such as the stimulation of haustoria formation by "good" hosts. Overall, being physically close to the host plant seems to have both advantages and disadvantages for *R. minor*, the balance between which is presumably dynamic, being affected by host/parasite nutritional status and changing through time as the host–parasite system matures, but also being influenced by prevailing environmental conditions and constraints. Nitrogen fertilization, for example, may have negative effects on hemiparasites, by increasing competition for light. Under the experimental conditions used, increases in seedling mortality during the early phases of the parasite's life cycle are contrasted with the apparent benefits of being close to the host plant in later life, suggesting that, in dense stands of plants *R. minor* may not establish, except perhaps in the presence of grazing animals. Conversely, in sparsely populated environments, *R. minor* plants which do germinate may either fail to attach to a host, or may derive little benefit from attaching to a distant host.

Often, as above, abstraction of nutrients and carbon by the parasite results in decreased host growth and fecundity (Adetimirin *et al.*, 2000; Scholes and Press, 2008), although this is not always the case. In Section IIC and D, we will explore the transfer of nutrients and C between the host and the parasite, specifically in terms of whether the parasites act as either predominantly phloem or predominantly xylem feeders.

C. RESOURCE ACQUISITION BY XYLEM-FEEDING PLANTS

1. Facultative hemiparasites are primarily parasitic for N

Recent studies have shown that predominantly xylem-feeding parasites can abstract significant quantities of N from "good" hosts, with *R. minor* abstracting 17% of a ^{15}N tracer taken up by *Cynosurus cristatus*, while these values were only 2.5 and 0.2% when *R. minor* was grown on *Leucanthemum vulgare* and *Plantago lanceolata*, respectively (Cameron and Seel, 2007).

Tennakoon *et al.* (1997) demonstrated a far higher flux than this, 56% of N-fixed by 18-month-old *A. littorea* plants was abstracted by *O. phyllanthi*. Both *O. phyllanthi* and *R. minor* are predominantly xylem-feeding root hemiparasites, deriving mineral nutrients and some C from the host, but able to complete their life cycles and photosynthesize without a host. As *R. minor* and *O. phyllanthi* have direct xylem–xylem connections with their hosts, solute transfer is facilitated by elevated transpiration, often twice or more that of the host plant, as the parasite effectively leaves its stomata open (Cernusak *et al.*, 2004; Jiang *et al.*, 2003), drawing a continuous water column from host-to-parasite via cohesion. This apparent lack of stomatal regulation is not found in every hemiparasite, for example *Melampyrum arvense* exhibits "normal" stomatal behavior, opening in the light and closing in the dark. At very high ABA concentrations, *R. minor* stomata will close, confirming that the ability to sense and act upon ABA functions within these plants, albeit with a greatly reduced sensitivity—ABA concentrations within *R. minor* were 50 times those of their host barley plants with no appreciable effects on stomatal aperture—perhaps a result of cytokinin transfer from the host (Jiang *et al.*, 2003). In *Striga*, potassium accumulates in the parasite leaves, acting as a cellular osmotica, and is known to be important in guard cell function (Smith and Stewart, 1990).

The very similar mineral nutrient composition of *R. minor* xylem sap, compared with host xylem sap suggests a lack of active ion transport mechanisms with mineral nutrients, water, hormones, and C transferred into the parasite by mass flow of xylem sap through small pores known as oscula across the haustorial bridge (Jiang *et al.*, 2003, 2004b). Further evidence of the nonselective nature of N abstraction from host plants was recently presented by Pate and Bell (2000), who demonstrated that the ^{15}N excess of the root hemiparasite *Parentucellia viscosa* is largely determined by the ^{15}N excess of the host plant. In addition to transpiration driven resource acquisition, *R. minor* further decreases its water potential relative to its host, through the synthesis of a variety of solutes, such as sugars and sugar alcohols (especially mannitol). This decrease in water potential may increase mass flow rates, especially under conditions where transpiration rates are particularly high, and may also be important in maintaining high growth rates in the parasite relative to the host.

As the parasite xylem sap is host derived, the composition of the host sap will have major effects on the parasites growth habit. For example, some species reduce nitrate preferentially in the root, while in others nitrate reduction is preferentially conducted in the shoot (Andrews, 1986). *R. minor* parasitizing plants which primarily assimilate nitrogen in their roots will be supplied with amino acids, while *R. minor* parasitizing a primarily shoot assimilating plant will be supplied with nitrate, which will have to be reduced by the parasite with commensurate carbon—costs both in terms of the direct energy and carbon costs, but also other indirect costs, such as those involved in maintaining ionic balance (Raven, 1983; Raven and Smith, 1976). Similarly, environmental factors, such as soil type, mean annual precipitation and temperature, and vegetation type are likely to affect the nutrients available to the host roots, which will in turn have effects on the vascular sap composition. In most natural ecosystems ammonium, which cannot be xylem transported, is in greater supply than nitrate (Addiscott *et al.*, 1991), which may shift the host xylem sap composition toward amino acids.

Both Press *et al.* (1987) and Tennakoon and Pate, (1996) estimated that the hemiparasitic plants *Striga* and *O. phyllanthi* could access approximately 35% of their respective host's nonstructural C, while Tennakoon *et al.* (1997) quantified the abstraction of C by *O. phyllanthi* at approximately 27% of recent photosynthate, accounting for approximately 40% of the C used in new parasite biomass. Measurements of $\delta^{13}C$ levels in the facultative hemiparasite *P. viscosa* and its host plants suggest a low water use-efficiency for the parasite, which would correspond with high stomatal aperture allowing high photosynthetic rates (Pate and Bell, 2000). Jiang *et al.* (2004b) estimated that *R. minor* plants were able to utilize approximately 20% of their barley (*Hordeum vulgare*) host's solutes, which is strikingly similar to their earlier estimates that *R. minor* abstracts approximately 18% of their host barley plants' water (Fig. 4; Jiang *et al.*, 2003). Jiang *et al.* (2004b) also noted decreases in the range of 10–18% for the uptake of N and other nutrients by parasitized plants, relative to unparasitized plants (Fig. 4), coupled with decreases in leaf nutrient concentrations, which may suggest that decreases in leaf nutrient status or photosynthetic rate were having significant effects on nutrient uptake. Of course the opposite argument could also be made, that decreased uptake was affecting leaf nutrient concentrations—undoubtedly this is true, yet is probably a secondary effect rather than being directly causal. A recent experiment, where the *R. minor*–*P. pratensis* association was grown at either ambient (350 ppm) or elevated (650 ppm) CO_2 levels showed that parasitism had a proportionately similar negative effect on host performance despite the photosynthetic gains resulting from the elevated CO_2 treatment (Hwangbo *et al.*, 2003). Parasite biomass was significantly increased, which contributed to an overall increase in the system (plant + parasite) biomass, although this seemed to be mainly the result of increased parasite photosynthesis, and a slightly (yet nonsignificantly) higher N content in the parasite at elevated CO_2. This ambivalence toward the higher rates of host photosynthesis seems to be further evidence that *R. minor* is primarily parasitic for nitrogen and other mineral nutrients

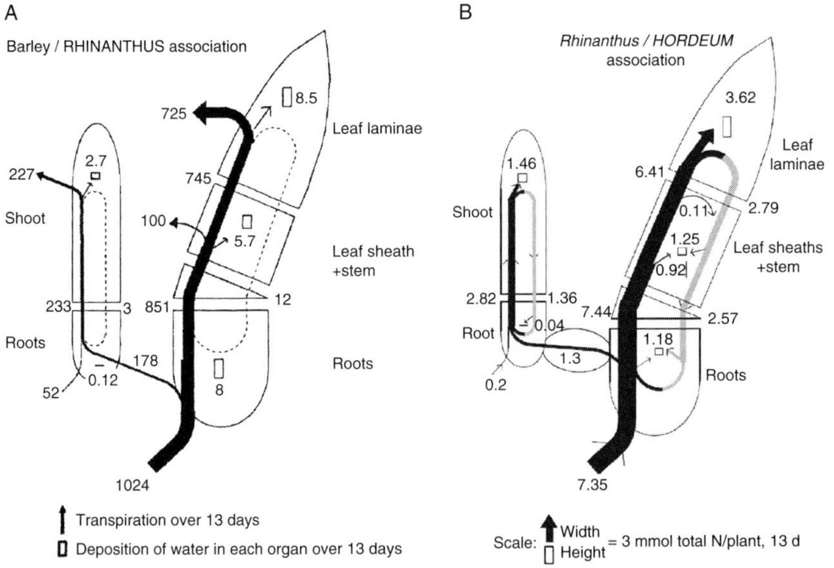

Fig. 4. Transfer of water (A) and nitrogen (B) from *Hordeum vulgare* to *Rhinanthus minor*. Black bars represent the xylem flows, while the grey/dashed lines represent the phloem flow. Bar widths are indicative of the flows, while the values show the transpiration or N flux in ml water or mmol N, respectively. Figures are taken from Jiang *et al.* (2003, 2004b).

rather than for carbon, and is consistent with the estimate of Jiang *et al.* (2004b) that *R. minor* obtains only about 10% of its C from its host, considerably less than many other parasites although a full carbon budget remains absent from the current literature. As *R. minor* is a xylem-feeding parasite, which can cause the host plant to become N deficient, increased host photosynthesis under elevated CO_2 conditions may be preferentially allocated to root growth or N uptake as the plant attempts to alleviate the physiological N stress imposed by the parasite. Indeed, in unparasitized tobacco, elevated CO_2 leads to both an increase in N uptake and an increase in root mass (Kruse *et al.*, 2002) while a recent paper (Cramer *et al.*, 2008), demonstrated that N limited un-parasitized *Ehrharta calycina* plants exhibited higher transpiration rates than N replete controls, and that increased transpiration had some effect in relieving the nutrient limitation. It seems likely that the increased stomatal aperture noted by Cramer *et al.* was a result of C deficit, caused by the increased energetic costs of N uptake from the environment, which manifested in reduced relative growth rates despite significantly higher photosynthetic rates. In a forum accompanying Cramer *et al.*'s paper, it was noted that other stresses, such as low irradiance (energy limitation) can have similar effects in increasing nutrient uptake (Raven, 2008). Of course in N stress caused by parasitism, increased host N uptake may prove to be counter-productive, acting to increase the N availability for the parasite, which does not incur the C cost of N uptake, although this may be dependant upon the size and architecture of the host root system and the attachment points of the parasite. This interesting hypothesis remains untested and requires further investigation.

2. Obligate hemiparasites are parasitic for both N and C

S. hermonthica is a xylem-feeding obligate parasite of many economically important species, causing significant human suffering in sub-Saharan Africa. Germination is reliant on chemical signals from the host plant, specifically xenognosins, such as strigolactones and SXSg, to which the parasite responds (Estabrook and Yoder, 1998; Hauck *et al.*, 1992). Germination of *Striga* in response to host root exudates is dependent upon the distance of the seed from the host root and may be constrained by stimulant diffusion. For example, Fate *et al.* (1990) showed that *Striga asiatica* did not germinate more than 1 cm away from the host (sorghum) root. Moreover, *Striga* radical growth is rhizotropic with respect to the host root which the parasite is able to locate by tracking the concentration gradient of the germination stimulant exuded from the host root, allowing the parasite to maximize its chances of successful attachment (Fate *et al.*, 1990). Over the first three to four weeks after seed germination, the young *Striga* plant remains below ground, relying entirely on the host plant for nutrition, with the parasite's carbon isotope signature closely following that of its host (Press *et al.*, 1987). Much of the damage sustained by the host happens during this below ground phase, when the parasite is both difficult to see, and virtually impossible to treat using conventional chemical treatments. As the *Striga* plants mature, they become less dependant upon their host for reduced carbon, with the isotope signatures of the C_3 *Striga* becoming more "C_3 plant-like" relative to its *Sorghum* hosts' C_4 isotope signature (Press *et al.*, 1987). However, it is worth noting that mature *Striga* plants, while photosynthetically competent, have very low photosynthetic rates, fixing barely enough excess carbon over the day to cover night-time respiration, and still derive a proportion, roughly one-quarter to one-third, of their carbon from the host, although this figure must be 100% when the parasite is underground (Press *et al.*, 1987). It is the combination of these three factors, the host necessity for germination, the dependence upon host photosynthate during below-ground development, and the low photosynthetic rate of the mature plant that renders *Striga* an obligate hemiparasite. *Striga* is known to have several interesting effects on the host plant, notably causing a decrease in host plant transpiration rate, apparently as a result of increased xylem ABA concentrations (Taylor and Seel, 1998; Taylor *et al.*, 1996), and, recent evidence suggesting, fundamentally altering the composition of the host plants xylem sap, with large increases in individual amino acid concentrations, particularly asparagine (Pageau *et al.*, 2003), although the mechanism underlying these changes remains elusive.

3. Host quality for xylem-feeding hemiparasites

Legumes are generally accepted to be "good" hosts for parasitic plants, although some legumes have been noted to be poor hosts, as for the *Euphrasia–Trifolium dubium* association (Yeo, 1964), or the slight decrease in mean fecundity noted in *Orthocarpus* grown on *Trifolium repens* compared with autotrophic growth (Atsatt and Strong, 1970). Despite these few exceptions, the reasons for generally high quality of leguminous hosts, and the effects of parasitism on the legume host remain unclear, although the assumption has been that N supplied by the legume as amino acids (since ammonia fixed by the symbiotic bacteria must be reduced to amino acids before transport in the host xylem) allows more rapid growth, both due to an increase in the absolute amount of amino acids available, as legumes typically exhibit high tissue-N concentrations, and the reduced energetic requirements of amino acid use for growth (Seel and Press, 1993). The phloem of leguminous lupin plants (*Lupinus albus*) contains a significantly higher concentration, on the order of five times more, of amino acids than either their own xylem (Parsons and Baker, 1996), or either the phloem or xylem of the non-leguminous plant *Ricinus communis* (Peuke *et al.*, 2001). Moreover, studies have revealed extensive penetration of *Vicia cracca* host roots by *R. minor* (Cameron *et al.*, 2006), yet more recent work suggests that the majority of vascular connections are with younger roots, upstream of the N fixing root nodules (Jiang *et al.*, 2008a), and also showed that nodulated faba bean plants relying on fixed N were inferior hosts to un-nodulated bean plants receiving exogenous N. It is unclear whether nodulation represents an active benefit for legume-parasitizing *R. minor* when all hosts are subjected to identical conditions, and perhaps the relative "poorness" of the nodulated plants in this experiment reflects the high energetic costs of symbiotic N fixation. Cameron *et al.* (2006) demonstrated that the biomass of two grass species used as hosts was significantly reduced by parasitism, although this was not the case for the legume used, *V. cracca*. Although N concentrations in the tissues were not measured, it would be reasonable to suspect that the parasite sequestered significant quantities of N from the graminoid hosts, while the legume may have been less affected. The question arises, therefore, of why some legume species may be parasitized yet apparently suffer few ill effects? As yet, this question has not been sufficiently addressed, and may simply reflect a greater degree of plasticity in the N concentrations required for growth by legumes, or may suggest that parasites attaching upstream of the root nodules may have little effect on the availability of N for growth, feeding on only N taken up by the fine roots. This may be some form of explanation for the rather similar biomass achieved by the parasite attached to graminoid and leguminous hosts, yet requires further investigation.

Rates of ^{15}N transfer between various hosts and their *R. minor* parasites were investigated by Cameron and Seel (2007), with *P. lanceolata* exhibiting the lowest rates of nutrient transfer, and *C. cristatus* the highest. A highly significant relationship between N sequestration by the parasite from the host and parasite growth rates was noted. Reductions in host biomass as a result of parasite attachment are not so easily explained however. The "missing biomass" when *R. minor* parasitizes one of the grass species, compared with the biomass attained by each species independently grown cannot be accounted for by simple mass-flow kinetics (Cameron *et al.*, 2008a). Rather, parasite attachment has been shown to have wide-ranging effects upon plant physiology and photosynthetic characteristics. When *P. lanceolata* was infected by *R. minor* a small yet significant decrease in host Fv/Fm was noted, which was coupled with a similarly small yet significant decrease in ΦPSII at low light intensities (Cameron *et al.*, 2008). However, when *Phleum bertolonii* was infected by the parasite, no significant decreases in host Fv/Fm were noted, yet ΦPSII decreased at all light intensities up to 750 μmol photons m^{-2} s^{-1}. Higher rates of decrease of ΦPSII than Fv/Fm may denote reductions in the quantity of electron transport chain proteins, and an increased proportion of PSII reaction centers being temporarily deactivated by light stress under steady state photosynthetic conditions. As ΦPSII is a relatively reliable indicator of photosynthesis, this may have significant effects on host photosynthesis. In *P. bertolonii*, these effects may be temporally localized to periods of low light, or when the host is growing in low light, perhaps understory, conditions, while in *P. lanceolata*, it would appear that decreases in photosynthetic rates would occur at all light intensities, and perhaps all environments.

4. *Other elements*

In a controlled experiment, unattached *R. minor* plants were supplied with exogenous N, P, and K, with significant biomass increases only resulting from P addition. This suggests that *R. minor* in the absence of a host may be P limited (Seel *et al.*, 1993). Indeed, another recent study showed that the increase in N, K, and P in *R. minor* connected to a *H. vulgare* host, relative to single plants, were 18-fold for N, 28-fold for K, and 42-fold for P

(Jiang *et al.*, 2004b). This is presumably due to the low bioavailability of phosphate in the soil, coupled with the absence of mycorrhizas in *Rhinanthus*, the comparatively low rooting volume of *R. minor*, and the high demand for P by organisms. Thus, the evolution of parasitism in *R. minor* may have been a mechanism for alleviating P deficiency in ancestral plants, although conversely the P stress seen in modern *R. minor* plants may be a secondary result of an evolutionary root reduction facilitated by parasitism.

Recently, Jiang *et al.* (2008b) used the incremental accumulation models of nutrient flux between host plants and their associated parasites (Pate *et al.*, 1979) to model boron (B) fluxes between barley plants and their *R. minor* parasites. Theory suggests that B, which is important for cell division and other processes, such as flowering, forms complexes with polyols such as mannitol in the phloem. Mannitol accumulating plants, such as *R. minor* should exhibit increased concentrations of B compared to control plants. Indeed, this seems to the case, yet the increases were mainly localized to lateral buds and inflorescences, both phloem fed, with the *R. minor* and barley leaf material being relatively similar. Perhaps most interestingly, high exogenous nitrate supplies led to significant reductions in the concentrations of mannitol and B, apparently due to the upregulation of mannitol dehydrogenase (Jiang *et al.*, 2008b).

5. Shoot: Root ratio

Tennakoon *et al.* (1997) noted a significant increase in host root mass relative to control plants, as a result of infection of *A. littorea* by *O. phyllanthi*. While the root mass was greatly increased in parasitized plants (67%), shoot mass of the host plant increased to a much smaller extent (21%); this led to an effective halving of the host shoot: root ratio in parasitized plants relative to unparasitized plants. In nitrogen terms, the same patterns were seen, although the extents were even greater. While the shoot N content was effective unchanged over the 4 month experimental period, the root N content increased by 75%, and was significantly higher than the unparasitized control plants (Tennakoon *et al.*, 1997). Similar patterns of a shift in shoot: root ratio have been noted in other studies of hemiparasitic plants, for example, in the *R. minor–H. vulgare* association, parasitism led to a 32% decrease in host leaf mass increase and a 52% decrease in leaf sheath and stem material biomass accumulation, while the reduction in root growth was only 20% relative to unparasitized controls over the measurement period (Jiang *et al.*, 2004b).

The reasons for these relative increases in host root mass in *Acacia* are unclear, yet it must be a result of either direct upregulation in root production, perhaps mediated by some parasite derived growth factor, or, more likely, a decrease in root growth repression. Nitrate accumulation in tobacco leaves has been shown to repress new root production (Scheible *et al.*, 1997), while it is well known that plant nitrogen deficiency, in this case caused by parasitism, can lead to localized root proliferation, especially in nutrient rich patches (Forde, 2002; Robinson *et al.*, 1994). Thus, a decrease in leaf nitrate concentration due to parasitism may fundamentally alter the patterns of root growth. Whatever the mechanism, decreased host shoot: root ratio in hemiparasites presumably benefits the parasite in terms of nutrient and water acquisition.

6. Effects of hormones

While the predominant theory is that increased nutrient supply is responsible for the increased biomass of parasitic plants when attached to a host, other factors, such as plant hormones may also have implications for parasite growth (Jiang *et al.*, 2005). Unattached *R. minor* plants have significantly lower concentrations of cytokinins in their tissues than host barley plants, with leaves exhibiting cytokinin and nitrogen deficiency symptoms (Jiang *et al.*, 2003). When attached to a barley host plant, cytokinin levels in *R. minor* plants approximately doubled, 70% of which were derived from the host plant. Meanwhile, host root production of cytokinins fell. As well as a decrease in zeatin production in host roots, zeatin utilization in the host leaves also fell, although the mechanisms behind these changes are unknown. Likewise, ABA levels were low in unattached parasites, while attached plants had much higher ABA levels, which were not derived from the host but were synthesized in the parasite roots (Jiang *et al.*, 2004a). The specific function of this elevated ABA in parasitic plants is unknown. It has however been clearly demonstrated that ABA significantly increases plasma membrane permeability (Jiang *et al.*, 2004a; Stillwell and Hester, 1984) and also plays a role in the regulation of aquaporins (Wan *et al.*, 2004). This is potentially significant for xylem-feeding parasites that lack vascular continuity, such as *Santalum album*, but instead rely on interfacial parenchyma because elevated ABA in the haustoria and roots may increase membrane permeability and hence facilitate water tapping. In the case of xylem-feeders with vascular continuity such as *R. minor*, there are no or fewer membranes between host and parasite through which water and xylem solutes must pass; thus, in these parasites, it is unlikely that ABA facilitates solute and water tapping (Jiang *et al.*, 2004a). One thing is clear, despite relatively high tissue ABA concentrations, the parasites' stomata are wide open, suggesting that the guard cells are insensitive to the increased ABA levels and stomatal regulation is not their function. Likewise, the physiological reason for the increase in ABA is unknown, and it may either be hormonally regulated itself, or a result of greater C or N availability for the parasite roots, or the greater water flows through the roots.

Jiang *et al.* (2007) grew *R. minor* plants on barley host plants, before removing the stem of the host plant (HFA—host free attached) and compared them with parasites which had never attached to a host (NP—non-parasitizing), and to plants which remained connected to their host. Unsurprisingly, the HFA parasites grew to a size intermediate to either of the other treatments—larger than the NP plants, smaller than the attached control plants. However, attachment to the host plant led to several interesting phenomena in the HFA plants. First, the HFA plants became able to develop a more "normal" root system, more like their barley host plants root system, than either the attached or unattached controls (NP). Secondly, the HFA plants exhibited "normal" stomatal responses, neither constantly open, like the attached parasites, nor constantly closed like the unattached controls. Rather they opened their stomata during the day, and closed their stomata at night. Unfortunately, ABA levels were not measured. Finally, the cytokinin levels of the HFA plants became more like those of non-parasitic plants, with HFA *R. minor* roots gaining the ability to produce zeatins (Jiang *et al.*, 2007). It seems likely that the host roots supplied cytokinins and nutrients to the parasitizing *R. minor* plants' leaves, which allowed major increases in leaf photosynthesis and metabolism, apparently facilitating the production of root growth hormones such as auxin and cytokinin in the leaves. When the host stem was excised, the host roots slowly died in the absence of a photoassimilate supply, since carbohydrate is not transferred from the parasite to the host. The subsequent reduction in nutrient supply from the senescing host roots seems to have led to the production of *R. minor* roots, facilitated by the now competent leaves. Attached *R. minor* plants do not produce roots normally, presumably due to root growth signal repression by foliar nitrate concentrations (Scheible *et al.*, 1997). The only things remaining to be explained are the depressed growth rate of HFA plants relative to parasitizing control plants, and the newly functioning stomata. First, it may be unwise to make a straight comparison between these two systems, *R. minor* plants parasitizing whole plants derive a significant advantage, in terms of being able to derive the all the benefits of a large root system while incurring none of the carbon—cost. That said, there may be ancillary effects also, for example, when connected to the host, *R. minor* derived cytokinins from the host roots. These have been implicated in the constant openness of parasitizing plants stomata—the deficiency of which would also explain the tight closure of the stomata preinfection. When the host root dies however, the *R. minor* plants must rely on their own endogenous production of cytokinins, which may very well be lower than the barley host plants. Thus, the stomata may have become more sensitive to ABA signals, also now produced by the new roots. Although further research on this is desperately required, we might suggest that the reductions in stomatal aperture, water and nutrient availability, not to mention the increased costs of nutrient foraging and assimilation faced by HFA plants relative to parasitizing control plants, explain the reduction in growth rate.

D. RESOURCE ACQUISITION BY PHLOEM-FEEDING PLANTS

1. Obligate parasites are parasitic for both N and C

Phloem-feeding parasites abstract their nutrition predominantly from the phloem of their host plant, as the name implies. Although phloem-feeding parasites typically retain a xylem connection, they derive the majority of their C and N requirements from the host plants phloem and are often classified as obligate holoparasites. *Orobanche* spp. growing on *T. pratense* sequesters 10–30% of N uptake by the host plant, although this represented 73.6% of N taken up by the individual root(s) parasitized (Kawachi *et al.*, 2008). This is a graphic demonstration that parasitism has severe effects on the individual roots parasitized, while N uptake was negligible in the roots that were not directly parasitized over the 90-min experiment. Interestingly, in Kawachi *et al.*, (2008) the host plants were fed ^{13}N nitrate, which cannot be transported in the phloem, yet, *Orobanche*, a phloem-feeding parasite, was apparently still able to sequester significant amounts of the ^{13}N tracer supplied. The reasons for this are unknown, although seem most likely to indicate significant N assimilation during the experimental period. In *Orobanche* parasitized tobacco plants, over 95% of N assimilated by the parasite is phloem derived, and thus amino acids, rather than N taken up directly from the soil (Hibberd *et al.*, 1999). As an achlorophyllous holoparasite *Orobanche* is completely unable to photosynthesize by itself, and is therefore unable to survive and complete its life cycle without a host. Nitrogen assimilation is an energy and carbon dependent process, and without photosynthesis to supply the necessary reducing potential and C-skeletons, utilizing host N assimilation products is an energy efficient way of providing N for growth. Transfer of nutrients and carbon from the host plant is not facilitated by xylem-based mass flow, as in the xylem-feeding hemiparasites, since transpiration rates are very much lower in *Orobanche* than its host, even considering relative mass (Cernusak *et al.*, 2004). *Orobanche* derives approximately 0.2% of its C supply from its host's xylem, while the amount of N measured in the xylem sap of the host could supply only 5% of that accumulated by the parasite (Hibberd *et al.*, 1999). Parasite xylem sap contained concentrations of N equivalent to 15% of the N accumulated over the experiment. Likewise, potassium, magnesium, sodium, and sulfur also seem to be provided via the phloem. Conversely, calcium seems to have been mainly supplied via the host's xylem. Furthermore, measurements of transpiration rates and xylem cross-sectional area, which was 10-fold less than its tobacco host, suggests that water-flow in *Orobanche* may be strongly limited by xylem conductivity. It remains unclear, however, whether this is the evolutionary reason for, or simply a result of, *Orobanche*'s phloem-feeding habit (Hibberd *et al.*, 1999).

In phloem conduits, organ to organ transfer of compounds is facilitated by the development of a concentration gradient between organs (Minchin and Lacointe, 2005), or in this case between the phloem conduits immediately adjacent to the haustoria and the phloem in other parts of the plant. Phloem, however, is prone to loss of function if disturbed *in vivo* (Minchin and Thorpe, 2003), so unlike xylem-feeding plants which often penetrate the host xylem cells, phloem-feeding plants do not penetrate the phloem conduits, but have specialized transfer mechanisms, such as interfacial solute transfer cells, which are highly hyphenated to increase their surface area, allowing more efficient solute transfer (Dörr, 1972). *Orobanche* is able to attach sieve plate to sieve plate with *Vicia narbonensis* phloem (Dörr and Kollmann, 1995), which suggests that *Orobanche* may be able to stimulate the production of new phloem cells, to which it can attach.

2. *Effects of phloem parasitism on the host plant*

Amino acids are known to cycle between the above- and below-ground organs in plants (Hibberd and Jeschke, 2001; Simpson *et al.*, 1982), between the phloem and the xylem. Abstraction of amino acids from the phloem by *Orobanche* leads to a decrease in the xylem amino acid concentration, with the concentration of amino acids in the xylem sap of infected tobacco plants approximately 30% that of control plants. Parasitism of tobacco by *Orobanche* seems to have the net effect of increasing both transpiration (26%) and photosynthetic (21%) rates in host plants relative to control plants. Hibberd *et al.* (1999) suggest this increase in photosynthesis is a result of a delayed senescence of older leaves, although both the increased photosynthesis and the delayed senescence may be the result of decreased leaf carbohydrate concentrations, since feedback repression of photosynthesis and stomatal closure often occur when carbohydrates accumulate in leaves (Araya *et al.*, 2006; Nakano *et al.*, 2000), while protein biosynthesis in plant leaves is significantly affected by leaf carbohydrate levels (Krapp *et al.*, 1991, 1993; Tholen *et al.*, 2007).

N uptake by the host plants was stimulated by 27% on a per unit root mass basis by *Orobanche* parasitism. Since we know from Kawachi *et al.* (2008) that parasites derive a majority of their resources from the uptake by the root (s) they are attached to, a 27% increase over the whole root system may translate into a far higher upregulation of nutrient uptake in individually parasitized roots, or may be spread evenly over the whole root system. Whether *Orobanche* releases chemical cues promoting N uptake, or whether this represents an upregulation of N uptake as a result of host plant deficiency, or is simply a result of the increased mass flow of water into the plant as a result of the increased host transpiration rates is debatable. Host xylem amino acid concentrations are significantly lower in parasitized than unparasitized plants, suggesting that the primary site of N reduction may shift from the roots to the leaves in parasitized plants, in addition to the decrease in the cyclic flow of amino acids between the phloem and xylem.

Although parasitism of tobacco by *Orobanche* led to an increase in host photosynthesis, the rate of C uptake by the parasite from the host's root system was higher than the increases in photosynthetic rate, leading to decreases in the amounts of carbon allocated to host roots and shoots. Although an 84% increase in C flux to the roots was noted, 73% of the total C flux to the roots was sequestered by the parasite (Fig. 5). When these figures are reconciled against each other, the C supply to the host root system of parasitized plants is approximately 49% of the C supply to unparasitized plants. Parasitism of tobacco by *Orobanche* did not lead to an overall decrease in the biomass of the host/parasite system, compared to unparasitized tobacco plants (Hibberd *et al.*, 1998); however, after day 50 increases in system biomass were almost exclusively attributable to increases in parasite biomass. Significantly, the relative growth rate of *Orobanche* on 50–73 day old tobacco plants (inoculated 24 days after planting with *Orobanche* seeds) exceeded the RGR of either the infected or uninfected host. Hibberd *et al.* (1998) attributed the lack of difference in system biomass, despite the reduced growth of the photosynthetic host plant, to an increase in the photosynthetic rates of old host leaves, illustrated by the fact that old leaves of infected tobacco plants contained 49% more protein and 94% more Rubisco than the old leaves of control plants. In the tobacco—*Orobanche* system, tobacco did not suffer notable decreases in leaf area as a result of parasitism, rather, *Orobanche* gained biomass at the expense of non-photosynthetic tobacco biomass—roots and stems. Hibberd *et al.* (1999) note that this increase in shoot photosynthesis is converse to the situation in *A. littorea* parasitized by the hemiparasitic *O. phyllanthi*, in which the photosynthetic rate decreases (Tennakoon *et al.*, 1997). Since hemiparasites such as *O. phyllanthi* only derive a relatively small proportion of their C from their hosts, yet derive significant quantities of N, this may be a result of a downregulation of photosynthesis in plants parasitized by hemiparasites as a result of decreased leaf N concentration, while the converse is true in holoparasites, where the parasites appear to maintain their hosts in a C deficient state while having relatively little effect on the transport of

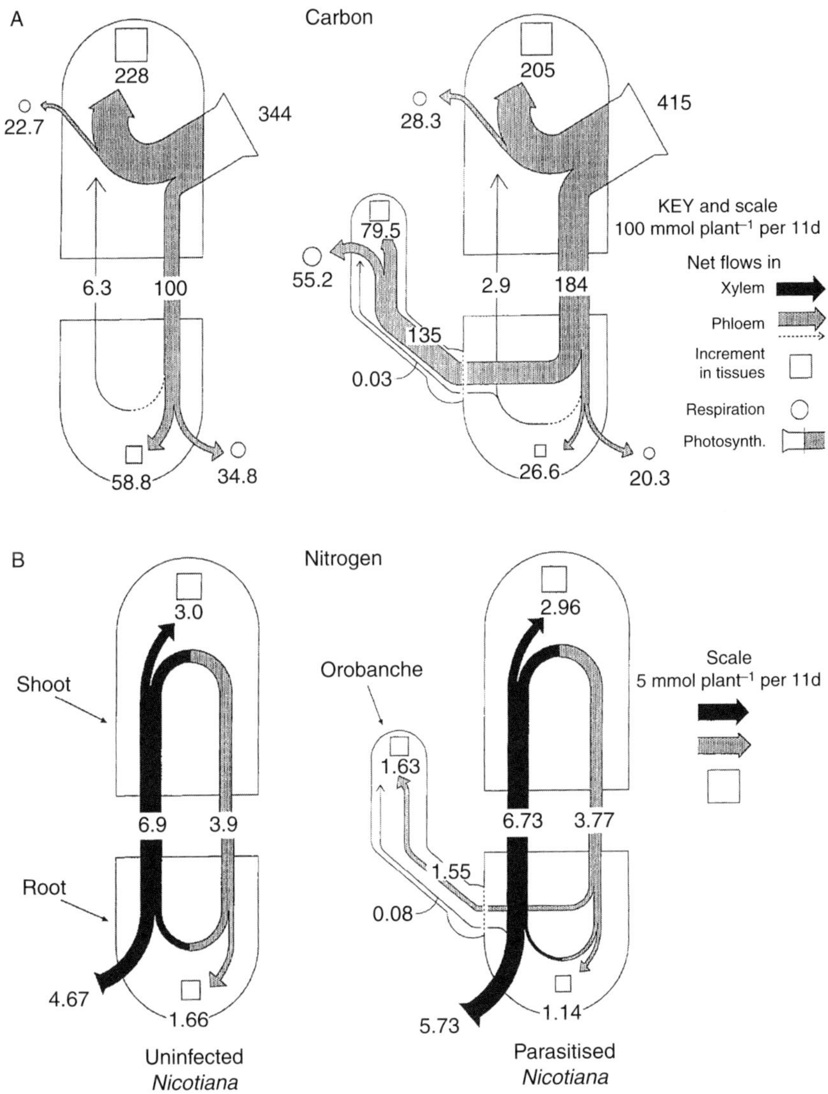

Fig. 5. Transfer of C and N from a tobacco host to an *Orobanche* parasite. Bar details are as Fig. 3, except the units are as shown in the diagrams. Figures are reproduced with permission from Hibberd *et al.* (1999).

N to the leaves. In concordance with this, the effects of multiple infection of tobacco plants by *Orobanche* are not cumulative, rather the parasites compete among themselves for resources (Hibberd *et al.*, 1998), suggesting that the size of the labile pool of resources is the limiting factor for holoparasite growth. Conversely, we would expect the effect of multiple infection by hemiparasites to be cumulative, as each additional parasite reduces the number of uninfected roots supplying water and nutrients to the host stem.

Similar to Hwangbo *et al.*'s (2003) CO_2 enrichment study discussed above, which was conducted to explore the relationship between the xylem-feeding hemiparasite *R. minor* and its host, *P. pratensis*, Dale and Press (1998) conducted a study exploring the relationship between the holoparasitic *Orobanche minor* and its host *T. repens* under CO_2 enrichment. In contrast with the *R. minor–P. pratensis* association where the shoot to root ratio decreased, in the *O. minor–T. repens* association host root biomass was negatively affected, while shoot growth was less affected, increasing the shoot to root ratio. The high CO_2 treatment had some restorative effect in parasitized plants, with the elevated CO_2 parasitized plants showing a smaller decrease in shoot mass than the control CO_2 parasitized plants. Atmospheric CO_2 partial pressure itself had little effect on unparasitized plants, however, with both the ambient and elevated CO_2 plants achieving similar biomass in the absence of parasites implying that even under ambient CO_2 unparasitized plants were not C-limited. In Dale and Press's study the *T. repens–O. minor* association supported a lower biomass than uninfected *T. repens* plants. As phloem-feeding parasites abstract amino acids and carbon from their host and invest it in non-photosynthetic biomass, this may in itself, be sufficient explanation. However, in Hibberd *et al.*'s studies on the *Orobanche*—tobacco association this suppression of biomass accumulation was not noted, which was primarily attributed to the maintenance of host photosynthetic tissues. Infection had a significant negative effect on leaf area in *T. repens*, the reasons for which are not clear, yet it may be related to the energetic costs of symbiotic nitrogen fixation in *T. repens*, which were not experienced by tobacco. Significantly, roots above the site of *O. minor* attachment were significantly thicker in infected than control plants, while roots below the site of infection were significantly thinner in infected plants, pointing to a large withdrawal of C from the host at the site of attachment, and perhaps increased C-flux to the roots. N fixation is energetically demanding and, coupled with the fact that ammonium is toxic to the plant and must be reduced to amino acids in the root, this has obvious energetic implications (Raven and Smith, 1976), and we can assume there would be a major reduction in plant available N. In non-N fixing plants this decrease in root available C may be of lesser importance, since nitrate can be transported to the leaves before being reduced in many species (Andrews, 1986), vastly reducing the energetic demands of the root system. In non-N-fixing plants, reductions in the volume of soil explored may be of greater importance than the availability of C in the roots *per se*, both in ammonium scavenging, since ammonia is relatively immobile in the soil solution compared with nitrate, and in competing effectively with other species for soil nitrogen (Hodge *et al.*, 1999; vanVuuren *et al.*, 1996). That said, *T. repens* grows poorly in fertile soils, where it is out competed by grasses. Thus, *T. repens* parasitism presumably only takes place in nutrient poor environments, which surely exacerbates the requirement for N fixation. Significant work remains to be done confirming these patterns and exploring the postulated differences between phloem and xylem-feeding plants.

Interestingly, infection led to a (calculated) transitory increase in RGR in *T. repens*, an observation which has been noted in other studies (Barker *et al.*, 1996), although this observation does not have widespread backing. The reason behind this increase in RGR would interesting to understand, yet, perhaps due to a lack of evidence leading to further investigation, has not been studied. Speculatively, it may be a hormonal response, or may be due to the alleviation of some stress suffered by the uninfected plant.

Unfortunately, the majority of these experiments, both in xylem- and phloem-feeding parasites, rely on a single method—mass increase—to estimate fluxes, and we require new approaches to confirm these patterns. For example, the use of short lived tracers, such as ^{13}N or ^{11}C can be used to visualize and quantify fluxes in real time (Kawachi *et al.*, 2008, Minchin and Thorpe, 2003), while ^{14}C and ^{15}N have been used successfully to model fluxes in longer term studies (Cameron and Seel, 2007; Carvalho *et al.*, 2006). Finally, natural abundance stable isotope studies have shown great promise (Cernusak *et al.*, 2004; Press *et al.*, 1987) and will be instrumental in improving our understanding of host–parasite biology.

III. PARASITE DEVELOPMENT AND HOST DEFENSE MECHANISMS

Although the evidence for host resistance to parasitism has long been controversial and patchy, the last 10 years has seen a great increase in the amount of attention given to this interesting and important subject. Although many questions remain, the basic mechanisms are becoming more clearly understood, and can largely be split into two main groups; host resistance and host tolerance. Host resistance can further be split into three main groups; suppression of parasite germination, prevention of parasite attachment, and host phytochemical defenses. While germination suppression and the phytochemical defenses are poorly understood, increasing amounts of evidence have accumulated regarding the physical and chemical defenses against the attachment of the parasite to the host in recent years.

A. HOST RESISTANCE

1. Parasite germination and its suppression

Obligate parasites, such as *Striga* and *Orobanche*, typically require germination cues supplied by the host plant to germinate. Facultative parasites, conversely, do not require host factors to germinate (Shen *et al.*, 2006). Prior to germination *Orobanche* requires a specific preconditioning period, which has been linked with the production of gibberellins and cAMP, which seem to be prerequisites for germination (Uematsu *et al.*, 2007). Xenognosins represent the primary chemical class of germination factors, specifically strigolacetones, SXSg, and resorcinol (Estabrook and Yoder, 1998; Pierce *et al.*, 2003). Some plant varieties promote lower rates of parasite germination than others, which has been suggested to be a result of reduced production of germination signal molecules. For example, some *Arabidopsis* (Goldwasser and Yoder, 2001) and *Pisum* (Perez-De-Luque *et al.*, 2005a) varieties induce the germination of fewer *Orobanche* seeds under controlled conditions than others. The mechanism by which this is affected is unknown, although it may be due to reduced stimulant production, the production of chemically discrete stimulant isoforms with altered properties, or potentially the *de novo* production of germination suppression factors. Recent work has shown that the amino acid methionine was both able to almost completely inhibit the germination of *Orobanche ramosa* seeds, and lead to severe reductions in the number of tubercles noted on infected tomato roots (Vurro *et al.*, 2006). Since it appears distance from the host to the seed was a major factor, we might postulate that some plants' germination stimulants are less diffusible than others, or that they degrade faster. Pierce *et al.* (2003) grew 10 varieties of maize plants, mainly parasitism tolerant (4.1–35.1% grain reduction) but also with parasitism susceptible varieties as controls (cf. 81.2% yield reduction). Aliquots of root exudate from each variety were tested for their ability to germinate parasite seeds, with most varieties yielding germination percentages between 55 and 86%, although two varieties were significantly lower than this, only around 20–26% of *Striga* seeds germinated. These varieties both supported only low parasite biomasses, with their own biomass being only slightly compromised by parasitism. Significantly, in every variety, the host above-ground biomass was more significantly affected than the below-ground biomass, which correlates with our discussion of the effects of xylem-feeding parasites on host shoot to root ratio earlier. Likewise, a recent study explored the potential for *Striga* resistance in a wild maize relative, *Tripsacum dactyloides*, finding that significantly fewer, approximately 38%, *Striga* seeds germinated in response to *T. dactyloides* root exudates than *Z. mays* root exudates

(Gurney *et al.*, 2003). Low parasite germination stimulation is genetically controlled and has been bred in *Sorghum* (Haussmann *et al.*, 2001). Finally, the possible interaction of other species should not be ignored—seven toxic chemicals produced by *Fusarium* fungi were able to completely suppress *O. ramosa* seed germination (Zonno and Vurro, 2002), which may have ecological implications.

Facultative parasites, conversely, do not require germination stimulants, although it is unknown whether germination stimulants may increase germination percentage or rate.

2. Parasite attachment and its suppression

a. Suppression of haustoria initiation and formation. Obligate parasite species, for example *Striga,* require specific haustorial initiation factors (HIFs), such as xenognosins, to initiate haustorial formation (Estabrook and Yoder, 1998), while facultative parasites tend to rely instead on low molecular weight secondary metabolites released during normal growth by the host (Yoder, 2001). The initial step in haustorium production is a major shift in cellular growth patterns, with cell expansion, and to a lesser extent cell division, promoted. Root elongational growth declines to nearly zero in *Triphysaria* during this phase (Yoder, 1999), which may be related to a decline in hydrogen peroxide concentrations in the root tip in response to the availability of xenognosins (Keyes *et al.*, 2007). Furthermore, the proto-haustorial cells exhibit a major upregulation in root hair production, all over the now swollen region. The mechanism of action of these HIFs is unclear, yet it is known that the parasite secretes H_2O_2 to the environment which, in a pH dependant reaction, breaks down host cell wall phenolic compounds into benzoquinones, particularly DMBQ (Keyes *et al.*, 2001; Kim *et al.*, 1998). These benzoquinones accumulate in the parasite cells, which causes haustorial formation. Accumulation appears to be necessary and is reversible if the benzoquinone supply is discontinued; this seems to be a mechanism to prevent "false positives," haustoria formed in the absence of the confirmed presence of a root (Keyes *et al.*, 2001, 2007). A large number of parasite genes are upregulated (Matvienko *et al.*, 2001), with 137 distinct transcripts identified, some of which were involved in quinine detoxification, while others appeared to be active in haustorial formation. Recent work has implicated cytokinins, IAA (indole-3-acetic acid) and ethylene in the formation of *Triphysaria* haustoria, suggesting that HIFs may serve to upregulate root hormone production. Whether parasite hormone production could be blocked, or inhibitors released by the host plant remains to be seen. Cytokinins appear to be important in haustorial formation (Keyes *et al.*, 2000) while Tomilov *et al.* (2005) showed that IAA at high concentrations suppressed haustorial initiation. 1-napthalyene acetic acid (NAA) has also been shown to block the formation of primary haustoria in *Striga* (Keyes *et al.*, 2000), and IAA/NAA overproduction may represent another potential defense mechanism for host plants. It is unclear whether *Striga* secondary haustoria are affected in the same way. Some varietal diversity exists in parasite attachment success between *Orobanche* and its host, *Pisum*, suggesting that there may be varietial variation in HIF production or efficacy (Perez-De-Luque *et al.*, 2005a). Likewise, variation exists between *Triphysaria* species in the induction efficiency of haustoria by certain chemicals. When three species of *Triphysaria* were assayed, 90% of *T. versicolor* individuals produced haustoria in the presence of HIFs, while this figure was 40% for *T. pusilla* and only 10% for *T. eriantha* (Jamison and Yoder, 2001), indeed differences were even noted in haustorial initiation between plants within each species. This suggests that haustorial development in each of these three species may be triggered by different HIFs or by different isoforms of specific HIFs. Presumably host root exudates are specific not only to differing species, but also differing individuals, which may represent a form of host selection by the parasite, or a defensive mechanism by the host.

Orobanche resistant purple vetch produces higher concentrations of germination stimulants than control vetch plants, yet these parasites failed to attach (Goldwasser *et al.*, 1997). The specific reasons for this lack of attachment, or the overexpression of the germination factors, are unknown yet may have the net effect of promoting the growth of the *Vicia* nonhost plants, by causing the parasitism and suppression of other species. Two species of vetch, *V. cracca* and *Vicia villosa* subsp. *varia*, are known to produce the allelochemical cyanamide, with *V. cracca* producing it at 10 times the concentration of *V. villosa* (Kamo *et al.*, 2008). It is worth noting, however, that other species, specifically facultative hemiparasites such as *R. minor*, do not require HIFs for haustoria formation, and will attempt to form haustoria upon contact with twigs, stones, or other debris in the soil. During experimental culture, *Rhiananthus* has even been noted trying to form haustoria on the plastic container it was being grown in (Cameron *et al.*, 2005). This may be indicative of some type of mechanical stimulation of haustoria induction in *R. minor*.

Interestingly, parasites show virtually no self attachment, despite hautoria production in the absence of a host being relatively common, with 13–67% of *Triphysaria* plants forming autohaustoria (Yoder, 1997). Autohaustoria, haustoria which spontaneously develop even in the absence of a host, behavior is apparently genetically controlled, with F1 progeny significantly more likely to exhibit autohaustoria if their parents exhibited autohaustoria behavior. This lack of self-parasitism may suggest that either parasites excrete some kind of inhibitory substance—perhaps in response to H_2O_2—or have cell walls which either do not produce haustoria-inducing benzoquinones in the presence of H_2O_2. This might be resolved by testing the effects of benzoquinines purified from the same or different species. When the parasite was grown in the vicinity of an *Arabidopsis* host, significant increases in the proportion of parasites producing haustoria were noted, as would be expected, along with an increase in the rate of haustorial production. Conversely, when two parasites of the same species were grown together, no increases in haustorial production rates or numbers were noted. Between these two extremes, two parasitic plants of the same genus, but different species, produced an intermediate response, where approximately twice the number of haustoria was produced compared with the negative, autohaustorial control plants (Yoder, 1997). While nonhost repression of haustorial induction prevents self-parasitism, allowing the plants to parasitize only host species, it suggests that root derived signals can act as promoters or repressors of haustorial production, which may have some role in host defense. Certainly, it appears that parasites themselves are able to prevent parasitism of their own roots or those of closely related species.

b. Host resistance to attachment. Initial parasite attachment is facilitated by proto-haustorial root hairs, which have a rich hemicellulose coat that binds to the root surface. Subsequent attachment to the host is facilitated by the secretion of mucilaginous substances. This mucilage seems to have some enzymatic activity, and may be responsible for loosening the cell to cell connections between host root cells (Losner-Goshen *et al.*, 1998; Olivier *et al.*, 1991), with vascular penetration effected by a combination of intrusive growth and cell wall digestion (Nagar *et al.*, 1984). As some host varieties are able to block parasite penetration to a greater or lesser degree, this suggests genetic variability on both the part of the host and the parasite in this interaction (Goldwasser *et al.*, 2000b).

Cameron *et al.* (2006) investigated, using histological techniques, electron microscopy and FTIR microscopy the apparent differential host susceptibility to parasitism by *R. minor*. The two grasses in the study were extensively parasitized and had substantial reductions in biomass accumulation as a result. The legume studied, *V. cracca*, supported a high parasite biomass, yet was not significantly negatively affected by the parasite. Likewise, recent work by Cameron and Seel (2007) explored the responses of a grass and two forbs (non-grass herbs) to parasitism by *R. minor*, showing resistance mechanisms in the forbs, and again an apparent lack of resistance mechanisms in the grass, *C. cristatus* (Fig. 3). The two forbs, *Leucanthenum vulgare* and *P. lanceolata,* both effectively prevented the parasite from gaining access to their vascular bundles, while *C. cristatus* either lacks the ability, or it is not energetically favorable, to do so. However, the resistance mechanisms differed between *L. vulgare* and *P. lanceolata* with the former lignifying the host–parasite boundary, and the latter employing both lignification and also undergoing widespread host cell necrosis at the host–parasite interface (Fig. 3; Cameron and Seel, 2007; Cameron *et al.*, 2006). It is unclear whether, although seems likely that, xylem vessel occlusion has a role in parasite resistance. More recent evidence has suggested that *P. lanceolata* may, as well as using apparent "avoidance" mechanisms, preventing the parasite haustoria from attaching to the host root, also use toxic chemicals, which act to disrupt *R. minor*'s photosynthetic capacity (Cameron *et al.*, 2008). Studies of the hypersensitive response in lettuce show that the generation of ROS, which have a putative direct role in microbial defense, increases, while the protective mechanisms to deal with ROS remain at background levels (Bestwick *et al.*, 2001). Indeed, it seems plausible that host root cell death is a by-product of the production of bioactive compounds such as reactive oxygen species (ROS) or phytoalexins, rather than a direct protection mechanism itself. These noted patterns were confirmed in a recent paper investigating the anatomy of *R. minor* parasitizing two grasses, *P. bertolonii* and *H. vulgare*, a legume, *V. cracca,* and two forbs, *L. vulgare* and *P. lanceolata* (Rümer *et al.*, 2007). While the good hosts allowed the production of functional, mature haustoria, the poor host, *P. lanceolata*, only allowed the production of fragmented, damaged haustoria. In *P. bertolonii*, significant lignification was noted, while in *L. vulgare* suberization of the endodermis appeared to be important, yet despite these defenses, the parasite remained able to penetrate the host stele. Although it is common to think of parasitism as an "all or nothing" situation, partial resistance may have functional importance also, perhaps by reducing the rate or efficiency of transfer from the host to the parasite, which in turn may have effects on both host and parasite growth and fecundity. Lignification was far less prominent in the legume, *V. cracca*, while *P. lanceolata* again exhibited both lignification and a pronounced hypersensitive response, with massive cell death noted around the site of attempted attachment. This correlation between the hypersentive response and the failure of *R. minor* to produce functional haustoria may be indicative of direct action by phytotoxins and ROS excreted by *P. lanceolata*. Interestingly, the hyaline body, the tissue responsible for haustorial metabolism, of the haustoria attached to the legume, *V. cracca*, was smaller than the hyaline body of the haustoria attached to *P. bertolonii*, which may reflect a lower requirement for N reduction or the post-abstraction reduction of nitrogen, or may be a toxic effect of cyanamide produced by the legume (Kamo *et al.*, 2008).

Similar host defense mechanisms have been noted in the interaction between *Orobanche crenata* and a variety of faba bean which shows some resistance to parasite infection (Perez-De-Luque *et al.*, 2005b). In resistant plants, significant lignification of host vascular tissue was noted, which included the occlusion of xylem conduits by an unknown, apparently host derived, substance. That the xylem was occluded seems somewhat strange, since *Orobanche* is predominantly a phloem-feeding parasite. The function of xylem occlusion is, therefore, a quandary, but may simply represent some form of physical barrier to further haustorial penetration.

While a parasite may be able to attach to a host's root, this is no guarantee that the endophyte will be able to successfully penetrate the host stele. Putative *S. hermonthica* resistance has been noted in the Nipponbare variety of rice, which seems to be neither a chemical defense, nor a hypersensitive response, nor attributable to lignification (Gurney *et al.*, 2006). In Nipponbare rice, *S. hermonthica* was unable to penetrate the host endodermis and gain access to its vascular tissue. This lack of attachment led to the death of the *S. hermonthica*, with 49% of young *S. hermonthica* plants either being dead or showing signs of necrosis 21 days after infection. The physiological reason for the apparent inability of *S. hermonthica* to penetrate the Nipponbare endodermis is unknown; however, it was heritable, and Gurney *et al.* were able to identify QTL associated with resistance, which suggests that we may in the future be able to successfully breed *S. hermonthica* resistance in rice. Since parasite penetration into the host root is via the intercellular interface, rather than directly through cells, it suggests that this endodermal barrier may be comprised of some currently uncharacterized suberin or other chemical responsible for endodermal impenetrability, or indigestibility. Recent work by Swarbrick *et al.* (2008) has characterized the global patterns of gene expression in Nipponbare and a *Striga* susceptible rice variety, IAC 165. In both varieties, large numbers of genes were downregulated, 1749 in IAC 165 and 1295 in Nipponbare, with 653 in common between the two. Likewise, a smaller number of genes were upregulated, 358 in Nipponbare and 330 in IAC 165, with 148 in common. The majority of downregulated genes were of unknown function, yet genes involved in cellular communication and metabolism were most strongly affected, both positively and negatively, in both varieties, while genes involved in cell rescue and defense responses appeared to be upregulated only. Nipponbare was generally less affected, with lower numbers of genes showing altered behaviors in response to parasitism than in IAC 165.

Endodermal defenses were also noted in *Vicia atropurpurea* (purple vetch), but not in common vetch, *Vicia sativa* (Goldwasser *et al.*, 2000a). However, in resistant vetch species a build up of a reddish-brown secretion, possibly some form of suberization, was noted, which was absent in the Nipponbare rice.

In lettuce, marigold and cowpea, *S. asiatica* was able to attach to the root surface, yet in all cases was able to traverse less than 25% of the host root's cortex (Hood *et al.*, 1998). The reason for the parasites inability to attach to the host's stele is unknown, yet has been postulated to be a result of the production of cytotoxic compounds. Since neither lettuce nor marigold are common host species for any *Striga* species, their response may be a "generic" response to invading organisms, which ironically the *parasite* has not evolved resistance to! Cowpea, conversely, is a host plant for *S. gesnerioides*, and it seems likely that *S. asiatica* is repulsed by the host's defense responses to *S. gesnerioides*. Sunflower (*Helianthus* spp) is postulated to use both the endodermal defenses exhibited by purple vetch and the apparent cytotoxin defenses noted in lettuce, although the evidence for cytotoxins in sunflower was based on cell necrosis, and remains provisional (Labrousse *et al.*, 2001).

c. Post-attachment phytochemical defenses to infection. *S. hermonthica* causes significant reductions in host biomass and yield when attached to *Zea mays* hosts. However, *S. hermonthica* in the presence of the *Z. mays* wild relative *T. dactyloides* produced fewer root tubercules and leaf primordia (Gurney *et al.*, 2003). Parasite growth rates were likewise host specific, with *T. dactyloides* supporting parasite biomasses two to three orders of magnitude lower than *Z. mays*. Parasitism, or at least attempted parasitism, by *S. hermonthica* had effects on host biomass accumulation and partitioning, and again this was host specific. *T. dactyloides* was relatively unaffected by parasitism—if anything its growth was actually promoted by *S. hermonthica* infection, while *Z. mays* plants typically suffered a 45% reduction in biomass over the experimental period, with, as we would expect from a xylem-feeding parasite, shoot and leaf material most affected, and root biomass least affected. *S. hermonthica* was able to successfully attach to the host *Z. mays* plants roots, with successful penetration by the endophyte, and clearly defined tissue differentiation, including large a hyaline body—an organelle and lipid rich tissue which appears to be important in haustorial metabolism, including N metabolism (Gurney *et al.*, 2003). However, while haustoria were able to form on *T. dactyloides roots*, they were approximately 50% smaller, little tissue differentiation was evident, and no hyaline body was present. The host root also seemed to be less damaged by infection in *T. dactyloides* than in *Z. mays*, with a *Z. mays–T. dactyloides* hybrid midway between the two extremes. *T. dactyliodes* effects were more widespread within the parasite, since secondary haustoria attempting to attach to *Z. mays*, after the plant has attached to *T. dactyliodes*, showed similar symptoms, including poor tissue differentiation and a lack of a hyaline body associated with attachment to *T. dactyliodes*. While there was some lignification evident in the system, it is clear that *T. dactyliodes* effects its parasite resistance after a xylem–xylem connection has been formed. The fact that attachment to *T. dactyloides* by the parasite was able to prevent successful parasite attachment to other species strongly suggests that *T. dactyloides* produces some type of xylem-borne chemical defense against *S. hermonthica*, which Gurney *et al.* (2003) postulated may act on the formation of the hyaline body, and thus parasite energetics. Chemical defenses against parasitism have been demonstrated in other studies, notably *S. hermonthica* infections of maize could be kept under check by intercropping with the allelopathic *Desmodium uncinitum* (Khan *et al.*, 2007). Likewise, *Striga gesneriodes*, of which there are a minimum of seven distinct races, each with its own specific virulence and host range, resistance has been noted in cowpea (*Vigna unguiculata*) populations. While *S. gesneriodes* was able to form haustoria, they did not develop properly, lacking clear tissue differentiation, suggesting that some cowpea varieties are able to produce chemical defenses which significantly retard haustorial development (Timko *et al.*, 2007). Significant progress has been made in understanding the genetic basis of this resistance, which can be localized to a small number of QTL positions, suggesting that resistance is either coded by a single allele or by a small complex of tightly grouped genes (Ouedraogo *et al.*, 2001, 2002). While parasitism of *Medicago truncatula* by *O. crenata* led to the upregulation of up to 81 presumably defense genes in the host, more than 30% of which with no known function (Die *et al.*, 2007).

In *Pisum* varieties, significant numbers of post-attachment *Orobanche* parasites underwent tubercle necrosis (Perez-De-Luque *et al.*, 2005a). Various hypotheses were forwarded for this phenomenon, such as the occlusion of parasite vascular tissue (presumably phloem) by host secretions, and the production of toxic chemicals. Recent work in this area has included QTL mapping, genomic, and proteomic studies. Two *Orobanche* resistance QTL have been discovered, explaining 21% of the noted variation in pea susceptibility (Valderrama *et al.*, 2004), with specific pathogenic and wound response genes upregulated as a result of parasitism (Griffitts *et al.*, 2004). Twenty-eight percent of the proteins showing a differential response to parasitism could be identified (Castillejo *et al.*, 2004). Although these percentages seem low, general genotype-specific patterns could be noted from the data, such as the apparent downregulation of metabolic genes while defense compound genes, including peroxidases, were upregulated (Castillejo *et al.*, 2004). Meanwhile, infection by certain *Rhizobium* strains was able to significantly reduce *Orobanche* success, both by preventing infection and the subsequent necrosis of tubercles (Mabrouk *et al.*, 2007a–c).

Secondary defense compounds may have a strong influence in host–parasite interactions. For example, nicotine is produced in tobacco roots and is transported in the xylem to the leaves where it accumulates (Kaplan *et al.*, 2008). Tobacco parasites thus tend to be phloem feeders, since nicotine levels are low in the phloem while xylem-feeding hemiparasites would undoubtedly accumulate nicotine to very high, presumably toxic levels. Tissue damage tends to upregulate the nicotine genes (Kaplan *et al.*, 2008), which may further render the host unpalatable to herbivores and phytophages such as aphids, with the parasite effectively using the host defense mechanisms to reduce competition. We might imagine xylem-feeding hemiparasites evolving resistance to the host toxins and acting in a similar manner, accumulating secondary metabolites in their leaves, although this is currently speculative.

B. HOST TOLERANCE

Although this is far less well understood than host resistance, some varieties of maize appear to allow parasite germination, and the establishment of functional haustoria in large numbers, yet are apparently unaffected in terms of biomass accumulation or height increase (Pierce *et al.*, 2003). Other potential differences, for example in host N concentration or the quality of grain produced by the host plants were not measured, so it is hard to tell what "real" effects the parasitism may have had. Indeed, a recent paper looking at the effects of N source on *R. minor* parasitism in *Vicia faba* (Jiang *et al.*, 2008a) showed significant differences in parasite growth between high (N supplied) and low N (N-fixation only) treatments, yet no differences were apparent in the host growth rates. Tissue analysis showed that the parasitized plants grown at low N had lower tissue N concentration and supported a lower parasite biomass, suggesting that while host N concentration was unimportant in determining the host growth rate within this system, it was important in determining the parasite's growth rate. These legumes might, on the basis of biomass alone, be considered to be parasitism "tolerant," yet the reduced N concentrations presumably have downstream effects on plant growth or fecundity.

IV. ECOLOGICAL IMPLICATIONS OF PARASITE–HOST PHYSIOLOGY

A. HOST RANGE

As a result of the above detailed factors, hemiparasites would be expected to have a significantly greater host range than holoparasites (Fig. 6); however, truthfully the full host range for most species remains unclear. Pate and Bell

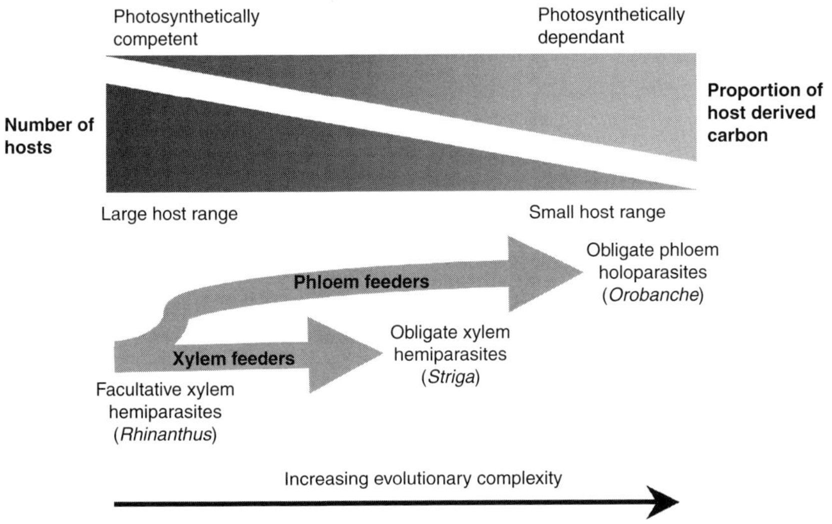

Fig. 6. Relationship between parasite functional type, the proportion of its C derived from its host and the number of species they are able to successfully parasitize. Note that the split arrow showing the differences between the evolutionary pathways of phloem and xylem feeders is not meant to indicate a monophyletic origin, rather that any given parasite, upon evolving to become a facultative parasite has two potential pathways it might follow—to remain a xylem feeder, or to become a phloem feeder.

(2000) noted that the facultative hemiparasite *P. viscosa*, a species introduced into Australia, was able to parasitize all of the 27 species tested, comprising of both indigenous and introduced species. While hemiparasites, such as *R. minor* which has a reported host range of in excess of 20 species, or *Triphysaria*, a hemiparasitic member of the Orobanchaceae (ex-Scrophuariaceae), which has a host range of at least 25 species in 18 different families, have relatively large host ranges, many holoparasites only parasitize one or a small range of species, for example *Orobanche hederae* can only parasitize *Hedera helix* (ivy). However, within the last 4 years Moroccan populations of *Orobanche foetida* have been noted infecting common vetch (*V. sativa*), a hitherto nonhost species, suggesting that genetic variability exists, even in phloem-feeding holoparasites, which allows them to adapt to the local range of potential host species (Patto *et al.*, 2008). Some *Orobanche* species are able to parasitize a relatively large number of species; recent work investigating the host range of *O. ramosa* found many species could be parasitized, for example 35 of the 55 weedy species investigated were susceptible, although relatively few appeared able to complete their life cycles (Qasem and Foy, 2007). As hemiparasites do not require a host to complete their life cycle, due to their ability to conduct all required biochemical reactions, there has been little apparent evolutionary pressure to acquire one or a few specific hosts; however, we cannot reiterate enough that it should not be concluded that all hosts are equal for hemiparasites, and significant differences in host quality are readily apparent as described at length above and by many other authors (Atsatt and Strong, 1970; Cameron *et al.*, 2006). For hemiparasites, parasitism represents an "easy" method of increasing their fecundity, may be required for normal growth (Jiang *et al.*, 2007), and their requirements of mineral nutrients and water can be met by a range of hosts. Conversely, holoparasites tend to have a limited host range, due to their far greater reliance on their host plants, which must meet all the parasites' metabolic demands.

Parasite evolution is envisaged to go through three distinct stages (Searcy, 1970; Searcy and MacInnis, 1970). The first stage is characterized by the attachment of the parasite to the host. The second stage involves the reduction in the amount of genetic material, as genes which are unnecessary for the parasites new lifestyle are selected against—such as the inability of *Rhinanthus* to grow normally, unless facilitated by a host, or the loss of photosynthetic capacity in *Orobanche*. The final stage is one of increasing genetic complexity, as the parasite becomes specialized to an individual host species—for example, the necessity for germination stimulants and HIFs.

The evolution of the haustorium, and thus the parasitic mode of life, has long been contentious. Atsatt (1966, quoted from Kuijt, 1969) speculated that root parasitism may have arisen by some form of root grafting; however, this was summarily dismissed by Kuijt (1969), as true root grafts are vanishingly rare in herbaceous plants. Furthermore, Kuijt makes the point that root grafting is a phenomenon of the secondary growth of older roots. Björkman (1960), based upon isotope labeling studies showing ^{14}C being transferred between plants via a fungal partner, suggested that parasitism may have arisen through a mycorrhizal association between the proto-host and the proto-parasite. This phenomenon of plant parasitism of fungi, termed mycoheterotrophy, has certainly evolved in at least 400 species across several plant families, most notably in the Orchidaceae (Leake, 1994) but also in a wide range of taxonomic groups including both "lower" and "higher" plants (Leake *et al.*, 2008) with up to 10% of all plants relying on fungus-derived carbon at some point in their life cycle (Cameron *et al.*, 2008b). Moreover, such tripartite symbioses whereby mycoheterotrophic plants steal carbon and mineral nutrients from mycorrhizal fungi which in turn form mutualistic mycorrhizas with other plants (so-called epi-parasitic plants) have also been clearly demonstrated (McKendrick *et al.*, 2000). There is however no evidence that mycoheterotrophy is the evolutionary origin of haustorial parasitism, indeed this seems unlikely especially given that mycoheterotrophy is more prevalent in the monocotyledonous plants than in dicotyledonous plants, in direct contrast with the haustorial parasites which have exclusively evolved in the dicotyledonous plants (Cameron and Leake, 2007). Furthermore, we would add that if Björkman's hypothesis is correct, we would expect to see the overwhelming majority of parasitic species evolving into holoparasites, and we would not expect to see xylem–xylem continuity in these species. Our preferred hypothesis for the evolution of parasitic plants revolves around the production of root hairs. We can imagine our proto-parasite and its host growing in the same environment, with their roots growing in close proximity to each other. All plants exude a variety of carbon-based compounds to the soil, including acid phosphatases, amino acids, and various soluble carbohydrates. Let us assume that our proto-parasite is able to produce root hairs in response to this; not unreasonable, considering this response is seen in many species when their roots happen across a nutrient rich patch (Hodge *et al.*, 1999). Proto-parasites which are able to maximize their contact with their donor plant, perhaps by encapsulating the host root, may have had a significant advantage over those which did not respond to plant exudates, or those individuals which produced few hairs. Finally, we may consider that some species evolved the trick of damaging their host roots to stimulate the release of exudates and other chemicals—graduating from the plant equivalent of petty theft to breaking and entering.

However parasitism *did* evolve, the physiological evidence suggests that the first parasites were hemiparasitic, and either xylem-feeding or a mixture of xylem and phloem feeding. Some parasites appear to have been able to access the host's phloem contents, and evolved into the phloem-feeding holoparasites (Fig. 6). However, the phloem-feeding holoparasites still retain features of their xylem-feeding ancestry, such as the largely functionally irrelevant xylem of *Orobanche*, or the functionally useless but still transcribed *rbc*L gene, encoding the large subunit of Rubisco (Benharrat *et al.*, 2000; Lusson *et al.*, 1998). Weber (1987) considered the evolution of primary and secondary haustoria in parasitic members of the Orobanchaceae and Scrophulariaceae, concluding that secondary haustoria probably evolved first, later evolving into the more complex, terminal primary haustoria. Ancestral phloem feeders could derive all their energetic and nutritional requirements parasitically, and did not require photosynthesis for C acquisition. Selective pressures would have tended to favor phloem-feeding parasites which, rather than investing resources in photosynthesis (Yoder, 1999), invested them in reproduction. As phloem feeders lost their autotrophic abilities (dePamphilis *et al.*, 1997; Wickett *et al.*, 2008; Wolfe *et al.*, 1992), they would be pressured into an evolutionary arms race to evolve complex mechanisms allowing them to sense other species, and to synchronize their life cycles. This complexity, however, resulted in greater host specificity in phloem-feeding parasites, due to their greater evolutionary specialization (Fig. 6). In this way, *Striga* is particularly interesting, in that it is an obligate hemiparasite, with a holoparasitic underground stage, and a hemiparasitic above-ground stage (with the exception of one virtually holoparasitic species, *Striga gesnerioides*, which has chlorophyll, but lacks leaves). Indeed, *Striga* may be the closest to holoparasitism possible in a xylem-feeding angiosperm. This honor may be shared in the gymnosperms by their single parasitic representative, *P. usta*, which is xylem feeding for water and nitrogen, but, as determined by isotopic discrimination studies, derives its carbon mycoheterotrophically, via a fungus common to both the host and the parasite. It seems clear that the parasite is unable to derive sufficient carbon directly parasitically from the host xylem, and may be effectively phloem parasitic by proxy. Indeed, in common with phloem-feeding parasites, *P. usta* demonstrates many of the secondary adaptations to holoparasitism, such as a severe reduction in chloroplast number, and the inability to photosynthesize independently (Feild and Brodribb, 2005).

Clearly, some parasites have the ability to choose their host based upon chemical signals in the environment. Differential germination, haustoria induction as well as presumably directional root growth may all play parts in host selection by plants. However, the relative importance of each of these factors varies between groups and species, with germination responses being more important in obligate parasites, while post-germination responses appear to be more important in facultative parasites. The flip side of the coin is that host defense mechanisms also play an important role in determining host range, with various species exhibiting responses such as lignification, phytotoxins, root, and tissue necrosis as well as various forms of attachment suppression, as noted earlier. The combination of the parasite's ability to penetrate the host stele and the host's ability to resist that attack which determines whether a potential host falls within the host range of the parasite.

Many parasite species are "r-strategists" and produce a large number of seeds, for example, *Striga* can produce up to 20 000 seeds, although they are very small. *Orobanche* is similarly fecund, producing 15 000 or more seeds. Due to the small size of *Striga* seeds, the seedling can only survive a short time, approximately 5 days, in the soil in the absence of a host. The evolution of complex germination signaling between the host and the parasite presumably allowed *Striga* plants to reduce the size of their seed, since the probability of host plant attachment would be very much higher in the presence of such mechanisms, in turn allowing *Striga* to become much more fecund. This fecundity, along with the longevity of *Striga* seeds, which can survive in a dormant state for up to 20 years, represent major issues for the control of *Striga* in agricultural settings, as it is impossible to completely eradicate them from the soil seed bank using methods such as soil fumigation. Facultative parasites, conversely, require a larger seed reserve, to provide energy for shoot emergence from the soil. However, although the r-strategist classification is common to most parasites, some parasites exploit other ecological niches, for example *Melampyrum sylvaticum*, which is classified as an annual k-strategist (Dalrymple, 2007).

B. ECOLOGICAL IMPLICATIONS OF PLANT PARASITISM

In the nineteenth century parasitic plants were often regarded simply as botanical curiosities; however, in recent times their importance in agricultural (Parker and Riches, 1993) and natural (Gibson and Watkinson, 1992) ecosystems has been more widely recognized. Over the past 30 years the majority of the research in this field has focused on parasitic weeds. Parasitic plants have become weeds in a range of agricultural systems; *Orobanche* species have become a pest of crops such as sunflowers and tomatoes in Mediterranean areas (e.g., Jorrin *et al.*, 1996) where the parasite has become economically significant. However, the most publicized of the parasitic weeds are species of the *Striga* genus that are obligate root hemiparasites of tropical grasses and form a significant threat to agriculture in sub-Saharan Africa (Parker and Riches, 1993). As such, recent work has focused on understanding host–parasite relations between these economically important species and their range of hosts (e.g., Goldwasser *et al.*, 2000a; Gurney *et al.*, 2003) while the interactions between other non-economically significant parasitic plants have been relatively under researched in comparison. This is despite parasitic plants also being significant components of natural and seminatural ecosystems around the globe. As a consequence their potential role in the functioning of these systems is poorly understood. A number of studies have however shown that parasitic plants represent a potent force for the regulation of the structure and dynamics of the communities they inhabit (Davies *et al.*, 1997; Gibson and Watkinson, 1992; Marvier, 1998; Pennings and Callaway, 1996). However, as would be expected, there is a great deal of variability in the response of the community to the parasite and debate as to the mechanisms through which the parasites effect changes in communities (Cameron *et al.*, 2005).

Host plants generally suffer suppressed biomass accumulation and reproductive output (Davies and Graves, 2000; Seel and Press, 1996) as a result of infestation by parasitic plants. Parasitic plants can thus influence communities if they parasitize and hence suppress the growth and competitive ability of dominant or indeed subdominant plant species (Callaway and Pennings, 1998). Interspecific competition between two or more species can be mediated by the parasite suppressing the dominate species to a greater extent than a subdominant with poor competitive ability, hence parasitic plants can facilitate coexistence (Smith, 2000). Consequently, they not only affect the growth of individual hosts, but may also influence the outcome of competitive interactions between host species (Gibson and Watkinson, 1991; Matthies, 1996). For example, Marvier (1998) showed that competition with grasses constrained the growth of the forb *Hypochaeris glabra.* However, when the root hemiparasite *Triphysaria* was present then the forb was freed from competition with the grasses which were suppressed by the parasite. Additionally, Callaway and Pennings (1998) and Pennings and Callaway (1996) showed that the shoot holoparasite, *Cuscuta salina*, can influence the structure of salt marsh communities by preferentially parasitizing *Salicornia virginica*, allowing the less preferred *Limonium californium* and *Frankenia salina* to proliferate (Pennings *et al.*, 1996).

R. minor is a good example of a parasitic plant widely distributed throughout natural and seminatural grasslands in northern temperate regions. *R. minor* is an annual root hemiparasite and, as discussed above, has the potential to undertake independent photosynthesis but is generally reliant on a host for the majority of its mineral nutrition. *R. minor* has a wide host range (Gibson and Watkinson, 1989) and significantly affects the biomass of at least some of the hosts to which it attaches (Davies and Graves, 2000). Indeed, the wide host range of *R. minor* suggests it must in almost all environments have access to potential hosts, which may have allowed the evolutionary reduction in its root system. Critically though, the impact of *R. minor* is not consistent across all host species (Cameron *et al.*, 2006, 2005; Davies *et al.*, 1997). Broadly grasses are damaged to the greatest extent while legumes and, particularly, forbs are left undamaged (Hwangbo, 2000; Joshi *et al.*, 2000). Moreover, this general trend was quantified in a recent meta-analysis by Ameloot *et al.* (2005) who showed that, on average, infestation by *R. minor* resulted in a 56% suppression of the grass biomass but an increase of 16% in the forb biomass, a trend observed in both the long and short term (Ameloot *et al.*, 2006). Consequently *R. minor* can induce shifts in the balance of a community, in favor of the forbs at the expense of the grasses (Davies *et al.*, 1997). Host suppression in the community by *R. minor* can initially act to reduce diversity (Gibson and Watkinson, 1991). However, in the long term *R. minor* can promote diversity as at the end of the season when this parasite dies the gap in the vegetation that is created provides a regeneration niche for other species to colonize (Joshi *et al.*, 2000) thus further enhancing botanical diversity. The presence of a parasitic plant can allow the coexistence of two plants that would not exist together in other circumstances (Smith, 2000). Conversely, if rare species are preferentially parasitized by a parasitic plant then diversity can also be reduced (Pennings and Callaway, 1996).

Gibson and Watkinson (1992) suggested that parasitic plants select or rather preferentially parasitize some species over others; however, the mechanism through which "preferential parasitism" acts is unclear and likely to be species specific. Choice experiments with *Cuscuta pentagona* (Runyon *et al.*, 2006), *Cuscuta campestris* (Koch *et al.*, 2004), and *C. salina* (Pennings and Callaway, 1996) have shown selective foraging for the most "rewarding" hosts by these stem holoparasites. Additionally, *Striga* and *Orobanche* species are able to exert some selectivity over the hosts they utilize by germinating in response to specific germination stimulants released into the rhizosphere by plant roots (Yoder, 2001).

Gibson and Watkinson (1989, 1991, 1992) have suggested that host choice by *R. minor* may account for its effects in the community although they report that *R. minor* chooses *P. lanceolata* as a preferred host yet subsequent studies have shown that this at this association may even be fatal for the parasite (Cameron and Seel, 2007; Cameron *et al.*, 2006). Additionally *R. minor* does not require a germination stimulant (Vallance, 1952) thus the element of choice at this stage is not present. Choice therefore seems an unlikely explanation for *R. minor*'s observed effects in the community. Vanhulst *et al.* (1987) have suggested that host resistance may play a part in the community level effects of *R. minor* and other hemiparasites, and this argument would intuitively seem to be stronger than host choice. Recent evidence, detailed above, supports this hypothesis suggesting that the differential resistance capacity of the forbs (that are resistant) and the grasses (that are not) in fact drives parasite-induced shifts in host community structure rather than selective parasitism (Cameron and Seel, 2007; Cameron *et al.*, 2006, 2005; Rümer *et al.*, 2007; Westbury and Dunnett, 2007). Moreover, such physiological studies are beginning to provide the link between field observations of the parasitic plant-induced shifts in host community structure and the mechanisms that underpin these shifts. In spite of these recent breakthroughs, the variable effect of the parasite on host community structure and dynamics remains highly unpredictable when studied in field situations (Ameloot *et al.*, 2005; Cameron *et al.*, 2005). Further research is now required to understand how biotic and abiotic factors interact with parasitic plants to determine the magnitude and polarity of any parasitic plant-induced effects on host community structure and dynamics.

V. CONCLUSIONS

In conclusion, while parasites are conventionally split into hemiparasites and holoparasites, obligate and facultative, we have split them more broadly into two groups; xylem feeders and phloem feeders, with the xylem feeders then split again as obligate or facultative. This grouping has allowed us to understand, on the basis of the available carbon and nutrients available to any given parasite, their behavior, evolution, and life cycle, along with aspects of their ecology. Xylem feeders, which are primarily parasitic for N and other mineral nutrients and water, are largely capable of supplying their own photosynthetic needs. As such, host stems tend to become N deficient and C replete, shifting the shoot: root ratio toward increased root biomass, at the expense of shoot growth. Phloem-feeding plants are typically wholly dependent upon their host for all their growth needs and are often achlorophyllous. The host stems are generally N replete as root to shoot transport of N is relatively unaffected, yet overall the plant may be C depleted, favoring an increase in shoot: root ratio, as the parasite utilizes root-bound C before it can reach the root apical meristems. Facultative xylem-feeding plants tend to interact with their hosts post-germination, when the parasite has emerged from the soil, while obligate parasites; both xylem and phloem feeding interact from before seed germination, often involving germination and haustorial initiation factors. Resistance to facultative parasites is largely based around physical and chemical defenses, such as lignification and allelopathy, while resistance to obligate parasites is mainly directed toward preventing parasite seed germination and haustoria attachment, although physical and phytochemical defenses are often also involved. The genetics of host resistance to parasitism is currently at an easy stage and is proving to be a complex study, with many different partial resistance loci being located. Large numbers of genes are affected when a host is attacked by a parasite, many of which we are currently unable to identify functions for. The subterranean habit of many parasites, for example *Striga* lives underground for 4–6 weeks before emergence, along with the large production of robust seeds capable of living in the soil for many years makes control by conventional means difficult, with crop rotation, the use of "catch crops" and intercropping currently the most effective control mechanisms. On the other hand it is clear that hemiparasites can have positive influences in promoting botanical diversity in natural ecosystems due to their differential deleterious effects, although the effects of holoparasites on natural and seminatural ecosystems remains only poorly understood.

Overall, studies of plant parasitism remain in an embryonic state, with even basic information such as host to parasite N and C fluxes only available for a small range of parasites, and even then only incompletely. However, significant progress has been made over recent years, which will undoubtedly yield great advances in both our understanding of both these interesting plants, and in the development of strategies to control these potent agricultural pests.

ACKNOWLEDGMENTS

LJI is supported by a Japan Society for the Promotion of Science postdoctoral fellowship (Award number: PO8097) and DDC is supported by a Natural Environment Research Council (UK) independent fellowship (Award number: NE/E014070/1). We would like to thank our two anonymous referees for their helpful comments.

REFERENCES

Addiscott, T. M., Whitmore, A. P. and Powlson, D. S. (1991). "Farming, fertilizers and the nitrate problem." CAB International, Oxon.

Adetimirin, V. O., Aken'ova, M. E. and Kim, S. K. (2000). Effects of *Striga hermonthica* on yield components in maize. *Journal of Agricultural Science* **135**, 185–191.

Ameloot, E., Verheyen, K. and Hermy, M. (2005). Meta-analysis of standing crop reduction by *Rhinanthus* spp. and its effect on vegetation structure. *Folia Geobotanica* **40**, 289–310.

Ameloot, E., Verheyen, K., Bakker, J. P., De Vries, Y. and Hermy, M. (2006). Long-term dynamics of the hemiparasite *Rhinanthus angustifolius* and its relationship with vegetation structure. *Journal of Vegetation Science* **17**, 637–646.

Andrews, M. (1986). The partitioning of nitrate assimilation between root and shoot of higher plants. *Plant, Cell and Environment* **9**, 511–519.

Araya, T., Noguchi, K. and Terashima, I. (2006). Effects of carbohydrate accumulation on photosynthesis differ between sink and source leaves of *Phaseolus vulgaris*. *Plant and Cell Physiology* **47**, 644–652.

Atsatt, P. R. (1966). The population biology of closely related species of *Orthocarpus*, PhD Thesis, University of California, Los Angeles.

Atsatt, P. R. and Strong, D. R. (1970). Population biology of annual grassland hemiparasites. Host environment. *Evolution* **24**, 278.

Barker, E. R., Press, M. C., Scholes, J. D. and Quick, W. P. (1996). Interactions between the parasitic angiosperm *Orobanche aegyptiaca* and its tomato host: Growth and biomass allocation. *New Phytologist* **133**, 637–642.

Barkman, T. J., McNeal, J. R., Lim, S. H., Coat, G., Croom, H. B., Young, N. D. and dePamphilis, C. W. (2007). Mitochondrial DNA suggests at least 11 origins of parasitism in angiosperms and reveals genomic chimerism in parasitic plants. *BMC Evolutionary Biology* 7, 248.

Benharrat, H., Delavault, P., Theodet, C., Figureau, C. and Thalouarn, P. (2000). RbcL plastid pseudogene as a tool for *Orobanche* (subsection minores) identification. *Plant Biology* **2**, 34–39.

Bestwick, C. S., Adam, A. L., Puri, N. and Mansfield, J. W. (2001). Characterisation of and changes to pro- and anti-oxidant enzyme activities during the hypersensitive reaction in lettuce (*Lactuca sativa* L.). *Plant Science* **161**, 497–506.

Björkman, E. (1960). *Monotropa hypopitys* L., an epiparasite on tree roots. *Physiologia plantarum* **13**, 308–327.

Callaway, R. M. and Pennings, S. C. (1998). Impact of a parasitic plant on the zonation of two salt marsh perennials. *Oecologia* **114**, 100–105.

Cameron, D. D. and Leake, J. R. (2007). A different kind of parasitic plant: A brief history of mycoheterotrophy and epi-parasitism. *Haustorium* **50**, 4–6.

Cameron, D. D. and Seel, W. E. (2007). Functional anatomy of haustoria formed by *Rhinanthus minor*: Linking evidence from histology and isotope tracing. *New Phytologist* **174**, 412–419.

Cameron, D. D., Hwangbo, J. K., Keith, A. M., Geniez, J. M., Kraushaar, D., Rowntree, J. and Seel, W. E. (2005). Interactions between the hemiparasitic angiosperm *Rhinanthus minor* and its hosts: From the cell to the ecosystem. *Folia Geobotanica* **40**, 217–229.

Cameron, D. D., Coats, A. M. and Seel, W. E. (2006). Differential resistance among host and non-host species underlies the variable success of the hemi-parasitic plant *Rhinanthus minor*. *Annals of Botany* **98**, 1289–1299.

Cameron, D. D., Geniez, J. M., Seel, W. E. and Irving, L. J. (2008a). Suppression of host photosynthesis by the parasitic plant *Rhinanthus minor*. *Annals of Botany* **101**, 573–578.

Cameron, D. D., Johnson, I., Read, D. J. and Leake, J. R. (2008b). Giving and receiving: Measuring the full cost of mycorrhizas in a model orchid, *Goodyera repens*. *New Phytologist* **180**, 176–184.

Carvalho, D. D., Irving, L. J., Carnevalli, R. J., Hodgson, J. and Matthew, C. (2006). Distribution of current photosynthate in two guinea grass (*Panicum maximum* Jacq.) cultivars. *Journal of Experimental Botany* **57**(9), 2015–2024.

Castillejo, M. A., Amiour, N., Dumas-Gaudot, E., Rubiales, D. and Jorrin, J. V. (2004). A proteomic approach to studying plant response to crenate broomrape (*Orobanche crenata*) in pea (*Pisum sativum*). *Phytochemistry* **65**, 1817–1828.

Cernusak, L. A., Pate, J. S. and Farquhar, G. D. (2004). Oxygen and carbon isotope composition of parasitic plants and their hosts in southwestern Australia. *Oecologia* **139**, 199–213.

Christensen, N. M., Dörr, I., Hansen, M., van der Kooij, T. A. W. and Schulz, A. (2003). Development of *Cuscuta* species on a partially incompatible host: Induction of xylem transfer cells. *Protoplasma* **220**, 131–142.

Cramer, M. D., Hoffmann, V. and Verboom, G. A. (2008). Nutrient availability moderates transpiration in *Ehrharta calycina*. *New Phytologist* **179**, 1048–1057.

Dale, H. and Press, M. C. (1998). Elevated atmospheric CO_2 influences the interaction between the parasitic angiosperm *Orobanche minor* and its host *Trifolium repens*. *New Phytologist* **140**, 65–73.

Dalrymple, S. E. (2007). Biological flora of the British Isles: *Melampyrum sylvaticum*. *Journal of Ecology* **95**, 583–597.

Davies, D. M. and Graves, J. D. (2000). The impact of phosphorus on interactions of the hemiparasitic angiosperm *Rhinanthus minor* and its host *Lolium perenne*. *Oecologia* **124**, 100–106.

Davies, D. M., Graves, J. D., Elias, C. O. and Williams, P. J. (1997). The impact of *Rhinanthus* spp. on sward productivity and composition: Implications for the restoration of species-rich grasslands. *Biological Conservation* **82**, 87–93.

dePamphilis, C. W., Young, N. D. and Wolfe, A. D. (1997). Evolution of plastid gene rps2 in a lineage of hemiparasitic and holoparasitic plants: Many losses of photosynthesis and complex patterns of rate variation. *Proceedings of the National Academy of Sciences of the United States of America* **94**, 7367–7372.

Die, J. V., Dita, M. A., Krajinski, F., Gonzalez-Verdejo, C. I., Rubiales, D., Moreno, M. T. and Roman, B. (2007). Identification by suppression subtractive hybridization and expression analysis of *Medicago truncatula* putative defence genes in response to *Orobanche crenata* parasitization. *Physiological and Molecular Plant Pathology* **70**, 49–59.

Dörr, I. (1972). Contact of *Cuscuta* hyphae with sieve tubes of its host plants. *Protoplasma* **75**, 167–187.

Dörr, I. (1997). How *Striga* parasitizes its host: A TEM and SEM study. *Annals of Botany* **79**, 463–472.

Dörr, I. and Kollmann, R. (1995). Symplasmic sieve element continuity between *Orobanche* and its host. *Botanica Acta* **108**, 47–55.

Estabrook, E. M. and Yoder, J. I. (1998). Plant-plant communications: Rhizosphere signaling between parasitic angiosperms and their hosts. *Plant Physiology* **116**, 1–7.

Fate, G., Chang, M. and Lynn, D.G (1990). Control of germination in *Striga asiatica* – Chemistry of spatial definition. *Plant Physiology* **93**, 201–207.

Feild, T. S. and Brodribb, T. J. (2005). A unique mode of parasitism in the conifer coral tree *Parasitaxus ustus* (Podocarpaceae). *Plant Cell and Environment* **28**, 1316–1325.

Forde, B. G. (2002). The role of long-distance signalling in plant responses to nitrate and other nutrients. *Journal of Experimental Botany* **53**, 39–43.

Gibson, C. C. and Watkinson, A. R. (1989). The host range and selectivity of a parasitic plant – *Rhinanthus minor* L. *Oecologia* **78**, 401–406.

Gibson, C. C. and Watkinson, A. R. (1991). Host selectivity and the mediation of competition by the root hemiparasite *Rhinanthus minor*. *Oecologia* **86**, 81–87.

Gibson, C. C. and Watkinson, A. R. (1992). The role of the hemiparasitic annual *Rhinanthus minor* in determining grassland community structure. *Oecologia* **89**, 62–68.

Goldwasser, Y. and Yoder, J. I. (2001). Differential induction of *Orobanche* seed germination by *Arabidopsis thaliana*. *Plant Science* **160**, 951–959.

Goldwasser, Y., Kleifeld, Y., Plakhine, D. and Rubin, B. (1997). Variation in vetch (*Vicia* spp.) response to *Orobanche aegyptiaca*. *Weed Science* **45**, 756–762.

Goldwasser, Y., Plakhine, D., Kleifeld, Y., Zamski, E. and Rubin, B. (2000a). The differential susceptibility of vetch (*Vicia* spp.) to *Orobanche aegyptiaca*: Anatomical studies. *Annals of Botany* **85**, 257–262.

Goldwasser, Y., Plakhine, D. and Yoder, J. I. (2000b). *Arabidopsis thaliana* susceptibility to *Orobanche* spp. *Weed Science* **48**, 342–346.

Griffitts, A. A., Cramer, C. L. and Westwood, J. H. (2004). Host gene expression in response to Egyptian broomrape (*Orobanche aegyptiaca*). *Weed Science* **52**, 697–703.

Gurney, A. L., Grimanelli, D., Kanampiu, F., Hoisington, D., Scholes, J. D. and Press, M. C. (2003). Novel sources of resistance to *Striga hermonthica* in *Tripsacum dactyloides*, a wild relative of maize. *New Phytologist* **160**, 557–568.

Gurney, A. L., Slate, J., Press, M. C. and Scholes, J. D. (2006). A novel form of resistance in rice to the angiosperm parasite *Striga hermonthica*. *New Phytologist* **169**, 199–208.

Hauck, C., Muller, S. and Schildknecht, H. (1992). A germination stimulant for parasitic flowering plants from *Sorghum bicolor*, a genuine host plant. *Journal of Plant Physiology* **139**, 474–478.

Haussmann, B. I. G., Hess, D. E., Reddy, B. V. S., Mukuru, S. Z., Kayentao, M., Welz, H. G. and Geiger, H. H. (2001). Quantitative-genetic parameters of *Sorghum* growth under *Striga* infestation in Mali and Kenya. *Plant Breeding* **120**, 49–56.

Hibberd, J. M. and Jeschke, W. D. (2001). Solute flux into parasitic plants. *Journal of Experimental Botany* **52**, 2043–2049.

Hibberd, J. M., Quick, W. P., Press, M. C. and Scholes, J. D. (1998). Can source-sink relations explain responses of tobacco to infection by the root holoparasitic angiosperm *Orobanche cernua*? *Plant Cell and Environment* **21**, 333–340.

Hibberd, J. M., Quick, W. P., Press, M. C., Scholes, J. D. and Jeschke, W. D. (1999). Solute fluxes from tobacco to the parasitic angiosperm *Orobanche cernua* and the influence of infection on host carbon and nitrogen relations. *Plant Cell and Environment* **22**, 937–947.

Hodge, A., Robinson, D., Griffiths, B. S. and Fitter, A. H. (1999). Why plants bother: Root proliferation results in increased nitrogen capture from an organic patch when two grasses compete. *Plant Cell and Environment* **22**, 811–820.

Hood, M. E., Condon, J. M., Timko, M. P. and Riopel, J. L. (1998). Primary haustorial development of *Striga asiatica* on host and nonhost species. *Phytopathology* **88**, 70–75.

Hwangbo, J. K. (2000). Interactions between a facultative annual root hemiparasite, Rhinanthus minor L. and its hosts. PhD Thesis, University of Aberdeen.

Hwangbo, J. K., Seel, W. E. and Woodin, S. J. (2003). Short-term exposure to elevated atmospheric CO_2 benefits the growth of a facultative annual root hemiparasite, *Rhinanthus minor* (L.), more than that of its host, *Poa pratensis* (L.). *Journal of Experimental Botany* **54**, 1951–1955.

Jamison, D. S. and Yoder, J. I. (2001). Heritable variation in quinone-induced haustorium development in the parasitic plant *Triphysaria*. *Plant Physiology* **125**, 1870–1879.

Jiang, F., Jeschke, W. D. and Hartung, W. (2003). Water flows in the parasitic association *Rhinanthus minor – Hordeum vulgare*. *Journal of Experimental Botany* **54**, 1985–1993.

Jiang, F., Jeschke, W. D. and Hartung, W. (2004a). Abscisic acid (ABA) flows from *Hordeum vulgare* to the hemiparasite *Rhinanthus minor* and the influence of infection on host and parasite abscisic acid relations. *Journal of Experimental Botany* **55**, 2323–2329.

Jiang, F., Jeschke, W. D. and Hartung, W. (2004b). Solute flows from *Hordeum vulgare* to the hemiparasite *Rhinanthus minor* and the influence of infection on host and parasite nutrient relations. *Functional Plant Biology* **31**, 633–643.

Jiang, F., Veselova, S., Veselov, D., Kudoyarova, G., Jeschke, W. D. and Hartung, W. (2005). Cytokinin flows from *Hordeum vulgare* to the hemiparasite *Rhinanthus minor* and the influence of infection on host and parasite cytokinins relations. *Functional Plant Biology* **32**, 619–629.

Jiang, F., Timergalina, L., Kudoyarova, G., Jeschke, W. D. and Hartung, W. (2007). Growth and development of the facultative root hemiparasite *Rhinanthus minor* after removal of its host. *Functional Plant Biology* **34**, 237–245.

Jiang, F., Jeschke, W. D., Hartung, W. and Cameron, D. D. (2008a). Does legume nitrogen fixation underpin host quality for the hemiparasitic plant *Rhinanthus minor*? *Journal of Experimental Botany* **59**, 917–925.

Jiang, F., Jeschke, W. D., Harung, W. and Cameron, D. D. (2008b). Mobility of boron-polyol complexes in the hemiparasitic association between *Rhianathus minor* and *Hordeum vulgare*: The effects of nitrogen nutrition. *Physiologia Plantarum* **134**, 13–21.

Jorrin, J., De Ruck, E., Serghini, K., Perez de Luque, A. and Munoz-Garcia, J. (1996). Biochemical aspects of the parasitism of sunflower by *Orobanche*. Advances in parasitic plant research. *In* "Proceedings of the 6th international parasitic weed symposium. Cordoba, Spain" (J. I. D. Berner, D. M. Joel, L. J. Mussleman and C, Parker, eds.), pp. 559–565.

Joshi, J., Matthies, D. and Schmid, B. (2000). Root hemiparasites and plant diversity in experimental grassland communities. *Journal of Ecology* **88**, 634–644.

Kamo, T., Endo, M., Sato, M., Kasahera, R., Yamaya, H., Hiradate, S., Fujii, Y., Hirai, N. and Hirai, N. (2008). Limited distribution of natural cyanamide in higher plants: Occurrence in *Vicia villosa* subsp varia *V. cracca* and *Robinia pseudo-acacia*. *Phytochemistry* **69**, 1166–1172.

Kaplan, I., Halitschke, R., Kessler, A., Rehill, B. J., Sardanelli, S. and Denno, R. F. (2008). Physiological integration of roots and shoots in plant defense strategies links above- and belowground herbivory. *Ecology Letters* **11**, 841–851.

Kawachi, N., Fujimaki, S., Sakamoto, K., Ishioka, N. S., Matsuhashi, S. and Sekimoto, H. (2008). Analysis of NO_3 interception of the parasitic angiosperm *Orobanche* spp. Using a positron-emitting tracer imaging system and (NO_3^-)-^{13}N: A new method for the visualization and quantitative analysis of the NO_3 interception ratio. *Soil Science and Plant Nutrition* **54**, 408–416.

Keith, A. M., Cameron, D. D. and Seel, W. E. (2004). Spatial interactions between the hemiparasitic angiosperm *Rhinanthus minor* and its host are species-specific. *Functional Ecology* **18**, 435–442.

Keyes, W. J., O'Malley, R. C., Kim, D. and Lynn, D. G. (2000). Signaling organogenesis in parasitic angiosperms, Xenognosin generation, perception, and response. *Journal of Plant Growth Regulation* **19**, 217–231.

Keyes, W. J., Taylor, J. V., Apkarian, R. P. and Lynn, D. G. (2001). Dancing together. Social controls in parasitic plant development. *Plant Physiology* **127**, 1508–1512.

Keyes, W. J., Palmer, A. G., Erbil, W. K., Taylor, J. V., Apkarian, R. P., Weeks, E. R. and Lynn, D. G. (2007). Sernagenesis and the parasitic angiosperm *Striga asiatica*. *Plant Journal* **51**, 707–716.

Khan, Z. R., Midega, C. A. O., Hassanali, A., Pickett, J. A. and Wadhams, L. J. (2007). Assessment of different legumes for the control of *Striga hermonthica* in maize and sorghum. *Crop Science* **47**, 730–736.

Kim, D. J., Kocz, R., Boone, L., Keyes, W. J. and Lynn, D. G. (1998). On becoming a parasite: Evaluating the role of wall oxidases in parasitic plant development. *Chemistry & Biology* **5**, 103–117.

Koch, A. M., Binder, C. and Sanders, I. R. (2004). Does the generalist parasitic plant *Cuscuta campestris* selectively forage in heterogeneous plant communities? *New Phytologist* **162**, 147–155.

Krapp, A., Quick, W. P. and Stitt, M. (1991). Ribulose-1,5-bisphosphate carboxylase-oxygenase, other Calvin-cycle enzymes, and chlorophyll decrease when glucose is supplied to mature spinach leaves via the transpiration stream. *Planta* **186**, 58–69.

Krapp, A., Hofmann, B., Schafer, C. and Stitt, M. (1993). Regulation of the expression of rbcS and other photosynthetic genes by carbohydrates – A mechanism for the sink regulation of photosynthesis. *Plant Journal* **3**, 817–828.

Kruse, J., Hetzger, I., Hansch, R., Mendel, R. R., Walch-Liu, P., Engels, C. and Rennenberg, H. (2002). Elevated pCO_2 favours nitrate reduction in the roots of wild-type tobacco (*Nicotiana tabacum* cv. Gat.) and significantly alters n-metabolism in transformants lacking functional nitrate reductase in the roots. *Journal of Experimental Botany* **53**, 2351–2367.

Kuijt, J. (1969). "The biology of parasitic flowering plants". University of California Press, Berkeley.

Labrousse, P., Arnaud, M. C., Serieys, H., Berville, A. and Thalouarn, P. (2001). Several mechanisms are involved in resistance of *Helianthus* to *Orobanche cumana* Wallr. *Annals of Botany* **88**, 859–868.

Leake, J. R. (1994). The biology of myco-heterotrophic (saprophytic) plants. *New Phytologist* **127**, 171–216.

Leake, J. R., Cameron, D. D. and Beerling, D. J. (2008). Fungal fidelity in the myco-heterotroph-to-autotroph lifecycle of Lycopodiaceae: A case of parental nurture? *New Phytologist* **177**, 572–576.

Losner-Goshen, D., Portnoy, V. H., Mayer, A. M. and Joel, D. M. (1998). Pectolytic activity by the haustorium of the parasitic plant *Orobanche* (Orobanchaceae) in host roots. *Annals of Botany* **81**, 319–326.

Lusson, N. A., Delavault, P. M. and Thalouarn, P. A. (1998). The rbcL gene from the non-photosynthetic parasite *Lathraea clandestina* is not transcribed by a plastid encoded RNA polymerase. *Current Genetics* **34**, 212–215.

Mabrouk, Y., Simier, P., Delavault, P., Delgrange, S., Sifi, B., Zourgui, L. and Belhadj, O. (2007a). Molecular and biochemical mechanisms of defense induced in pea by *Rhizobium leguminosarum* against *Orobanche crenata*. *Weed Research* **47**, 452–460.

Mabrouk, Y., Zourgui, L., Sifi, B. and Belhadj, O. (2007b). The potential of *Rhizobium* strains for biological control of *Orobanche crenata*. *Biologia* **62**, 139–143.

Mabrouk, Y., Zourgui, L., Sifi, B., Delavault, P., Simier, P. and Belhadj, O. (2007c). Some compatible *Rhizobium leguminosarum* strains in peas decrease infections when parasitised by *Orobanche crenata*. *Weed Research* **47**, 44–53.

Marvier, M. A. (1998). Parasite impacts on host communities: Plant parasitism in a California coastal prairie. *Ecology* **79**, 2616–2623.

Matthies, D. (1996). Interactions between the root hemiparasite *Melampyrum arvense* and mixtures of host plants: Heterotrophic benefit and parasite-mediated competition. *Oikos* **75**, 118–124.

Matvienko, M., Torres, M. J. and Yoder, J. I. (2001). Transcriptional responses in the hemiparasitic plant *Triphysaria versicolor* to host plant signals. *Plant Physiology* **127**, 272–282.

McKendrick, S. L., Leake, J. R., Taylor, D. L. and Read, D. J. (2000). Symbiotic germination and development of myco-heterotrophic plants in nature: Ontogeny of *Corallorhiza trifida* and characterization of its mycorrhizal fungi. *New Phytologist* **145**, 523–537.

Minchin, P. E. H. and Lacointe, A. (2005). New understanding on phloem physiology and possible consequences for modelling long-distance carbon transport. *New Phytologist* **166**, 771–779.

Minchin, P. E. H. and Thorpe, M. R. (2003). Using the short-lived isotope ^{11}C in mechanistic studies of photosynthate transport. *Functional Plant Biology* **30**, 831–841.

Nagar, R., Singh, M. and Sanwal, G. G. (1984). Cell-wall degrading enzymes in *Cuscuta reflexa* and its hosts. *Journal of Experimental Botany* **35**, 1104–1112.

Nakano, H., Muramatsu, S., Makino, A. and Mae, T. (2000). Relationship between the suppression of photosynthesis and starch accumulation in the pod-removed bean. *Australian Journal of Plant Physiology* **27**, 167–173.

Olivier, A., Ramaiah, K. V. and Leroux, G. D. (1991). Selection of sorghum (*Sorghum bicolor* Moench) varieties resistant to the parasitic weed *Striga hermonthica* (Del) Benth. *Weed Research* **31**, 219–225.

Ouedraogo, J. T., Maheshwari, V., Berner, D. K., St-Pierre, C. A., Belzile, F. and Timko, M. P. (2001). Identification of AFLP markers linked to resistance of cowpea (*Vigna unguiculata*) to parasitism by *Striga gesnerioides*. *Theoretical and Applied Genetics* **102**, 1029–1036.

Ouedraogo, J. T., Tignegre, J. B., Timko, M. P. and Belzile, F. J. (2002). AFLP markers linked to resistance against *Striga gesnerioides* race 1 in cowpea (*Vigna unguiculata*). *Genome* **45**, 787–793.

Pageau, K., Simier, P., Le Bizec, B., Robins, R. J. and Fer, A. (2003). Characterization of nitrogen relationships between *Sorghum bicolor* and the root-hemiparasitic angiosperm *Striga hermonthica* (Del.) Benth. Using (KNO_3)-^{15}N as isotopic tracer. *Journal of Experimental Botany* **54**, 789–799.

Parker, C. and Riches, C. R. (1993). "Parasitic weeds of the world: Biology and control". CAB International, Wallingford, UK.

Parsons, R. and Baker, A. (1996). Cycling of amino compounds in symbiotic lupin. *Journal of Experimental Botany* **47**, 421–429.

Pate, J. S. and Bell, T. L. (2000). Host associations of the introduced annual root hemiparasite *Parentucellia viscosa* in agricultural and bushland settings in western Australia. *Annals of Botany* **85**, 203–213.

Pate, J. S., Layzell, D. B. and McNeil, D. L. (1979). Modeling the transport and utilization of carbon and nitrogen in a nodulated legume. *Plant Physiology* **63**, 730–737.

Patto, M. C. V., Diaz-Ruiz, R., Satovic, Z., Roman, B., Pujadas-Salva, A. J. and Rubiales, D. (2008). Genetic diversity of Moroccan populations of *Orobanche foetida*: Evolving from parasitising wild hosts to crop plants. *Weed Research* **48**, 179–186.

Pennings, S. C. and Callaway, R. M. (1996). Impact of a parasitic plant on the structure and dynamics of salt marsh vegetation. *Ecology* **77**, 1410–1419.

Perez-De-Luque, A., Jorrin, J., Cubero, J. I. and Rubiales, D. (2005a). *Orobanche crenata* resistance and avoidance in pea (*Pisum* spp.) operate at different developmental stages of the parasite. *Weed Research* **45**, 379–387.

Perez-De-Luque, A., Rubiales, D., Cubero, J. I., Press, M. C., Scholes, J., Yoneyama, K., Takeuchi, Y., Plakhine, D. and Joel, D. M. (2005b). Interaction between *Orobanche crenata* and its host legumes: Unsuccessful haustorial penetration and necrosis of the developing parasite. *Annals of Botany* **95**, 935–942.

Peuke, A. D., Rokitta, M., Zimmermann, U., Schreiber, L. and Haase, A. (2001). Simultaneous measurement of water flow velocity and solute transport in xylem and phloem of adult plants of *Ricinus communis* over a daily time course by nuclear magnetic resonance spectrometry. *Plant Cell and Environment* **24**, 491–503.

Pierce, S., Mbwaga, A. M., Press, M. C. and Scholes, J. D. (2003). Xenognosin production and tolerance to *Striga asiatica* infection of high-yielding maize cultivars. *Weed Research* **43**, 139–145.

Press, M. C., Shah, N., Tuohy, J. M. and Stewart, G. R. (1987). Carbon isotope ratios demonstrate carbon flux from C4 host to C3 parasite. *Plant Physiology* **85**, 1143–1145.

Qasem, J. R. and Foy, C. L. (2007). Screening studies on the host range of branched broomrape (*Orobanche ramosa*). *Journal of Horticultural Science & Biotechnology* **82**, 885–892.

Raven, J. A. (1983). Phytophages of xylem and phloem – a comparison of animal and plant sap-feeders. *Advances in Ecological Research* **13**, 135–234.

Raven, J. A. (2008). Transpiration; how many functions? *New Phytologist* **179**, 905–907.

Raven, J. A. and Smith, F. A. (1976). Nitrogen assimilation and transport in vascular land plants in relation to intracellular pH regulation. *New Phytologist* **76**, 415–431.

Robinson, D., Linehan, D. J. and Gordon, D. C. (1994). Capture of nitrate from soil by wheat in relation to root length, nitrogen inflow and availability. *New Phytologist* **128**, 297–305.

Rümer, S., Cameron, D. D., Wacker, R., Hartung, W. and Jiang, F. (2007). An anatomical study of the haustoria of *Rhinanthus minor* attached to roots of different hosts. *Flora* **202**, 194–200.

Runyon, J. B., Mescher, M. C. and De Moraes, C. M. (2006). Volatile chemical cues guide host location and host selection by parasitic plants. *Science* **313**, 1964–1967.

Scheible, W. R., Lauerer, M., Schulze, E. D., Caboche, M. and Stitt, M. (1997). Accumulation of nitrate in the shoot acts as a signal to regulate shoot-root allocation in tobacco. *Plant Journal* **11**, 671–691.

Scholes, J. D. and Press, M. C. (2008). *Striga* infestation of cereal crops – An unsolved problem in resource limited agriculture. *Current Opinion in Plant Biology* **11**, 180–186.

Searcy, D. G. (1970). Measurements by DNA hybridization in-vitro of genetic basis of parasitic reduction. *Evolution* **24**, 207.

Searcy, D. G. and MacInnis, A. J. (1970). Measurements of DNA renaturation of the genetic basis of parasitic reduction. *Evolution* **24**, 796–806.

Seel, W. E. and Press, M. C. (1993). Influence of the host on 3 sub-arctic annual facultative root hemiparasites. 1. Growth, mineral accumulation and above-ground dry-matter partitioning. *New Phytologist* **125**, 131–138.

Seel, W. E. and Press, M. C. (1996). Effects of repeated parasitism by *Rhinanthus minor* on the growth and photosynthesis of a perennial grass, *Poa alpina*. *New Phytologist* **134**, 495–502.

Seel, W. E., Parsons, A. N. and Press, M. C. (1993). Do inorganic solutes limit growth of the facultative hemiparasite *Rhinanthus minor* in the absence of a host. *New Phytologist* **124**, 283–289.

Shen, H., Ye, W., Hong, L., Huang, H., Wang, Z., Deng, X., Yang, Q. and Xu, Z. (2006). Progress in parasitic plant biology: Host selection and nutrient transfer. *Plant Biology* **8**, 175–185.

Simpson, R. J., Lambers, H. and Dalling, M. J. (1982). Translocation of nitrogen in a vegetative wheat plant (*Triticum aestivum*). *Physiologia Plantarum* **56**, 11–17.

Sinclair, W. T., Mill, R. R., Gardner, M. F., Woltz, P., Jaffre, T., Preston, J., Hollingsworth, M. L., Ponge, A. and Moller, M. (2002). Evolutionary relationships of the New Caledonian heterotrophic conifer, *Parasitaxus usta* (Podocarpaceae), inferred from chloroplast TRNL-F intron/spacer and nuclear rDNA ITS2 sequences. *Plant Systematics and Evolution* **233**, 79–104.

Smith, D. (2000). The population dynamics and community ecology of root hemiparasitic plants. *American Naturalist* **155**, 13–23.

Smith, S. and Stewart, G. R. (1990). Effect of potassium levels on the stomatal behavior of the hemiparasite *Striga hermonthica*. *Plant Physiology* **94**, 1472–1476.

Stillwell, W. and Hester, P. (1984). Abscisic acid increases membrane permeability by interacting with phosphatdylethanolamine. *Phytochemistry* **23**, 2187–2192.

Swarbrick, P. J., Huang, K., Liu, G., Slate, J., Press, M. C. and Scholes, J. D. (2008). Global patterns of gene expression in rice cultivars undergoing a susceptible or resistant interaction with the parasitic plant *Striga hermonthica*. *New Phytologist* **179**, 515–529.

Taylor, A. and Seel, W. E. (1998). Do *Striga hermonthica*-induced changes in soil matric potential cause the reduction in stomatal conductance and growth of infected maize plants? *New Phytologist* **138**, 67–73.

Taylor, A., Martin, J. and Seel, W. E. (1996). Physiology of the parasitic association between maize and witchweed (*Striga hermonthica*): Is ABA involved? *Journal of Experimental Botany* **47**, 1057–1065.

Tennakoon, K. U. and Cameron, D. D. (2006). The anatomy of *Santalum album* (sandalwood) haustoria. *Canadian Journal of Botany-Revue Canadienne De Botanique* **84**, 1608–1616.

Tennakoon, K. U. and Pate, J. S. (1996). Heterotrophic gain of carbon from hosts by the xylem-tapping root hemiparasite *Olax phyllanthi* (Olacaceae). *Oecologia* **105**, 369–376.

Tennakoon, K. U., Pate, J. S. and Fineran, B. A. (1997). Growth and partitioning of C and fixed N in the shrub legume *Acacia littorea* in the presence or absence of the root hemiparasite *Olax phyllanthi*. *Journal of Experimental Botany* **48**, 1047–1060.

Tennakoon, K. U., Bolin, J. F., Musselman, L. J. and Maass, E. (2007). Structural attributes of the hypogeous holoparasite *Hydnora triceps* (Drege & Meyer) (Hydnoraceae). *American Journal of Botany* **94**, 1439–1449.

Tholen, D., Pons, T. L., Voesenek, L. and Poorter, H. (2007). Ethylene insensitivity results in down-regulation of Rubisco expression and photosynthetic capacity in tobacco. *Plant Physiology* **144**, 1305–1315.

Timko, M. P., Gowda, B. S., Ouedraogo, J. and Ousmane, B. (2007). Molecular markers for analysis of resistance to *Striga gesnerioides* in cowpea. *In* "Towards ending the witch-hunt" (G. Ejeta and J. Gressell, eds.), pp. 115–128. World Scientific Publishing Co. Pte Ltd, Singapore.

Tomilov, A. A., Tomilova, N. B., Abdallah, I. and Yoder, J. I. (2005). Localized hormone fluxes and early haustorium development in the hemiparasitic plant *Triphysaria versicolor*. *Plant Physiology* **138**, 1469–1480.

Uematsu, K., Nakajima, M., Yamaguchi, I., Yoneyama, K. and Fukui, Y. (2007). Role of cAMP in gibberellin promotion of seed germination in *Orobanche minor* (Smith). *Journal of Plant Growth Regulation* **26**, 245–254.

Valderrama, M. R., Roman, B., Satovic, Z., Rubiales, D., Cubero, J. I. and Torres, A. M. (2004). Locating quantitative trait loci associated with *Orobanche crenata* resistance in pea. *Weed Research* **44**, 323–328.

Vallance, K. B. (1952). The germination of the seeds of *Rhinanthus crista-galli*. *Annals of Botany* **16**, 409–420.

Vanhulst, R., Shipley, B. and Theriault, A. (1987). Why is *Rhinanthus minor* (*Scrophulariaceae*) such a good invader. *Canadian Journal of Botany-Revue Canadienne de Botanique* **65**, 2373–2379.

vanVuuren, M. M. I., Robinson, D. and Griffiths, B. S. (1996). Nutrient inflow and root proliferation during the exploitation of a temporally and spatially discrete source of nitrogen in soil. *Plant and Soil* **178**, 185–192.

Vurro, M., Boari, A., Pilgeram, A. L. and Sands, D. C. (2006). Exogenous amino acids inhibit seed germination and tubercle formation by *Orobanche ramosa* (broomrape): Potential application for management of parasitic weeds. *Biological Control* **36**, 258–265.

Wan, X. C., Steudle, E. and Hartung, W. (2004). Gating of water channels (aquaporins) in cortical cells of young corn roots by mechanical stimuli (pressure pulses): Effects of ABA and of $HgCl_2$. *Journal of Experimental Botany* **55**, 411–422.

Watling, J. R. and Press, M. C. (2001). Impacts of infection by parasitic angiosperms on host photosynthesis. *Plant Biology* **3**, 244–250.

Weber, H. C. (1987). Evolution of the secondary haustoria to a primary haustorium in the parasitic Scrophulariaceae/Orobanchaceae. *Plant Systematics and Evolution* **156**, 127–131.

Westbury, D. B. and Dunnett, N. P. (2007). The impact of Rhinanthus minor in newly established meadows on a productive site. *Applied Vegetation Science* **10**, 121–129.

Wickett, N. J., Zhang, Y., Hansen, S. K., Roper, J. M., Kuehl, J. V., Plock, S. A., Wolf, P. G., dePamphilis, C. W., Boore, J. L. and Goffinet, B. (2008). Functional gene losses occur with minimal size reduction in the plastid genome of the parasitic liverwort *Aneura mirabilis*. *Molecular Biology and Evolution* **25**, 393–401.

Wolfe, K. H., Morden, C. W. and Palmer, J. D. (1992). Function and evolution of a minimal plastid genome from a nonphotosynthetic parasitic plant. *Proceedings of the National Academy of Sciences of the United States of America* **89**, 10648–10652.

Yeo, P. F. (1964). The growth of *Euphrasia* in cultivation. *Watsonia* **6**, 1–24.

Yoder, J. I. (1997). A species-specific recognition system directs haustorium development in the parasitic plant *Triphysaria* (Scrophulariaceae). *Planta* **202**, 407–413.

Yoder, J. I. (1999). Parasitic plant responses to host plant signals: A model for subterranean plant–plant interactions. *Current Opinion in Plant Biology* **2**, 65–70.

Yoder, J. I. (2001). Host-plant recognition by parasitic Scrophulariaceae. *Current Opinion in Plant Biology* **4**, 359–365.

Zonno, M. C. and Vurro, M. (2002). Inhibition of germination of *Orobanche ramosa* seeds by *Fusarium toxins*. *Phytoparasitica* **30**, 519–524.

Low Oxygen Signaling and Tolerance in Plants

FRANCESCO LICAUSI AND PIERDOMENICO PERATA

PlantLab, Scuola Superiore Sant'Anna, Piazza Martiri della Libertà 33, 56127 Pisa, Italy

Advances in Botanical Research, Vol. 50

0065-2296/09 $35.00
DOI: 10.1016/S0065-2296(08)00804-5

ABSTRACT

Plants often experience low-oxygen conditions, not only as an environmental stress condition but also as part of their normal developmental process. Oxygen deficiency signaling in the plant cell has been shown to involve reactive oxygen and nitrogen species, hormones, and calcium as secondary messengers, similarly to the low-oxygen signaling observed in other eukaryotic systems. However, in plants, evidences suggesting the existence on an oxygen sensor are scant. To date, research efforts have been aimed at understanding the strategies used by plants to low oxygen. Anatomical modifications, which encompass leaf elongation, adventitious rooting production and aerenchyma formation, can help the plants to avoid the stress consequent to low oxygen availability. On the other hand, metabolic adaptations enable the plant tissue to survive while experiencing oxygen deficiency, mainly coupling structural maintenance with energy saving. Application of the knowledge already available to crops may provide solutions for both agricultural and industrial processes involving low-oxygen conditions.

I. INTRODUCTION: PLANT CELLS DEALING WITH LOW OXYGEN

The Earth is the only planet in the solar system with an oxygen rich atmosphere. Evolving biological organisms have developed mechanisms that link efficient energy production with oxygen availability. This has made life on earth strictly dependent on oxygen not only for energy production but also for a number of different biochemical reactions. However, oxygen limitations are a normal part of the developmental process, especially in multicellular organisms which, as a consequence, have developed different solutions for oxygen transport.

Despite the fact that plants are able to produce oxygen in their photosynthetic tissues, non-green organs are extremely sensitive to low-oxygen concentrations. Although a well developed aerechyma can be effective in conducting oxygen in some organs, plants, unlike animals, lack efficient oxygen distribution systems (Van Dongen *et al.*, 2008).

While anoxia is usually described as a condition whereby Cytochrome *c* oxidase (COX) is impeded to donate electrons to oxygen, given its extremely low concentration, hypoxia has been defined as an expression of inadequate oxygen availability that hinders other oxidases, such as the Alternative oxidase (AOX) while COX retain a limited capacity (Igamberdiev and Hill, 2008). According to this definition, hypoxia is, therefore, independent of the actual concentration of oxygen, but rather depends on wide variety of factors such as the rate of respiration, porosity, and thickness of the tissue.

The most studied environmental condition associated with hypoxia is flooding, since gas diffusion in water is 10,000 times slower than in air (Armstrong, 1979). However submergence does not exactly coincide with low-oxygen stress, since the first condition imply a plethora of other factors, such as ethylene entrapment, CO_2 accumulation, reduction in light intensity and increased amounts of reduced compounds to toxic levels (Bailey-Serres and Voesenek, 2008). Underwater, photosynthetic tissues can still provide oxygen if light conditions permit, but competition from microorganisms is usually so strong in water saturated (waterlogged) soils that, often, a hypoxic or even anoxic condition is established. In crop plants especially, this can lead to a huge reduction in yield (Setter and Waters, 2003). However, flooding is not the only determinant for hypoxia in plant tissues: densely packed cell structures or tissues with high metabolic rates can suffer from hypoxia even when external oxygen concentrations are close to that of the atmosphere (21%). Both conditions occur to developing seeds, for which, an oxygen concentration between 2 and 10 μM has been reported (Borisjuk *et al.*, 2007; Van Dongen *et al.*, 2004; Vigeolas *et al.*, 2003).

Interestingly, plants growing under microgravity conditions also experience hypoxia. This is demonstrated by the increased expression of a reporter gene under the control of a promoter of a well-known anaerobiosis-induced gene, *alcohol dehydrogenase* (ADH), in transgenic Arabidopsis during spaceflights.

The ability to manipulate whole plants or plant organs under hypoxic or anoxic conditions can also be useful in industrial processes. Usually fruit, cut flowers, and vegetables are maintained in a controlled, low-oxygen atmosphere, to delay the metabolic processes, primarily respiration, and to prolong their shelf life. Physiological and molecular responses in fruit (oranges, avocados) have therefore been investigated (Kanellis *et al.*, 1991; Pasentsis *et al.*, 2007). One particular case of industrial interest in plant cell anaerobiosis is *Chlamydomonas reihnhardtii*. This green alga in anoxia uses protons as final electron acceptors, leading to hydrogen production (Melis and Happe, 2004). Unfortunately anoxic conditions are, as expected, extremely challenging for the cell. Therefore the study of this alga and its adaptive mechanisms to low oxygen is acquiring more and more interest (Mus *et al.*, 2007).

II. OXYGEN SENSING IN EUKARYOTS

To achieve rapid and highly tuneable responses, oxygen signaling should rely on one or more sensors capable of immediately perceiving when oxygen concentrations fall below a critical level and should be able to trigger different responses depending on the organ, tissue or cell type. A number of direct and indirect sensors probably represent the best solution to cope with a dynamic stress situation, such as a slow shift from hypoxia to complete anoxia through increasingly harsh hypoxic conditions, followed by a rapid re-oxygenation.

In mammals and yeast, where events involved in low-oxygen sensing have been extensively studied, the general sensing machinery seems to be conservatively split into different mechanisms that either directly measure the oxygen level within the cell or perceive variations in oxygen-dependent metabolites. In mammals, the hypoxia inducible factor (HIF) transcriptional complex plays a major role, since it is regulated at both a transcriptional and posttranscriptional level. Briefly, HIF is a heterodimer composed of HIF-α (hypoxia inducible), and HIF-β (constitutively expressed) subunits, all members of the bHLH-PAS (PER-ARNT-SIM) family of transcription factors (Wenger, 2002). The HIFα subunits can be hydroxylated in an oxygen dependent manner on two prolyl residues by prolyl-4-hydroxylases (PHDs) that requires molecular oxygen as co-substrate. Hydroxylated HIFα is directed to the 26S proteasome by the von Hippel–Lindau tumor suppressor protein (pVHL) that represents the recognition subunit of an E3 ubiquitin-protein ligase (Acker *et al.*, 2006). Moreover, prolyl hydroxylation of the alpha subunits inhibits binding of the co-activator protein p300 and the cAMP response element binding CREB binding protein, necessary for assembling the RNA polymerase II complex. Reduced availability of oxygen is immediately reflected in lower rates of HIF1-α hydroxylation and degradation, giving way to a transcriptional activator complex that is translocated to the nucleus where it can regulate downstream genes (Semenza, 2007).

In fission yeast *Saccharomyces cerevisiae*, SRN1, a basic helix–loop–helix leucine zipper, is responsible for the induction of those genes required for hypoxic growth (Todd *et al.*, 2006). In normoxic conditions SRN1 is bound to the membrane but, when the oxygen level decreases, this transcription factor is activated by proteolysis, releasing an N-terminal fragment, SRN1N, which is translocated to the nucleus. SRN1 activation by low oxygen is partially explainable by an inhibition of sterol synthesis (Goldstein *et al.*, 2006), but it is also mediated by hypoxic stabilization. In fact, in a similar way to mammals, SRN1 is rapidly degraded, under normoxic conditions, by

OFD1, a 2-oxoglutarate dioxygenase (PHD) which hydroxylates SRN1 at a specific prolyl residue (D'Angelo *et al.*, 2003; Hughes and Espenshade, 2008).

It is interesting to note that, apparently counter intuitively, hydroxylating enzymes acting as sensors are upregulated under low-oxygen conditions (Hughes and Espenshade, 2008). A possible explanation for this feedback mechanism is that the cell needs to rapidly adapt to normoxic conditions after re-oxygenation.

Hydroxylation and subsequent degradation of the transcriptional activator are not the only oxygen-sensing mechanisms that have been found in eukaryots. It has been suggested that reactive oxygen species (ROS) production via NADPH oxidase or mitochondrial electron transport chain (ETC) and reactive nitrogen species (RNS), via NO-synthase isoforms, act on oxygen signaling stabilizing HIF1-α (Appelhoff *et al.*, 2004). In fact ROS and RNS contribute to changing the cellular redox state and to imbalancing the iron population from Fe^{2+} to Fe^{3+}, which is no longer a co-factor for PHDs (Semenza, 2007).

Energy charge, AMP levels, calcium (Ca^{2+}) concentrations and ion channels have also been suggested as inter-dependent modulators of hypoxic signaling in different mammal cell types (Fähling, 2008).

III. OXYGEN SENSORS IN PLANTS

It is still unknown how plants do sense oxygen. Different hypotheses have been provided, involving direct oxygen binding or metabolism. However, while some of them have been excluded from the plethora of candidates for the oxygen sensing, others still need to be characterized. Indirect sensing, via signal compound(s) accumulated during oxygen shortage, is also likely to contribute to trigger or modulate the hypoxic response in the plant cell.

Gibbs and Greenway (2003) raised the question whether oxygen sensing occurs ubiquitously in all plant cells, or it is rather triggered by "anoxic cores," that is, portions of tissue which suffer of limitations in oxygen diffusion and therefore become anoxic while surrounding tissues are still provided with oxygen for respiration (Berry and Norris, 1949). An alternative hypothesis would state that all plant cells can perceive different degrees of oxygen deficiency. Evidences for this second hypothesis come from the recent literature, since the hypoxic response in Arabidopsis protoplasts (Baena-Gonzalez *et al.*, 2007) is largely overlapping with that of whole plants (Branco-Price *et al.*, 2005), and the extent of induction of anaerobic genes is inversely proportional to the environmental O_2 concentration in Arabidopsis roots (Van Dongen *et al.*, 2008). However both hypotheses imply the presence of a molecular sensor, that directly or indirectly, perceive the actual oxygen level in the cell.

A. DIRECT SENSING BY OXYGEN BINDING

Non-symbiotic hemoglobins were the first proteins to be put forward as oxygen sensors in plants (Appleby *et al.*, 1988). Early objections to this hypothesis were based on the high affinity of non-symbiotic hemoglobins for oxygen (Garrocho-Villegas and Arredondo-Peter, 2008). In fact, a low-oxygen response is taking place at oxygen concentrations that does not affect the binding of oxygen to hemoglobins. Nevertheless, non-symbiotic hemoglobins are induced under low-oxygen levels and have also been characterized as playing a major role in nitric oxide detoxification and hydrogen peroxide scavenging under hypoxia (Perazzolli *et al.*, 2004). In plants, no horthologs of the animal HIF1-α or the fungal SRN1 have been found so far. This is not surprising, since there is virtually no similarity between HIF1-α and SRN1. Also fungal and animal PHDs appear to be divergent, despite the conservativeness of the catalytic domain. Nevertheless, in plant genomes several members of the PHD family have been predicted. However so far, few have been characterized, and none have been investigated in terms of low-oxygen sensing.

Plant PHDs resemble HIF1-α PHD because they act as monomers instead of tetramers composed of two subunit types ($\alpha 2\beta 2$) such as collagen PHDs. In Arabidopsis, two of the six PHDs have been characterized so far. AtPHD-1 can hydroxylate collagen and HIF1-α-type peptides, together with plant prolyl-rich peptides (Hieta and Myllyharju, 2002). Prolyl-rich sequences in plants are characteristic of cell wall proteins that include proline-rich glycoproteins, extensins, and arabinogalactanoproteins. These proteins all require hydroxylation at the prolyl rich moiety to be glycosylated with arabino-oligosaccharides or larger arabinogalactan polysaccharides (Lamport, 1965; Pope, 1977). Interestingly, several of these proteins are upregulated by hypoxia and anoxia in Arabidopsis and rice (Lasanthi-Kudahettige *et al.*, 2007; Loreti *et al.*, 2005). At-PHD-2, the second Arabidopsis PHD to be characterized, is unable to hydroxylate HIF1-α-like substrates (Tiainen *et al.*, 2005). Several prolyl-4 hydroxylase are induced by hypoxia and anoxia under low-oxygen conditions, in rice and Arabidopsis (Branco-Price *et al.*, 2005; Lasanthi-Kudahettige *et al.*, 2007; Loreti *et al.*, 2005)

Even though oxygen sensing through O_2 dependent enzyme activities is conserved among animals and fungi, it may take place in a different way in plant cells. Microarray analyses of Arabidopsis roots exposed to different hypoxic conditions, from 8% to 1% O_2 (Van Dongen *et al.*, 2008), and of rice coleoptiles under anoxia (Lasanthi-Kudahettige *et al.*, 2007) to low oxygen revealed that several genes encoding for enzymes, whose activities rely on oxygen availability, are upregulated. For instance genes encoding for proteins of yet unknown function but with a high degree of similarity to the animal dioxygenase Cysteamil-dioxygenase (*At5g39890* and *At5g15120* in Arabidopsis, *Os01g09030* and *Os07g01090* in rice) are induced to a very large extent in both plant species. The highly conserved aminoacidic structure and anaerobic induction among plant species deserves investigation in terms of the role of these proteins in anaerobic signaling or tolerance.

B. INDIRECT OXYGEN SENSING

Hypoxia does not necessarily have to be sensed by the plant cell by directly measuring the oxygen concentration. As in other eukaryotes, other signal molecules could act as indirect oxygen signals.

1. Energy shortage

Good candidate molecules as signals of low oxygen are those whose concentration changes depending on a reduction in oxygen availability are ATP and ADP or the whole pool of adenosine nucleotide phosphates, which together make up the energy charge. Efforts to distinguish direct low-oxygen responses and adaptations due to changes in the energy charge have been performed using Arabidopsis, pea roots and potato tubers as models (Riewe *et al.*, 2008; Zabalza *et al.*, 2009).

Although a general energy deficit is usually associated with stress since efficient energy production, structural maintenance, and development are impaired, oxygen shortage is probably the stress condition most directly linked to a reduction in energy availability. Recently, two Arabidopsis protein kinases belonging to the SnRK superfamily KIN10 and KIN11 were reported to regulate the common transcriptional response to darkness and multiple stresses that bias energy availability (Baena-Gonzalez *et al.*, 2007). Both proteins are orthologs of yeast SNF1 and mammal AMPK, which are activated under hypoxia and ischemia and, once activated by increasing the AMP to ATP ratio, switch off energy consuming processes and activate catabolism. However, the transcriptome changes mediated by KIN10 only partially overlap with the anaerobic changes, suggesting that the SnRKs only partially contribute to the low-oxygen response.

2. Reactive oxygen and nitrogen species

ROS and, more recently RNS, have been proposed as also playing a role in oxygen sensing in plants. Two antithetical hypotheses deal with ROS production and low-oxygen signaling. According to the oldest and more intuitive hypothesis, a drop in oxygen levels is associated with a reduction in ROS synthesis. The second hypothesis, first formulated by Blokhina *et al.* (2001), holds that ROS is produced in the apoplast of both tolerant (*Iris pseudoacorus* and *Oryza sativa*) and intolerant (*Triticum aestivum* and *Iris germanica*) plant species under anoxic conditions. Further evidence for ROS production under hypoxic conditions has been provided by Santosa *et al.* (2007). They reported the emission of ethane, a volatile product deriving from membrane peroxidation, in rice seedlings subjected to flooding or a low-oxygen atmosphere. Subsequent re-oxygenation further increased the ethane emission but only for a short time (less than 2 h). A further link between ROS and low-oxygen signaling is strongly supported by H_2O_2 production by a NADPH oxydase as requirement for a controlled induction of *ADH* in Arabidopsis (Baxter-Burrell *et al.*, 2002). Treatment using diphenyleneiodonium (DPI), an inhibitor of H_2O_2 production via NADPH oxidase, was in fact able to inhibit the hypoxic increases in ADH activity in Arabidopsis seedlings. H_2O_2 accumulation under low oxygen was demonstrated to depend on a ROP (Rho of Plants) rheostat. ROPs are monomeric proteins capable of binding GTP. When bound to GTP, ROPs are active and can be inactivated by dephosphoryation of the guanosin nucleotide. Inactivation of RopGAP4 in *Arabidopsis thaliana* causes an increase in *ADH* induction under anoxia. A link between ROS and hypoxic signaling has also been shown by the fact that ZAT12 and ZAT10, two transcription factors involved in oxidative stress responses, are also induced by low-oxygen treatment (Davletova *et al.*, 2005; Rossel *et al.*, 2007). Moreover, overexpression of ZAT12 leads to the repression of some anaerobic genes during treatment at 4 °C. (Vogel *et al.*, 2005), such as *ADH*, *Sucrose synthase 4* (*SUS4*) and *Non-symbiotic Hemoglobin* 1 (*HB1*).

Nitric oxide (NO) has been recently shown to play a role in both signaling and tolerance to low oxygen. NO production from nitrite (NO_3^-), which probably occurs through the combined activities of rotenone-insensitive NAD(P)H dehydrogenases, mitochondrial complex III (ubiquinone: Cytochrome *c* reductase) and IV (COX), increases as a consequence of the transition from normoxia to hypoxia (Igamberdiev and Hill, 2008). This could be explained by an increased synthesis rate or by an increase in NO stability directly linked to a lack of oxygen. Molecular O_2 reacts quickly with NO to generate NO_2. Nitric oxide is then able to inhibit oxygen consumption by COX, therefore preventing the tissue from becoming anoxic (Geigenberger, 2003). Moreover NO has been shown to also inhibit oxygen consuming reactions involved in storage metabolism and to promote mitochondrial ROS production (Amirsadeghi *et al.*, 2006; Borisjuk *et al.*, 2007). NO can also exert its biological role through the nitrosylation of tyrosine and cysteine and transition metals. The activities of several plant proteins involved in transcription, the central metabolism, stress responses and innate immunity have so far been shown to be regulated by *S*-nitrosylation (Abat *et al.*, 2008; Grennan, 2007; Huber and Hardin, 2004; Lindermayr *et al.*, 2006; Serpa *et al.*, 2007).

3. pH as a low-oxygen signal

Another candidate signal for the plant cell to sense the shift from aerobic to hypoxic or anoxic condition may be represented by the cytoplasmic acidification that rapidly takes place under hypoxia concomitantly with a slower increase in external (or apoplastic) pH. Indeed, under low-oxygen conditions, production of lactate and protons leaking from the vacuole can significantly lower the cytoplasmic pH (reviewed by Magneschi and Perata (2009) and Perata and Alpi (1993)). pH changes have been suggested to act as signal from the root to the shoot during drought stress (Davies and Zhang, 1991), while cytosolic acidification may represent a precondition for gene activation in response to attacks by pathogens (He *et al.*, 1998). The evidence that pH decrease under oxygen deprivation with a rate that is not constant, but reaches a new set point after a few hours, would support its role in signaling the oxygen (or energy) shortage. Felle (2001), using microprobes, non-invasively inserted in the sub-stomatal cavity of leaves, showed that pH changes cannot be transferred on long distance. In fact, imposing acidification to the root system or to the cut petiole was not sufficient to cause rapid and significant pH changes in the leaves. However, it cannot be ruled out that the hypoxic acidification occurring in the roots triggers a secondary signal, able to be transported to the roots. However, as proposed by Felle (2001) changes in extracellular and intracellular pH are very common responses to a wide number of stress and stimuli, and pH alone can not be sufficient to signal a specific stimulus. pH should rather be characterized as having an additional signaling effect, synergistic with a specific low-oxygen signal.

IV. LOW-OXYGEN SIGNAL TRANSDUCTION IN PLANTS

The search for signaling components in low-oxygen plant responses has been carried out mostly using genetic approaches; although some reverse genetic approaches have provided encouraging results. Provided the relatively vast amount of data on the hypoxic response at the transcriptional level, the easiest starting points with respect to the search for signal transducer have been the functional analysis of induced proteins known to be involved in signaling in other conditions, or sharing homology with these. At the same time, analyses of the promoter sequences of well studied and uncharacterized genes was aimed at the identification of DNA elements required for their induction. These represent the ideal targets for transcription factors that play a major role in the transcriptional response to anaerobiosis in plants. Subsequent analyses on these transcription factors by means of transgenic approaches, involving either overexpression or silencing provided, in some cases, a clear demonstration of the relevance of specific transcription factors in species or variety specific manner. Analyses on miRNAs, whose expression changes during submergence stress, started to shed light at a further step in the regulation of gene expression under oxygen deprivation (Zhang *et al.*, 2008). A different, still fruitful, approach in the identification of the hypoxic signal transduction has been based on the analyses of molecules already proved to have a similar role in other stress conditions, as it happened for calcium and phosphorylation cascades.

Several whole transcriptome analyses on different plant species and different tissues under hypoxia and anoxia have led to the identification of a number of genes that make up primary anaerobic responses (Branco-Price *et al.*, 2005; Gonzali *et al.*, 2005; Klok *et al.*, 2002; Lasanthi-Kudahettige *et al.*, 2007; Mus *et al.*, 2007; Pasentsis *et al.*, 2007; Van Dongen *et al.*, 2008). Besides several genes involved in maintaining metabolisms and structures, a number of genes involved in signal transduction and gene transcription has also been identified.

A. TRANSCRIPTIONAL REGULATION OF HYPOXIC SIGNAL

1. Cis-acting elements

An extensive search for DNA elements responsible for gene induction has been carried out over the last 20 years. A first anaerobic response element (ARE) with a GGTTT core was subsequently rejected as a general element for hypoxic induction since it was also present in genes whose expression does not change under anaerobiosis (Russell and Sachs, 1989; Walker *et al.*, 1987). The Arabidopsis *ADH* promoter probably represents one of the most studied DNA sequences among the anaerobic genes. It has been shown to contain a general repressor element in its 5′ second half, while hypoxic regulation depends on the first 172 nucleotides that are upstream of the transcription-starting site, encompassing GT and GC boxes. The G box, which is usually associated with light-regulated genes, does not influence gene expression under hypoxia. It has also been reported that cold and ABA regulation occur via elements that are not required for hypoxic induction (Dolferus *et al.*, 1994).

A comparison of glycolytic enzyme promoters that are upregulated under low oxygen and flooding in different plant species has confirmed these elements, but has also identified five new ones (5′-AAACAAA-3′, 5′-AGCAGC-3′, 5′-TCATCAC-3′, 5′-GTTT(A/C/T)GCAA-3′, and 5′-TTCCCTGTT-3′), for which no binding protein has previously been reported (Mohanty *et al.*, 2005). In rice, *ADH* and *PDC* expression has also been shown to increase biphasically: an initial expression takes place after 2 h and a second one after 12 h of flooding stress. Both have been studied with regard to the posttranslational modifications occurring at the histone octamer associated with the 5′ sequence upstream the *ADH* gene. The first event was shown to be associated with a change of histone 3 (H3) from di- to tri-methylation while acetylation of H3 and H4 occurred at the time of the second inductive step. Both modifications were shown to increase the binding of RNA polymerase II in the genomic region, suggesting that *ADH* upregulation is mediated, at least partially, at a transcriptional level (Tsuji *et al.*, 2006).

2. Trans-acting elements

A number of transcription factors belonging to different protein families (AP2-ERF, LOB, WRKY, MYB) are coherently upregulated in rice and Arabidopsis under anoxia. So far only a few have been characterized with respect to hypoxia signaling and tolerance. AtMYB2, a member of the Myb family of transcription factors, has been shown to be able to bind the GC motif in the *ADH* promoter *in vitro* and to upregulate the GUS reporter gene expression under the control of Arabidopsis *ADH* promoter in protoplasts and leaves (Hoeren *et al.*, 1998). Interestingly, *MYB2* mRNA levels were shown to increase after the addition of cycloheximide, a protein synthesis inhibitor (Hoeren *et al.*, 1998). This observation led to the hypothesis that MYB2 represents a primary actor in anaerobic gene induction, since it is independent of transcription. Further analyses at the whole transcriptome level did not identify significant changes in mRNA levels for *MYB2*, under hypoxic conditions (Branco-Price *et al.*, 2005, 2008; Van Dongen *et al.*, 2008). Unfortunately, so far no *MYB2* knockout or silenced line has been characterized in terms of the ability to induce *ADH* and other anaerobic genes under low oxygen. Overexpression of *MYB2* in Arabidopsis plants, on the other hand, showed an increase in *ADH* induction, but only when exogenous ABA was applied. The same was shown in plants overexpressing another transcription factor, *MYC2*, whose mRNA level does not change under low oxygen and is involved in jasmonate signaling. Interestingly, plants overexpressing both transcription factors showed a constitutively enhanced expression of *ADH* and other anaerobic genes irrespectively of ABA supplementation (Abe *et al.*, 2003). This suggests that at least one member of either the Myb or the Myc family is required for hypoxic gene regulation. Other members of the Myb family have been reported to be induced under low oxygen in Arabidopsis, rice and wheat. Interestingly, *TaMYB1* expression, a MYB transcription factor from *T. aestivum*, was shown to be induced by low-oxygen treatment and further enhanced by light (Lee *et al.*, 2007). The activation of these transcription factors, however, may take place via posttranslational modification or interaction with other transcription factors. DNA binding of *MYB2* has in fact been demonstrated to be inhibited by *S*-nitrosylation of a conserved cystein residue (Serpa *et al.*, 2007). MYBLEU, another member of the Myb rice family, has been proposed to play a role in gene activation under low oxygen (Locatelli *et al.*, 2000). Heterologous expression of this transcription factor in Arabidopsis has been reported to lead to elongation in the primary roots and in the internodal region of the floral stem, together with an enhancement of tolerance under oxygen deficiency (Mattana *et al.*, 2007).

Among the transcription factors that are upregulated under low-oxygen conditions, the AP2-ethylene responsive factor (ERF) family seems to be the most conserved, with at least one member reported to be induced in nearly all the plant species considered (Lasanthi-Kudahettige *et al.*, 2007). All the low-oxygen ERF members belong to the same group, named either VII or B2. Interestingly, rice, which is one of the few crops able to germinate in the absence of oxygen and which often experiences flooding, is the plant with the biggest VII group. Moreover *Sub1a*, a variety specific ERF transcription factor, can confer tolerance to short-term flooding to intolerant varieties (Xu *et al.*, 2006). *Sub1a* overexpression in intolerant varieties was reported to induce a faster *ADH* gene induction and at the same time repress the induction of genes involved in energy-expensive processes such as cell expansion (Fukao *et al.*, 2006). The absence of *Sub1a* homologues in other plant species suggests that it represents a variety specific allele that acts on a more general regulation pathway. Other transcription factors belonging to the ERFVII group have been shown to play a role in ethylene signaling as part of the response to biotic and abiotic stresses (Jung *et al.*, 2007; Xu *et al.*, 2007; Youm *et al.*, 2008). Increased transcription of a gene does not necessarily result in accumulation of the encoded protein. Posttranscriptional mechanisms may also regulate gene expression by regulating the levels of a specific mRNA or its translational rate through the activity of microRNAs (miRNAs). They are approximately 21-nt-long, noncoding RNA produced by excision from a stem-loop precursor. Despite their critical role in mediating development and response to stresses (Jones-Rhoades and Bartel, 2004), only few analyses of miRNAs associated with low oxygen or flooding conditions have been published. Zhang *et al.* (2008) recently reported the change in expression of more than 100 miRNAs maize roots after 12, 24, and 36 h flooding. miRNAs which target sucrose degradation and carbohydrate breakdown were repressed after 24 and 36 h flooding. Surprisingly, one miRNA (osa-miR528-like) which regulates genes involved in ROS and acetaldehyde detoxyfication was upregulated. Among the upregulated miRNAs, zma-miR166 may regulate the root meristem cell differentiation under submergence stress in maize, since this miRNA can target transcription factors such as Rolled leaf 1 (Rdl1), a homologue of the Arabidopsis HD-ZIP. Moreover repression of zma-miR167 and zma-miR168, which target auxin responsive factors (ARFs) at the early submergence stage, was suggested to mediate the formation of adventitious rooting in the hypocotyl of submerged maize seedlings.

3. Transcriptional regulation of the anaerobic response in chlamydomonas
Low-oxygen signaling in *Chlamydomonas* cells requires a specific discussion. Among genes induced by anoxic treatment in *Chlamydomonas*, only those belonging to the general energy starvation response were upregulated. However no anoxia-regulated transcription factors in *Chlamydomonas* shared homology with those induced in multicellular plants. On the other hand, many of the enzymes involved in glycolisis, fermentation and stress responses were the same, indicating a conserved response from unicellular algae to higher plants (Mus *et al.*, 2007). Hypoxic responses in *Chamydomonas* have been reported to mirror that of copper deficiency in the induction of Cytochrome c_6 Cpx1 (encoding coprogen oxidase, a tetrapyrrole biosynthesis enzyme) and Crd1 (encoding a putative di-iron enzyme) (Eriksson *et al.*, 2004). Moreover treatments with $HgCl_2$, an inhibitor of the copper deficiency response, also blocked the hypoxic response. A physiological connection between copper- and oxygen-deficiency-induced gene expressions was therefore proposed (Moseley *et al.*, 2000; Quinn and Merchant, 1995). A common copper response element, called CuRe, whose core sequence is GTAC and is present in the promoter, has been shown to be necessary and sufficient to stimulate gene expression under copper deficiency. This element was shown to be necessary but not sufficient for low-oxygen induction, which also required the presence of a hypoxic response element (HyRE), which shares the same core sequence as the CuRe.

A trans-acting element capable of regulating copper deficiency responses has been associated with the copper response regulator locus (*CRR1*).

Mutation in the *crr1* locus leads to inhibition of low copper responses but only part of the low-oxygen response, since the hydrogenases gene was still induced under low oxygen. Cloning of the *crr1* locus led to the identification of a putative transcription factor with a plant-specific DNA-binding domain called SBP, ankyrin repeats, and a C-terminal Cys-rich region, which is similar to a Drosophila metallothionein. Since *crr1* only slightly increases with copper deficiency, it has been proposed that it acts as a positive regulator after being activated at the protein level (Kropat *et al.*, 2005).

B. OTHER ELEMENTS INVOLVED IN HYPOXIC SIGNALING

Signal cascades in stress physiology usually end up with the activation of transcription factors which, in turn, activate genes involved in adaptation and promote tolerance to the biological system. However, the path from the signal to the activation of transcription factors is mediated by several effectors that need to interact with each other and with components of the developmental program. A great number of proteins and small molecules that play this role have been extensively studied in relation to abiotic stresses different from anaerobiosis. The major components in abiotic stress signaling have been hypothesized as being phosphorylation/dephosphorylation events, variations in Ca^{2+} concentrations and ROS.

In mammalian cells, a decrease in oxygen levels activates voltage-gated plasma membrane channels triggering an influx of Ca^{2+}, while stimulating an efflux from the mitochondria (Fähling, 2008). It has been shown that ruthenium red (RR), an inhibitor of calcium flux from organelles, is sufficient to block the induction of *ADH1* and *SH1* (encoding a sucrose synthase isoform) in maize seedlings and reduced their ability to survive short time flooding. On the other hand, the supplementation of calcium prevented RR effects (Subbaiah *et al.*, 1994b). Other authors then found that Ca^{2+} is necessary for anaerobic induction of *ADH* in Arabidopsis and rice (Chung and Ferl, 1999; Tsuji *et al.*, 2000).

Taking advantage of fluorescence imaging and photometry of Ca^{2+} in maize suspension-cultured cells, Subbaiah *et al.* (1994a) demonstrated an immediate increase in calcium levels in cytosol, which was fully reversible a few seconds after re-oxygenation. This first increase in calcium level was followed by a second, more persisting one. The existence of a double, temporally distinct, Ca^{2+} increase in Arabidopsis seedlings was also reported by Sedbrook *et al.* (1996) who used a calcium sensitive luminescent protein Aequorin. Repeated anoxic treatment, with the addition of a calcium chelator EGTA and PM calcium channel blockers hampered the first spike in Ca^{2+} concentration.

The hypothesis that a calcium increase in the cytosol comes from mitochondria stemmed from the observation that these organelles are probably the first to be influenced when oxygen levels drop. Observations that Ca^{2+} accumulates in the cytoplasmic periphery of mitochondria using confocal fluorescence microscopy strongly enforced this hypothesis (Subbaiah *et al.*, 1998). The mechanism through which Ca^{2+} is released from mitochondria still needs elucidation. However, treatment with caffeine, which promotes calcium releases without membrane depolarization, induced *ADH* in normoxic conditions. Exposing caffeine treated cells to anoxia, leads to a further increase in calcium, suggesting that Ca^{2+} is released through an initial, caffeine sensitive and Na^{+} or H^{+}–Ca^{2+} antiport and, later, via mitochondrial membrane depolarization (Subbaiah *et al.*, 1998). Greenway and Gibbs (2003) proposed that a variation in pH of 0.5, which occurs a few minutes after the onset of anoxia, is sufficient for calcium to move against its electrochemical gradient using energy provided by an influx of protons in the mitochondria (Greenway and Gibbs, 2003; Gunter *et al.*, 1994).

Calcium was also shown to be involved in the induction of a cystein protease (calpaine), starting from the root apex and then spreading to the root axis in maize roots (Subbaiah *et al.*, 2000). Calcium is not considered to play a role on proteolytic activity in plant calpaines, since these proteins lack the calcium-binding domain IV (Margis and Margis-Pinheiro, 2003; Wang *et al.*, 2003). However, the induction of this protease has been proposed as being responsible for cell death occurring at the root tip and, in fact, de-tipping maize roots improved root tolerance to anoxic stress (Subbaiah *et al.*, 2000).

Oxygen consumption at the ETC is also probably Ca^{2+}-mediated. In potato tubers, calcium was shown to inhibit the AOX pathway mitochondria energized by NADH or succinate, but only when the cytochrome pathway was inhibited by cyanide (Mariano *et al.*, 2005). Moreover, in the absence of cyanide, calcium stimulates NO degradation via the activation of NADPH dehydrogenase, therefore preventing the inhibition of COX (complex IV) (De Oliveira *et al.*, 2008).

Despite the fact that mitogen-activated kinases (MAPKs) play a role in almost all signaling and developmental programs in plants (Colcombet and Hirt, 2008). Indeed, although these kinases have been shown to modulate hypoxic signaling in mammals (Fähling, 2008), little is known about their role with respect to oxygen deficit in plants. A minimal MAP kinase cascade consists of three levels: MAP3Ks, which phosphorylate MAP2Ks which, in turn, phosphorylates MAPKs which activate different signaling components (Colcombet and Hirt, 2008). Not many genes encoding for MAP kinases of phosphatases are upregulated under low-oxygen conditions in Arabidopsis

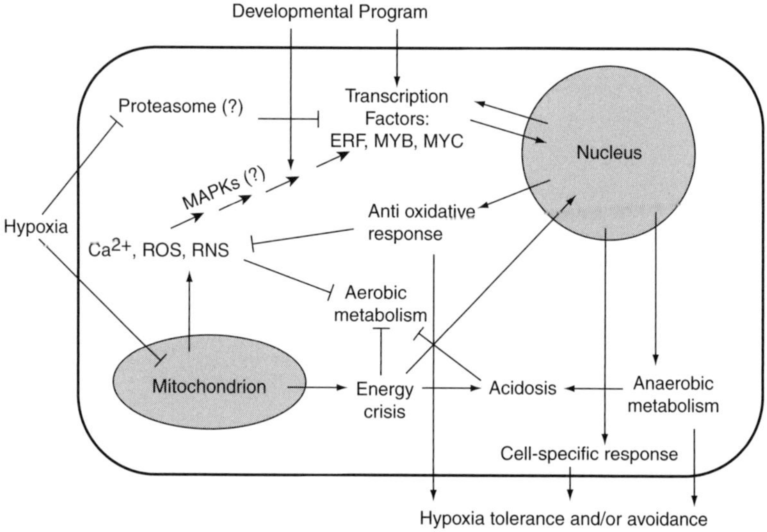

Fig. 1. Integrative view of the different pathways proposed to be involved in hypoxic signaling.

(Branco-Price *et al.*, 2008; Loreti *et al.*, 2005). This is however not surprising, since a rapid signal transduction should rely more on existing peptides rather than waiting for new components to be synthesized, especially when translation is impaired because of energy deficit. Two MAPKs are already upregulated after 2 h hypoxia in Arabidopsis: *MKK9* and *MKK11*. MKK9 was reported to stimulate ethylene synthesis and abiotic stress responses through the activation of MPK3 and MPK6 (Xu *et al.*, 2008). *MKK11*, together with *MYB2*, have been reported to be regulated by ABA in a ROP10 dependent manner (Xin *et al.*, 2005) (Fig. 1).

V. LOW-OXYGEN RELATED STRESSES: ENERGY DEFICITS AND CONSEQUENCES

The main cellular stress caused by low oxygen availability consists in a reduced respiration and, therefore, lower energy production, resulting in slower metabolic processes. Recently, studies in pea and Arabidopsis roots suggested that plant cell adjust respiration to oxygen availability to prevent, or at least postpone, anoxia (Zabalza *et al.*, 2009). When respiratory activity is reduced or blocked, energy production is limited to the glycolytic process. To avoid slowing in the glycolytic flux, plant cells activate fermentative pathways whose end products, if accumulated, could be detrimental to the plant survival. Especially, cytosolic acidification has been observed as a consequence of the energy deficit, although it is unclear whether this is a cause or a consequence of cell death under low-oxygen conditions. Moreover, return to normoxia following a hypoxic or anoxic condition also implies a new stress condition, because of the production or reactive oxygen and nitrogen species.

A. COX, AOX, AND IMPAIRED ENERGY PRODUCTION

In all aerobic organisms, a decrease in oxygen concentrations inside the cell results in reduced energy production via oxidative phosphorylation. This is because O_2 represents the ideal last electron acceptor in the mitochondrial electron transport chain (mETC) (Geigenberger, 2003).

Two types of mitochondrial terminal-oxidases play a role in the oxidative phosphorylation pathway: the cyanide-sensitive COX and cyanide-insensitive alternative oxidase (AOX). The COX complex (IV) consists of at least nine peptides, of which the three largest are mitochondrial encoded, and whose activity is linked to active ATP synthesis (Millar *et al.*, 2004). The AOX complex on the other hand, is a simple homodimer encoded by genomic genes and surprisingly its activity is not coupled to ATP production. The expression of *AOX1a* is instead increased in conditions that impair mitochondrial activity, such as abiotic stresses (Clifton *et al.*, 2005). AOX may play a role in avoiding oxidative stress and modulating metabolic flexibility in plants (Plaxton and Podestá, 2006; Watanabe *et al.*, 2008).

When oxygen levels drop, the COX function is reduced since its K_m for oxygen is in the order of 10 μM (Affourtit *et al.*, 2001; Millar *et al.*, 2004) and it is inhibited by NO (Dordas *et al.*, 2003). Zabalza *et al.* (2009) suggested that adaptation of respiration under hypoxia takes place at the COX level, since its activity in pea roots follows a biphasic trend, as the respiratory activity does, while AOX activity decreases just linearly. Under hypoxia, energy production within the cell is provided mainly by glycolysis, which requires constant NAD^+ regeneration to work efficiently (Perata and Alpi, 1993). The energy status of the cell can be measured by two parameters: adenylate energy charge (AEC), which indicate the proportion of high energy phosphate bonds in the nucleotide reserve pool (AEC = ATP + 0, 5ADP ATP + ADP + AMP), and ATP production, expressed as a percentage of sugar catabolism in the air. These two parameters are both necessary since AEC alone represents the energy status in the plant only until there is no further decrease in the nucleotide pool (Greenway and Gibbs, 2003). In complete anoxic conditions, energy production, based on rates of ethanol formation, can vary between 3 and 37.5% when compared to production in normoxic conditions. These variations depend on the glycolytic flux and the amount of polysaccharides (starch or sucrose) used as a starting source (Greenway and Gibbs, 2003). Several authors have reported an increased glycolytic ATP production, when conditions become completely anoxic, defined as a "Pasteur effect," but this is not enough to prevent the energy deficit (Gibbs *et al.*, 2000; Neal and Girton, 1955; Vartapetian, 1982). Moreover, slowing of glycolysis during prolonged (24–64 h) anoxic treatment has been proposed to occur in rice coleoptiles (Colmer *et al.*, 2001) and aged storage tissue of red beet (Zhang and Greenway, 1994), estimated by observing reduced ethanol accumulation.

B. DRAWBACKS OF METABOLIC ADAPTATIONS TO HYPOXIA

Continuous substrate availability and regeneration of oxidized NAD are required for an efficient glycolysis. This is achieved mainly through fermentative pathways: a short initial lactic and longer lasting ethanolic fermentation (Perata and Alpi, 1993). Lactic acid accumulation can be toxic to the plant tissues since its dissociation would contribute in decreasing cytosolic pH, while ethanol barely reaches dangerous concentrations in the cells since the membranes are permeable, provided a gradient in maintained (Davies, 1980). However, ethanol can probably reach high values in bulky tissues such as tubers, rhizomes or seeds sealed by the testa. Concentrations of around 70 umol g^{-1} fresh weight have been measured for *I. germanica* rhizomes treated for 16 days in anoxic conditions in a humid atmosphere (Monk *et al.*, 1984). Evidence from carrot protoplasts cultures, showed that ethanol toxicity is directly linked to acetaldehyde production. The addition of 4.15–5.35 mM ethanol to aerobic cultures delayed cell growth, however treatment with 4-methylpyrazole, an inhibitor of *ADH*, prevented the toxic effects of exogenous ethanol up to 40–80 mM, while toxic effects were obtained by directly adding acetaldehyde (Perata and Alpi, 1991; Perata *et al.*, 1984). Acetaldehyde, produced by ethanol oxidation, can also be a problem for the plant when aerobic conditions are restored since it may react with proteins and DNA (Perata *et al.*, 1992) or act as an electron donor for ROS production via xantine oxidase (Mustroph *et al.*, 2006). Acetaldehyde production upon re-oxygenation may represent a pathway for consuming H_2O_2 via catalase mediated ethanol peroxidation (Zuckermann *et al.*, 1997). Oxygen shortage also lead to a drop in cytoplasmic pH in almost all the plant systems studied so far, though this can vary among species, varieties, organs, and tissues (Felle, 2005, 2006). Initial cytoplasmic acidification has been proposed to be caused by the accumulation of organic acids, such as lactic acid (Vartapetian and Jackson, 1997), which dissociates rapidly. However,

Saint-Ges *et al.* (1991) showed that, in anoxically shocked maize root tips, a decrease in pH preceded a peak in lactate concentration. Gout *et al.* (2001) linked the initial drop in cytoplasmic pH to the hydrolysis of NTP to NMP, while Greenway and Gibbs (2003) suggested the strong inhibition of vacuolar H^+ ATPase as the cause of this phenomenon. H^+ influx together with K^+ efflux across the plasma membrane and tonoplast, and organic acids accumulation in the cytoplasm are the most likely causes of acidification when anaerobic conditions last more than few hours.

It is still not clear whether acidosis, in anoxic cells, is a cause or a consequence of cell death (Greenway and Gibbs, 2003; Roberts *et al.*, 1984; Tadege *et al.*, 1998; Xia and Roberts, 1994); however tolerance is strongly associated with a reduction in cytosolic acidification: in both excised shoots of anoxia tolerant rice (*O. sativa cv. Arborio*) and wheat (*T. aestivum cv. MEK*) a quick drop in cytosolic pH was observed within the first 30 min of anoxic treatment. However, while in rice this did not drop below pH 7.1, in wheat it reached values a pH value of 6.6 (Menegus *et al.*, 1991).

C. THE RE-OXYGENATION STRESS

A return to the aerobic state from low-oxygen conditions leads to further, mainly oxidative, stress possibly due to ROS and NOS production caused by mitochondria malfunctioning (Smirnoff, 1995). ROS and NOS can react with proteins, inhibiting their functions with nucleic acids, and with polyunsaturated lipids generating a peroxidation chain reaction, which leads to general membrane damage. In rice, ROS detoxifying enzymes show a reduced expression when coleoptile grows under anoxic conditions rather than in air (Lasanthi-Kudahettige *et al.*, 2007). However it has also been reported that flooded seedlings, when exposed to normoxia, restored catalase activity to a greater extent than aerobic controls, suggesting ROS production upon re-oxygenation (Biemelt *et al.*, 1998). The synthesis of catalase may represent a waste of energy under anoxia (Fukao and Bailey-Serres, 2004; Magneschi and Perata, 2009); however it is also possible that ROS scavenging is even detrimental in low oxygen, since it hampers a possible signal for regulation and adaptation.

VI. METABOLIC ADAPTATION TO ENERGY CRISIS

When plants, or some of their organs, cannot avoid hypoxic conditions, they try to metabolically adapt to cope with the stress.

As mentioned previously, energy production under anaerobic conditions mainly depends on the glycolytic pathway, which, in turn, requires the regeneration of NAD^+. As respiration is reduced, pyruvate is rapidly accumulated in concentrations similar to the K_m of pyruvate decarboxylase (PDC), whose affinity for the substrate is usually lower than that of pyruvate dehydrogenase (PDH) (Pronk *et al.*, 1996).

Therefore, accumulated pyruvate can be used as a substrate in fermentative pathways that regenerate NAD^+. Pyruvate has also been shown to be strongly linked to the ability of plants to adjust oxygen consumption by respiration (Zabalza *et al.*, 2009). A shift from pyruvate utilization by the TCA cycle to the fermentative pathways is associated with a decrease in *PDH* mRNA translation (Branco-Price *et al.*, 2005) and an upregulation of PDH kinase, which inactivates PDH (Marillia *et al.*, 2003).

Almost all plants studied so far showed an increase in ethanolic and, at lower levels, lactic fermentations under hypoxia and anoxia, although to different extents and with different effects on tolerance. Therefore, despite its conservativeness in nearly all the eukaryotes, fermentation alone is not necessarily able to confer resistance. A crucial point may rather be the efficiency in mobilizing the carbohydrate reserves during hypoxia. Other reactions, producing mainly alanine and γ-aminobutiryc acid (GABA), may also contribute to NAD^+ regeneration (Bailey-Serres and Voesenek, 2008). A further, recently suggested, role for the fermentative pathways is to regulate the pyruvate level within the cell, therefore modulating the respiration rate under hypoxia (Zabalza *et al.*, 2009). As for many other stress conditions, efficient metabolic adaptation to low oxygen requires time. Therefore a gradual acclimation from mild hypoxia such as O_2 levels at which respiration is reduced to half of the maximum rate to more extreme conditions, often has a positive effect on tolerance (Saglio *et al.*, 1988). The environmental relevance of this phenomenon with respect to flooding is questionable since it depends on the type of the soil and the temperature (Gibbs and Greenway, 2003). However it may be of great importance in case of slow hypoxia onset, such as the one caused by growth of bulky tissues or envelopes (Borisjuk *et al.*, 2007).

A. LACTATE SYNTHESIS AND ACCUMULATION

Lactic fermentation occurs through the reduction of pyruvate by lactate dehydrogenase (LDH) with the concomitant oxidation of NADH to NAD^+. A rapid activation of LDH has been observed in almost all plant species (e.g., *O. sativa*, *Zea mays*, *Solanum lycopersycum*, *Solanum tuberosum*) (Christopher and Good, 1996; Rivoal and Hanson, 1994; Rivoal *et al.*, 1991;

Sweetlove *et al.*, 2000). Glycolytic flux to lactate is quite low compared with that to ethanol in several plant species, with rates usually not higher than 1 μmol g^{-1} fresh weight h^{-1}. Lactic acid accumulation can be harmful for the cell, since it quickly dissociates lowering cytosolic pH. However lactate production may contribute to a posttranslational regulation of the fermentative pathways. According to this hypothesis, lactate-induced acidification would adjust pH to an optimal value for PDC activity, therefore channeling NAD regeneration toward ethanol synthesis via ADH (Davies, 1980). However, in maize roots, increased ethanolic fermentation is observed prior to cytosolic acidification (Saint-Ges *et al.*, 1991). However, in some species, such as in *Solanaceae*, anaerobic LDH isoforms may play a role in pyruvate regeneration during long lasting hypoxic conditions. In fact, in *Solanum lycopersicum*, hypoxia inducible LDH1 has been shown to possess greater activity toward pyruvate synthesis, whereas, constitutive LDH2 catalyzes the reaction in direction of lactate production (Germain *et al.*, 1997a). The observation of LDH-silenced potato tubers is in agreement with this hypothesis, since transgenic tubers contain twofold more lactate than wild types (Sweetlove *et al.*, 2000). Lactate is continuously produced in Arabidopsis tissues 2 h after the onset of hypoxia and its level rises further after 9 h. However, a high cytosolic accumulation of this metabolite is probably prevented by lactate extrusion via a hypoxia-inducible nodulin intrinsic protein NIP2;1 (Choi and Roberts, 2007). A similar function can be assumed for some pleiotropic drug resistance (PDR) type ATP-binding cassette (ABC) transporters, whose expression in rice is indeed regulated by lactate and other weak acids (Moons, 2008). Considering these premises, lactate production does not likely to play a major role in low-oxygen intolerance, since its production is restricted in time and the small amount of lactate produced during the first hours of hypoxia is easily expelled from the cell. Moreover, overexpression of LDH in Arabidopsis significantly increases root tip survival and root tip growth compared with the wild-type plants in non-hypoxically pre-treated tissues submitted to anoxic stress (Dolferus *et al.*, 2008). The observation that increased LDH activity also stimulates ethanolic fermentation led Dolferus *et al.* (2008) to hypothesize that lactic fermentation in Arabidopsis is either required to initiate or to favor the ethanolic fermentation.

B. ETHANOL PRODUCTION

Unlike lactic fermentation, ethanolic fermentation consists of two reactions: the conversion of pyruvate to ethanol proceeds through the coupled reactions of PDC and ADH. Of the two enzyme activities, PDC controls anaerobic sugar catabolism in several plant species, as shown with different transgenic approaches. Tobacco leaves overexpressing PDC produced 10–20-fold more ethanol than wild types under anoxia (Bucher *et al.*, 1994). Transgenic Arabidopsis overexpressing either *PDC1* or *PDC2* showed about 50–150% ethanol concentrations compared to wild types after 24 h in a 5% O_2 atmosphere, whereas *ADH* overexpression led to very small increases (10–20%) (Ismond *et al.*, 2003). In addition, overexpression of both *PDC* isoforms, but not *ADH*, enhanced survival rates under low-oxygen conditions (Ismond *et al.*, 2003). In general, except for *ADH* knockout mutants, where ADH activity is completely abolished, no correlation has been observed between ADH activity (measured *in vitro*) and ethanol production (Roberts *et al.,* 1989). This suggests that ADH activity is not crucial to anoxia tolerance in ideal situations such as those in laboratories. It can be hypothesized that maximum activity, on the other hand, is crucial in environmental conditions where plants experience different stress conditions at the same time, or when ADH is required after re-oxygenation, to utilize ethanol as a carbon source (Gibbs and Greenway, 2003).

C. OTHER PRODUCTS OF THE ANAEROBIC METABOLISM

Alanine and GABA are the other two products of the anaerobic metabolism (Bailey-Serres and Voesenek, 2008; Magneschi and Perata, 2009). Alanine is produced by the transfer of an amino group from glutamate to pyruvate, with the generation of α-ketoglutarate. The continuous removal of α-ketoglutarate contributes to $NAD(P)^+$ regeneration and hinders the reverse reaction toward alanine degradation. This can be achieved by the glutamine synthetase/glutamate synthase cycle GS-GOGAT (Reggiani *et al.*, 1988) or the glutamate dehydrogenase pathway (Fan *et al.*, 1997). The observation that NH_4^+ incorporation in the GS-GOGAT pathway would consume 1 ATP molecule per pyruvate molecule converted to alanine, whereas GDH does not consume ATP and regenerates 1 $NADP^+$ molecule seems to favor the glutamate dehydrogenase pathway (Gibbs and Greenway, 2003).

In fact, a modest increase in *GDH2* mRNA translation has been observed in Arabidopsis (Branco-Price *et al.*, 2008).

Glutamate is involved in several anaerobic metabolic reactions, as demonstrated by its decreased content after 2 h hypoxia in Arabidopsis (Branco-Price *et al.*, 2008). It can act as an amino group donor in aspartate transamination, with the production of the TCA intermediate oxalacetate, which is then converted to malate thus generating NAD^+. Aspartic-transaminase mRNA has been reported to be both induced and actively translated under hypoxia (Branco-Price *et al.*, 2008). Alternatively, glutamate can be decarboxylated by glutamate decarboxylase to generate γ-Aminobutyric acid (GABA), with a concomitant H^+ consumption and therefore counteracting the cytosolic acidification caused by lactic fermentation (Aurisano *et al.*, 1995). GABA transaminase (GABA-T) metabolizes GABA to succinic semialdehyde (SSA) coupling consumption of α-ketoglutarate and additional conversion of pyruvate to alanine (Breitkreuz *et al.*, 2003). SSA is possibly reduced to γ-hydroxybutyrate by a NADPH consuming reaction catalyzed by GHB dehydrogenase. None of these enzymes seem to be regulated at the transcriptional or translational level (Branco-Price *et al.*, 2008); therefore GABA accumulation ought to be explained by posttranslational activation of GDH or inactivation of GABA-T. Upon re-oxygenation SSA is probably converted to succinate, and then channeled to the TCA cycle by succinic semialdehyde dehydrogenase (SSADH) in a NAD^+-consuming reaction. SSADH activity has been reported to inhibit ROS formation (Bouché *et al.*, 2003). In agreement with this, while SSADH is induced under low oxygen, it is highly translated only upon re-oxygenation, when ROS are likely to be formed. A link between hypoxic accumulated GABA and secondary metabolism has been proposed by Liu and Castelfranco (1970), who hypothesized reactions in pea-seedlings involving ethanol incorporation with the formation of ethyl-β-glucoside, which are part the cell wall.

D. RESERVES MOBILIZATION TO FUEL THE GLYCOLYTIC FLUX

Provided that there is a group of efficient reactions that regenerates NAD^+, glycolisis coupled to the fermentative metabolism requires constant carbohydrate supplementation. In storage organs such as cereal grains, potato tubers, and *Acorus calamus* rhizomes, hexose provision requires efficient starch mobilization via starch degrading enzymes, such as endo-amylases, exo-amylases, debranching enzymes, and starch phosphorylase (Magneschi and Perata, 2009; Perata and Alpi, 1993; Smith *et al.*, 2005; Stitt, 1990). Starch phosphorylase may act on starch degradation in the late phases of low-oxygen stress, as demonstrated by its three- to sixfold increase in activation after 2 days of flooding treatments in rice. This enzyme could play a role in stress tolerance, since rice intolerant varieties did not show any increase in Starch phosphorylase (Das *et al.*, 2000). Rice seeds, but not wheat, barley, oat, and rye which are unable to germinate under anoxia, induce α-amylase genes in low-oxygen conditions (Guglielminetti *et al.*, 1995). α-Amylases in rice have been grouped into three subfamilies: AMY1 (A–B–C), AMY2 (A) and AMY3 (A–B–C–D–E–F) (Magneschi and Perata, 2009). Isoforms encoded by *AMY1A* are present both in aerobic and anaerobic seedlings, while *AMY3D* mRNA is anoxia-specific (Loreti *et al.*, 2003). This has been explained by a differential regulation at a transcriptional level: *AMY1A* is induced by gibberellins (GA) in air but not in anoxia, whereas *AMY3D* does not depend on GA since its promoter lacks the distinctive cis-acting element that confers GA-responsiveness (Morita *et al.*, 1998; Loreti *et al.*, 2003; Perata *et al.*, 1997). On the other hand, anaerobic induction of *AMY3D* is repressed by high sugar levels, suggesting that under aerobic conditions, GA-dependent activation of *AMY1A* maintains high sugar levels in the aleurone layer, thus preventing *AMY3D* induction. Sugar dependent regulation of *AMY3D* depends on cis-acting DNA elements contained in its promoter, named SRC (sugar repression core) (Chen *et al.*, 2006) and on transacting transcription factor *MYBS1* (Lu *et al.*, 2002). *MYBS1* is more expressed in anoxic rice coleoptiles compared to aerobic controls (Lasanthi-Kudahettige *et al.*, 2007). It has also been shown that both *AMY3D* and *MYBS1* are activated by the general regulator OsSnRK1 (Ismond *et al.*, 2003; Lu *et al.*, 2007), belonging to the same superfamily of the Arabidopsis KIN10 and KIN11, and already reported to play a role in sugar and energy depletion signaling (Baena-Gonzalez *et al.*, 2007).

During germination in rice seeds, starch-derived glucose is transferred to the scutellum, where sucrose synthesis takes place (Nomura *et al.*, 1969). Although this anabolic step is energy expensive, it may be required to transport carbon units to the growing tissues of the developing seedlings, thus providing substrate for glycolysis (Magneschi and Perata, 2009). Anoxic sucrose synthesis in rice seeds and Arabidopsis seedlings depends on the expression of sucrose-phosphate synthase and glucose-6-phosphate isomerase, but not on sucrose synthase (Bertani *et al.*, 1981; Guglielminetti *et al.*, 1995, 1999; Ricard *et al.*, 1991). In cereals unable to germinate under anoxia, no sucrose synthesis has been observed under anoxia (Guglielminetti *et al.*, 1999).

The supplementation of exogenous sucrose but not glucose improves survival under anoxia (Germain *et al.*, 1997a; Loreti *et al.*, 2005). Sucrose degradation, to provide hexose-6-phosphates for glycolysis, is achieved through two distinct pathways in plant cells: the sucrose synthase (Susy) pathway, which is bi-directional but favored in the catabolic direction, and the unidirectional invertase pathway. Under low-oxygen conditions, sucrose synthase isoforms are induced in almost all the plants examined, while invertase is repressed (Branco-Price *et al.*, 2008; Lasanthi-Kudahettige *et al.*, 2007). Degradation of sucrose to fructose 6-P and UDP-glucose by Susy only requires one molecule of pyrophosphate if the UTP produced is directly used by fructokinase to phosphorylate fructose or to regenerate ATP from ADP via NDP-kinase (Guglielminetti *et al.*, 1995). In contrast, the invertase pathway requires two ATP molecules per sucrose molecule degraded (Mustroph *et al.*, 2005). Therefore sucrose synthase isoforms are probably induced and Susy activity is stimulated while genes encoding for invertases are repressed. The advantage conferred by sucrose synthase under hypoxia was demonstrated by the reduced ATP levels observed in transgenic potato tubers overexpressing bacterial invertase, compared to wild types under 8% O_2 (Bologa *et al.*, 2003). The existence of more than a single hypoxia-inducible Susy isoform in several species suggests that functional redundancy is required to ensure low-oxygen tolerance in plant tissues. In fact, Arabidopsis mutants, whose hypoxia-inducible Susy genes (*SUS1* and *SUS4*) have been knocked-down, showed an increased susceptibility to root flooding (Bieniawska *et al.*, 2007). A sugar or energy starvation signal may mediate anaerobic upregulation of *SUS1* and *SUS4*, since sucrose supplementation was shown to reduce their induction (Loreti *et al.*, 2005). However, previous microarray analyses in sucrose starved Arabidopsis cells did not identify significant changes in *SUS1* mRNA transcription and translation, whereas *SUS4* transcription was strongly repressed (Contento *et al.*, 2004; Nicolai *et al.*, 2006). A non-sucrolytic role for anaerobic isoforms of the Susy family has also been proposed (Subbaiah *et al.*, 2006) following the observation of intra-mitochondrial localization for two Susy isoforms in maize root tips together with their interaction with voltage-dependent anion channel. Subbaiah *et al.* (2006) hypothesized that these proteins could modulate fluxes in and out of mitochondria in an anoxia- and tissue-specific manner (Subbaiah *et al.*, 2006).

Glucose and fructose need to be phosphorylated by hexokinases to be channeled to the glycolytic pathway. The anoxic induction of a fructokinase (OsFK2) in rice seedlings has been reported (Guglielminetti *et al.*, 2006), together with a hexokinase OsHXK7 whose induction is probably mediated by a sucrose starvation signaling pathway (Cho *et al.*, 2006; Lasanthi-Kudahettige *et al.*, 2007). Hexokinase activity exerts great control over the glycolytic pathway under anoxia in excised maize root tips (Bouny and Saglio, 1996) and tomato roots (Germain *et al.*, 1997b). Since sucrose degradation is presumed to mainly occur via a Susy pathway, glucose kinase(s) may also play a role in utilizing glucose produced either by invertase or derived from amylases activities. The generation of fructose-1,6bisP from fructose-6P can be catalyzed by the unidirectional phosphofructokinase PFK which uses ATP as a phosphate group donor, or via the bidirectional phosphofructokinase PFP which uses pyrophosphate (PP_i). Together with PDC, PFK may regulate carbon flux during anaerobiosis in several plant species (Faiz-ur-Rahman *et al.*, 1974; Gibbs *et al.*, 2000; Mohanty *et al.*, 1993). PFP is activated under anoxia in rice (Mertens *et al.*, 1990; Mohanty *et al.*, 1993) but the direction of the reaction it catalyzes is uncertain. The role of these substrate cycles between PFK and PFP is related to that of another cycle, mediated by pyruvate kinase (PK) and pyruvate orthophosphate dikinase (PPDK). Depending on PP_i consumption/production rates in the cell, these two cycles perhaps work in opposite directions. Interestingly, PPDK is also upregulated by low oxygen in rice at the mRNA and protein level (Lasanthi-Kudahettige *et al.*, 2007; Moons *et al.*, 1998).

E. MITOCHONDRIAL FUNCTION UNDER LOW-OXYGEN CONDITIONS

Under oxygen deprivation mitochondria do not necessarily stop functioning and, though with some changes in the enzyme composition, they maintain their ultrastructure even in anoxic conditions, as showed in rice and *Echinocloa* seedlings (Couee *et al.*, 1992; Fox and Kennedy, 1991). Nitrate has been shown to have a positive effect on mitochondria, and it has been proposed, but not proved, to take the place of oxygen as an alternative electron acceptor in the mtETC (Vartapetian *et al.*, 2003) or playing a role in NAD (P)H oxidation (Igamberdiev *et al.*, 2004). Further $NAD(P)^+$ regeneration is ensured by the reduction of nitrate to nitrite, catalyzed by nitrate reductase (Igamberdiev *et al.*, 2004), competing with production of ethanol. Interestingly, a reduction in anoxic cytoplasmic acidification has been obtained by supplementing maize root segments with nitrate (Libourel *et al.*, 2006). Supplementing nitrate to transgenic tobacco roots with reduced nitrate reductase (NR) levels, on the other hand, showed an increased lactate and ethanol production under anoxia, suggesting competition with the fermentative pathways (Stoimenova *et al.*, 2003). Nitrate reductase (NR) is significantly induced in Arabidopsis and rice under anoxia (Lasanthi-Kudahettige *et al.*, 2007; Loreti *et al.*, 2005) but not hypoxia (Branco-Price *et al.*, 2008). Further NR activation, at the posttranslational level, is mediated by pH acidification (Kaiser and Brendle-Behnisch, 1995). Nitrite dependent NAD (P)H oxidation is catalyzed via two Ca^{2+} dependent, rotenone-insensitive NAD(P)H dehydrogenases, which transfer electrons to the ubiquinone pool at the inner membrane. As mentioned previously Ca^{2+} releases from mitochondria are triggered within few minutes of anoxic imposition. Possible explanations for Ca^{2+} involve either the H^+–Ca^{2+} antiport, enhanced by pH acidification, or an adenilate kinase-dependent change in Mg^{2+} concentration which, in turn, modulates calmodulin and opens Ca^{2+} pores (Igamberdiev and Kleczkowski, 2003). Electron transfer from the ubiquinone pool to nitrate, associated with NO generation, has been proposed to be mediated by complex III (Cytochrome *b* reductase) and IV (COX) (Igamberdiev and Hill, 2008). AOX has also been proposed to mediate NO production from nitrite (Gupta *et al.*, 2005; Tischner *et al.*, 2004), however there is no strong evidence supporting this hypothesis (Igamberdiev and Hill, 2008). Nitrite to NO reduction in mitochondria from algae and higher plants has been observed under strict anaerobic conditions (Gupta *et al.*, 2005; Planchet and Kaiser, 2006), and may generate the proton motor force required for ATP synthesis (Stoimenova *et al.*, 2007). To prevent the inhibition of mtETC by NO accumulation, plants under hypoxic conditions have the advantage of non-symbiotic hemoglobins. The induction of members of this subfamily under low-oxygen conditions has been reported for a number of plant species (Lasanthi-Kudahettige *et al.*, 2007; Loreti *et al.*, 2005; Taylor *et al.*, 1994). Non-symbiotic hemoglobins have been proposed to react with NO and oxygenate it to regenerate nitrate, thus becoming an oxidized form of ferric Hb–Fe^{3+}. Ferric Hb–Fe^{3+} is reduced to ferrous Hb–Fe^{2+} by free ascorbate with the production of the strong oxidant monodehydroascorbate (MDHA). Moreover, ascorbate and MDHA can scavenge peroxynitrite, which is formed by the reactions of NO and superoxide and can modulate protein activity via nitrosylation (Barone *et al.*, 2003). Cytosolic MDHA reductase has been demonstrated (Igamberdiev *et al.*, 2006) to rapidly enhance ferric Hb^{2+} regeneration which then binds oxygen and starts the cycle again. Interestingly, a gene encoding MDHAR has been shown to be induced under low-oxygen conditions in Arabidopsis (Loreti *et al.*, 2005). MDHA can also be reconverted to ascorbate by the mitochondrial transport chain (Li *et al.*, 2002), as with succinate at the level of complex II (Szarka, 2007).

VII. DEALING WITH OXYGEN SHORTAGES: AVOIDANCE STRATEGIES

To avoid the energy crisis caused by oxygen deprivation, plants developed a number of constitutive or inducible tolerance strategies depending on their growing environment. Plant species that experience periodic and long-lasting flooding (e.g., *Rumex palustris* and *Ranunculus sceleratus*) and species able to germinate in flooded soils (e.g., *O. sativa* and *Potagemoton pectinaus*) have indeed evolved a number of strategies that enable photosynthetic tissues to reach the surface of the water and therefore provide oxygen to the organs remaining under water (Bailey-Serres and Voesenek, 2008; Voesenek *et al.*, 2006). The group of traits underlying these strategies, elongation, adventitious rooting, and aerenchyma formation, has been named low-oxygen escape syndrome (LOES; Pierik *et al.*, 2008). Among the three main strategies aimed to avoid low oxygen discussed in this section, fast elongation is quite controversial since it requires a considerable energy expense. Therefore, stimulation or inhibition of underwater elongation can both have adaptive value depending on the condition where the hypoxic environment is established. Interestingly, there is a consistent overlap in the signal molecules, mainly phytohormones, which regulate these processes.

A. LEAF GAS FILMS

In terrestrial plants, gas exchanges can also take place across the cuticola that, however, can represent an impediment to O_2 uptake and CO_2 release in the leaf during the night, while in the day the opposite (O_2 release and CO_2 uptake) is required to prevent restriction of photosynthesis (Colmer and Pederse, 2008). Some wetland species developed hydrophobic cuticles able to retain a layer of air when submerged. These gas films are also defined as plant plastrons for their similarity with gas layers of aquatic insects that enlarge gas–water interface between the tracheary elements and the water (Vogel, 2006). The leaf gas layers may act in a similar way facilitating CO_2 collection during the day and O_2 collection during the night. Moreover, Colmer and Peders (2008) speculated that the presence of gas layers could help leaves to keep the stomata open (Mommer *et al.*, 2005). So far, no evidence for this hypothesis has been provided. However, several reports described the significant contribution provided by leaf gas films to internal aeration in partially flooded deep water rice (Raskin and Kende, 1983), a tolerant rice type under complete submergence (Pederse *et al.*, 2009). The origin of water repellency has been intensively studied since it is relevant to agricultural spray-application processes (Wagner *et al.*, 2003). An extensive study on over 200 plant species revealed that a wide variety of morphologies serves as a basis for surface roughness that causes water repellency (Neinhuis and Barthlott, 1997). Trichomes, papillae-structured epidermal cells and the biochemical composition of waxes and wax crystals are supposed to be the major determinants of this property (Wagner *et al.*, 2003). Cuticular waxes are composed by a mixture of very-long chain fatty acids (VLCFA), embedded in cutin polymers (Kunst and Samuels, 2003). Understanding the genetic control of these properties would probably provide a useful tool toward the improvement of plant resistance to flooding (Pederse *et al.*, 2009), in a way similar to that provided by the discovery of the *Sub1a* gene (Xu *et al.*, 2006). Studies on cuticular wax biosynthesis in rice and Arabidopsis identified some genes encoding enzymes directly involved in VLCFA synthesis (Chen *et al.*, 2003; Greer *et al.*, 2007; Rowland *et al.*, 2006; Yu *et al.*, 2008) and a clade of AP2/EREBP transcription factors involved in the regulation this process (Aharoni *et al.*, 2004).

B. FAST ELONGATION

The fast growth of shoots under flooding conditions resembles in many aspects the shade avoidance syndrome (SAS), since (petiol and leaf) elongation is preceded by the imposition of a hyponastic habitus that enables the leaf blade to reach the water's surface in the shortest time possible (Cox *et al.*, 2003; Mommer *et al.*, 2006). The enhanced shoot elongation is known to be mediated by ethylene (Jackson, 2008; Vreeburg *et al.*, 2005). Ethylene accumulation within tissues under submergence has been reported in fast elongating species such as *Ranunculus scleratus* (Samarakoon and Horton, 1984) and *R. palustris* (Voesenek *et al.*, 1993) as well as in non-elongating species (Banga *et al.*, 1997). So far no clear demonstration has been provided as to whether submergence induced ethylene accumulation is just a consequence of gas entrapment or enhanced synthesis. In fact, most reports on anaerobic ethylene biosynthesis involve measurements after a hypoxic or anoxic treatment (Khan *et al.*, 1987; Peng *et al.*, 2001). Indeed, since ethylene synthesis is oxygen dependent, it would be reasonable to expect its inhibition under anoxia, although under hypoxia synthesis of this hormone could be possible or even enhanced. Indeed low-oxygen conditions strongly induce ACC synthase (*ACS*) and ACC oxidase (*ACO*) genes in several species (Peng *et al.*, 2005; Vriezen *et al.*, 1999; Zhou *et al.*, 2001) and O_2 for ethylene synthesis can be provided following transport from above-water tissues. Hypoxia also induces the ethylene *ETR2* receptor in Arabidopsis (Loreti *et al.*, 2005) and *OsERL1* in rice (Lasanthi-Kudahettige *et al.*, 2007). Since ethylene signaling acts through a negative regulation by the receptors, an increase in the levels of these proteins would hamper ethylene signal during hypoxia. Instead, the induction of ethylene sensors may be explained as a block for ethylene sensitivity when tissues return to normoxic conditions (Bailey-Serres and Voesenek, 2008). In partial agreement with this, a comparison of total and polysome-associated mRNA in *Arabidopsis thaliana* seedlings treated with 2 and 9 h hypoxia showed that an increase in the mRNA level for ETR2 was not associated with the same increase in translation. However, after 1 h of re-oxygenation, the association of this mRNA with polysomes further decreased instead of increasing as expected (Branco-Price *et al.*, 2008).

Ethylene promotes elongation also by enhancing ABA metabolism to phaseic acid while, at the same time, genes encoding ABA biosynthetic enzymes are repressed (Bailey-Serres and Voesenek, 2008; Benschop *et al.*, 2005). In rice, ABA breakdown is mediated by an ABA 8′-hydroxylase, whose expression depends on ethylene, while inhibition of ABA biosynthetic enzymes is not ethylene dependent (Saika *et al.*, 2007). A decrease in ABA levels results in an increase in gibberellins (GA) activity, either following conversion of inactive GA forms to active, as in *Rumex*, or resulting from increased cell sensitivity to GA, as in rice internodes (Bailey-Serres and Voesenek, 2008). Petiole and leaf elongation rely on cell expansion and division and both mechanisms are positively regulated by ethylene and GA (Gray, 2004). Cell wall-modifying enzymes, such as expansins and xyloglucan endotransglycosidases have been reported to be regulated at the transcriptional and posttranslational level by flooding and, in maize and rice, directly by ethylene or low-oxygen treatment (Lasanthi-Kudahettige *et al.*, 2007; Lee and Kende, 2002; Peschke and Sachs, 1994). In *R. palustris*, ethylene also triggers proton efflux into the apoplast, facilitating the actions of extensins (Vreeburg *et al.*, 2005). In rice, GA seems to regulate cell division though the upregulation of cyclin (*CycOs1*, *CycOs2*), cyclin dependent kinase and replication protein A1 (Lorbiecke and Sauter, 1998). Since all these processes require a lot of energy, a supply of carbohydrates deriving from starch or sucrose degradation and the efficient translocation of photosynthates is required. Therefore, shoot elongation provides only an ecological advantage when flooding lasts few days and the water level can be reached by at least the leaf tips plant, whereas it can be detrimental to plant survival in case of short period of flooding followed by re-aeration. In this perspective, the prevention of elongation triggered by submersion may provide a consistent acclimation value in several plant species living in environments characterized by frequent, though short-lasting, flooding events. For instance, some rice varieties belonging to the *indica* group, which can endure short-lasting submergence, have been shown to possess a specific allele, associated with the locus named *SUB1*, which is responsible for almost 70% of their flooding tolerance (Fukao *et al.*, 2006). Introgression of this locus via marked assisted selection in an intolerant indica variety turned it into flooding tolerant (Xu *et al.*, 2006). The *SUB1* tolerance alleles contain three genes encoding for ERF transcription factors, called *SUB1A*, *SUB1B*, and *SUB1C*. *SUB1C*, which is present in all varieties, is GA responsive and positively regulates the expression of several expansins. *SUB1A-1*, the allele present only in tolerant varieties, is sufficient alone to provide flooding tolerance. In fact, overexpression of this gene in a submergence-intolerant *japonica* variety conferred enhanced flooding tolerance (Fukao *et al.*, 2006). SUB1A-1, directly or indirectly, enhances the expression of the negative regulators of GA signaling SLR1 and SLRL1 (Fukao and Bailey-Serres, 2008). SLR1 may hinder the activity of the transcriptional activators, similarly to one of its Arabidopsis orthologs does with phytochrome interacting factor 3 and 4 (PIF3 and PIF4) (De Lucas *et al*, 2008) thus preventing induction of a subset of GA responsive genes, such as SUB1C.

Furthermore SLR1 also contains an N-terminal DELLA domain that mediates its GA-dependent degradation (Ueguchi-Tanaka *et al.*, 2007). On the contrary, SLRL1 does not possess any DELLA domain and therefore it is not degraded in response to GA. SUB1A is also able to restrict ethylene production under submergence, limiting the ethylene-mediated enhancement of GA-responsiveness (Fukao and Bailey-Serres, 2008) therefore enabling the plant to save energy by inducing a quiescent state until the water level decreases and normoxic conditions are restored (Xu *et al.*, 2006). Moreover, in some species such as *R. palustris*, the morphology of newly, underwater developed leaves changes quite dramatically, with a lower starch content, thinner cuticle and chloroplasts oriented toward the epidermis. These changes allow higher rates of CO_2 assimilation, lower CO_2 compensation points and facilitate O_2 transfers into the shoots, as shown in acclimated plants compared with non-acclimated ones (Mommer *et al.*, 2006).

C. LOW OXYGEN-INDUCED ADVENTITIOUS ROOTING

Flooding events often result in the partial submergence of the plant and some species have evolved the ability to produce new roots closer to the surface of the water. As adventitious rooting is a process associated with several productive applications, such as rootstock propagation or *in vitro* plant regeneration, this process has been extensively studied. However, there have been no conclusive studies explaining this phenomenon under hypoxic conditions. For instance, several studies have reported the involvement of almost all known plant hormones in this process, but it is still not clear whether hypoxia alone can induce adventitious roots production, or whether ethylene accumulation, high humidity or high CO_2 level is also required.

Adventitious root formation begins with an increase in cell proliferation from quiescent cells modulated by a number of stimuli, including auxin, peroxidase activity, and ABA (De Klerk *et al.*, 1999; Moncousin and Gaspar, 1983; Tari and Nagy, 1996). Rapid cell duplication generates a mass of metabolically active cells, called root primordium, whose structure resembles that of the root apical meristem (Malamy and Benfey, 1997). It is not known whether endogenous auxins are required for adventitious rooting under flooding conditions. However ABA levels have been reported to decrease in root tissues in hypoxic conditions (Smit *et al.*, 1990). After this phase, cell replication decelerates (Friedberg and Davidson, 1971; MacLeod and McLachlan, 1975) and the primordium elongates mainly by expansion, passing through the parental tissue and penetrating the epidermis (Malamy and Benfey, 1997). This requires the degradation of pericycle and epidermis cell walls to enable the adventitious root to emerge unwounded. Cell wall-modifying proteins such as expansins (Cho and Kende, 1997), subtilisin-like proteases (Neuteboom *et al.*, 1999), pectate lyases (Laskowski *et al.*, 2006), and endo-β-1,4-glycanases (Kimpara *et al.*, 2008) may play a role in this process. It is hard to observe the induction of genes encoding these enzymes using the methods currently available since adventitious rooting processes require strictly tissue-specific gene regulations. In addition, a modulation of their activity is also likely to occur at a posttranslational level (Yoshida and Komae, 2006). However, the induction of expansins in hypoxic and anoxic conditions has been reported both in rice and Arabidopsis (Branco-Price *et al.*, 2008; Lasanthi-Kudahettige *et al.*, 2007). It has also been demonstrated that underwater ethylene accumulation together with a reduction in ABA levels and a concomitant increases in active GA level shows a synergistic effect on epidermal cell death further facilitating the emergence of adventitious roots (Steffens *et al.*, 2006).

D. AERENCHYMA FORMATION

Adventitious rooting and leaf morphology modifications could not provide any real advantage to the flooded plant if oxygen, either produced by photosynthesis or taken up by the aerobic tissues, is not transported to the under water organs.

Gas flow through the intracellular spaces is probably able to provide enough oxygen to sustain root respiration and elongation in short roots (shorter than 3 cm) (Greenwood, 1967). Cell porosity is contributing considerably, as demonstrated by the correlation between increased fractional root porosity (FRP) and the ability to elongate (Justin and Armstrong, 1987). FRP depends on a number of parameters including cell packing, cortical cell configuration, ratio between porous and non-porous tissue and proportion of different configuration types. Cubic cell arrays were shown to have higher FRP when compared to hexagonal cell packing (Justin and Armstrong, 1987). As expected, the longer the root grows, the less oxygen is transported to the root tip, as demonstrated by Armstrong *et al.* (1983) on pea plants whose roots were growing on O_2 free medium. Oxygen levels reached values around 2.5% at the tip of pea roots 9–11 cm long. Moreover, lateral roots considerably reduced oxygen levels at the tip of the main root, where it reached values close to anoxia. Gas transport through submerged tissues is enhanced by the formation, in roots, hypocotyls and petioles, of the so called aerenchyma, a modification of parenchyma cells that generates enlarged air chambers, cavities or enlarged air spaces, which work as gas exchange channels (Chaffey, 2007; Evans, 2004). Aerenchyma can be formed through two different processes, defined as schizogeny, lysigeny, or schizo-lysigeny and, lately also, expansigeny (Drew *et al.*, 2000). The generation of schizogenous aerenchyma involves cell separation via cell wall reorganization, whereas lysigenous aerenchyma is produced as a consequence of programmed cell death (Campbell and Drew, 1983; Evans, 2004; Gunawardena *et al.*, 2001). Expansigeny, on the other hand, occurs via cell division and cell enlargement and is not followed by cell separations and cell collapse or death (Seago *et al.*, 2005). Aerenchyma formation as Arabidopsis was ignored until recently, when reports by Muhlenbock *et al.* (2007) revealed for the first time that 12-week-old plants showed hypoxia-induced lysigeny, although only in tissues where secondary growth was occurring. It had been previously proposed that failure in aerenchyma formation in young plants depended on the impossibility to entrap ethylene in lignified xylematic structures (Evans, 2004). Therefore, the developmental stage of the plants appears to play a role in their ability to form aerenchyma. Lysigenous aerenchyma formation in Arabidopsis seems to require a combination of light, H_2O_2 and ethylene signals, which affect stomatal conductance in leaves (Mateo *et al.*, 2004).

The induction of enzymatic activities involved in cell wall degradation (cellulases) and cell death in maize adventitious roots depends on ethylene production under hypoxia but not anoxia (He *et al.*, 1996). Moreover, these processes are inhibited in hypoxic roots by antagonists of inositol phospholipids, Ca^{2+}-calmodulin and protein kinases. On the other hand they are stimulated by the chemical activation of G-proteins, an increase in cytosolic Ca^{2+}, or an inhibition of protein phosphatases (He *et al.*, 1996).

From a genetic point of view, the induction of aerenchyma is subject to a tissue-specific program involving lesion stimulating disease 1 (LSD1), a zinc finger transcription regulator together with enhanced disease susceptibility 1 (EDS1) and phytoalexin deficient 4 (PAD4) , two lipase-like proteins. These three proteins can interact with each other and are all involved in responses to pathogen infection, photo-oxidation and leaf senescence (Muhlenbock *et al.*, 2007; Ochsenbein *et al.*, 2006; Parker *et al.*, 1996).

The expression of *LSD1*, *EDS1*, and *PAD4* is induced by an increase in CO_2 concentration, together with phosphorus depletion, which is also associated with flooded and hypoxic environments (Drew *et al.*, 1979; Topa and Cheeseman, 1994). A search on microarray databases showed that *LSD1* is also moderately upregulated by low-oxygen treatments in Arabidopsis young plants (4 and 7 days old) in low-oxygen conditions. The longitudinal O_2 transfer from shoots to the root apex can be enhanced by the formation, in adventitious roots, of an external cell layer characterized by low radial permeability to O_2, a barrier to radial O_2 loss (ROL) (Armstrong and Beckett, 1987; Colmer, 2003). While in some wetland plants this barrier to

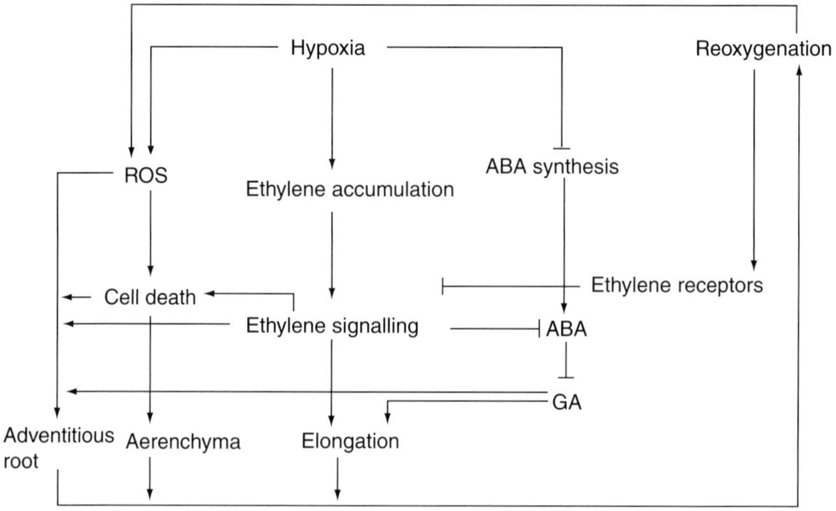

Fig. 2. Scheme of LOES signaling (modified from Bailey-Serres and Voesenek, 2008) involving cross-talk between phytohormons and ROS.

ROL is constitutively present in adventitious roots (e.g., Juncus effusus, Echinochloa crus-galli, Schoenoplectus validus) (McDonald *et al.*, 2001, 2002; Visser *et al.*, 2000) in others, for example, in O. sativa and some wild Hordeum species, the ROL barrier is induced by hypoxia (Colmer *et al.*, 1998; Garthwaite *et al.*, 2003; Jackson and Colmer, 2005) (Fig. 2).

VIII. FUNCTIONAL MAINTENANCE OF THE CELL AND ENERGY SAVING

Cell survival under oxygen deprivation relies on the ability to maintain a minimal functionality that minimizes energy costs and sustains membrane integrity and cellular compartmentation. To cope with the energy crisis, ATP-expensive processes such as protein synthesis are strongly reduced, while the limited amount of ATP produced is used to counteract the detrimental cytosolic acidification by fuelling the H^+ transport to the vacuole. Presence of anaerobic proteins preceding an anoxic stress improves the tolerance of the tissues, as shown by experiments on hypoxic acclimation. Surprisingly, also heat acclimation can improve anoxia tolerance, probably because of the induction of proteins, such as the heat shock proteins (HSP) that respond to both stimuli.

A. ADAPTATION OF THE TRANSLATIONAL MACHINERY TO THE ENERGY SHORTAGE

Despite the possibility of producing energy under low-oxygen conditions, plant tolerance also relies on the ability to reduce energy requirements for the maintenance of minimal cell functions and prevention of cell death. A decrease in transcription of most of the aerobic proteins associated with different aerobic cell functions has in fact been reported in rice and Arabidopsis (Branco-Price *et al.*, 2008; Lasanthi-Kudahettige *et al.*, 2007). In Arabidopsis, a reduction in the association of mRNA encoding for aerobic proteins with polysomes has already been reported after 2 h of hypoxic stress, which reverses within 1 h of re-oxygenation. A general analysis of genes known to be involved in cytosolic or organellar translation revealed an average threefold repression of these proteins under hypoxia, suggesting an energy saving mechanism similar to that observed for mild dehydration stress (Kawaguchi and Bailey-Serres, 2005; Kawaguchi *et al.*, 2004) and sucrose starvation (Nicolai *et al.*, 2006). A bioinformatic analysis was conducted to find out whether differential ribosome loading under hypoxia could be caused by specific RNA motifs in the 5′ or 3′untranslated regions (TRL). General features such as short UTR sequences (75–250 nt) and low GC content positively correlate with polysome association during hypoxia (Branco-Price *et al.*, 2005). The UTR sequences spanning the *ADH* coding sequences have been extensively studied with respect to their ability to regulate mRNA translation. It was reported that a 17 nt element in the 5′UTR of *ADH* mRNA was able to increase its translation rate without affecting its stability under hypoxia and heat stress (Bailey-Serres and Dawe, 1996; Mardanova *et al.*, 2007). mRNA translation in eukaryotes involves the assembly of the translation initiation complex on the mRNA through an interaction of the eIF4E subunit with the 5′m^7GpppN cap. The translation initiation complex then recruits the small ribosomal subunit to scan the mRNA for a favorable AUG start codon. The amount of eIF4E can be limiting for translation and hypoxia has been shown to reduce transcription and translation of the eIF4E isoforms. When cap-dependent protein synthesis is impaired, translation has been proposed to be achieved by a recruitment of the initiator complex at an internal ribosome entry site (IRES) (Baird *et al.*, 2006; Komar and Hatzoglou, 2005; Macejak and Sarnow, 1991). *ADH* mRNA contains IRESs but they have been proved to contribute only marginally to overall translation in tobacco cells (Mardanova *et al.*, 2008). However these experiments were done in aerobic cells, when there was no decrease in eIF4E levels. Further investigation, under anaerobic conditions would probably help toward an understanding of the importance of IRES in *ADH* translation during hypoxia.

B. CONTROL OF PH ACIDIFICATION DURING OXYGEN DEPRIVATION

Despite its possible role as a signal in hypoxic cells, cytoplasm acidosis could be extremely threatening for cell viability if not efficiently controlled. Greenway and Gibbs (2003) described the possible involvement of a biochemical pH stat to mitigate acidification. In other words, a set of biochemical reactions that compensate for the synthesis of weak organic acids. A mechanism that has been suggested to play this role includes the decarboxylation of acids, supported by the formation of γ-aminobutiric acid via glutamate decarboxylation (Drew, 1997). This relates to the consumption of nitrate and the accumulation of cations, whose paradigm is putresceine biosynthesis from arginine in rice (Reggiani *et al.*, 1989). The importance of putresceine in rice coleoptile has been shown by the use of [α-difluoromethylarginine (DFMA) an inhibitor of putresceine synthesis, which was able to prevent anoxic coleoptiles elongation (Reggiani *et al.*, 1989) while putresceine supplementation increased survival of wheat roots under anoxia (Reggiani *et al.*, 1990). The formation of NH_4^+ and its retention within the cell may also play a role in pH maintenance: high levels of NH_4^+ have been reported in hypoxic tomatoes (Horchani *et al.*, 2008) and NH_4^+ efflux from rice seedlings was also observed (Menegus *et al.*, 1993). A general strategy of anoxia tolerant plants to reduce energy costs to compensate for the proton leakage into the cytoplasm is also to increase vacuolar pH (Greenway and Gibbs, 2003), by decreasing the loading of undissociated acids in the tonoplast. A further reduction of proton leakage from vacuoles into the cytosol in tolerant plants is probably due to a switch from vacuolar ATPase to PP_iase to pump H^+ back into the vacuole (Brauer *et al.*, 1992). The inhibition of H^+–ATPase possibly depends on low ATP availability (Saint-Ges *et al.*, 1991) but this is still under discussion (Greenway and Gibbs, 2003), while the induction of H^+–PP_iase has been reported at the protein and mRNA level in rice seedlings (Carystinos *et al.*, 1995). A drop in cytoplasmic pH may also play a role in low-oxygen adaptation: Webster *et al.* (1991) related the inhibition of translation under anoxia to cytoplasmic acidification, demonstrating *in vitro* that translation is blocked by low pH values. Greenway and Gibbs (2003) suggested that pH is a fine tuner of ethanolic fermentation, and Tournaire-Roux *et al.* (2003) demonstrated that anoxic aquaporin closure depends on decreasing pH.

C. HYPOXIC AND HEAT TREATMENTS LEAD TO ACCLIMATION TO ANOXIA

Mild hypoxic pre-treatments appear to improve tolerance to subsequent anoxic treatment in Arabidopsis, rice, wheat, and tomatoes (Ellis and Setter, 1999; Ellis *et al.*, 1999; Saglio *et al.*, 1988; Waters *et al.*, 1991). This acclimation seems to require protein synthesis since the addition of the translation inhibitor cycloheximide prevented the hypoxic acclimation. Possible explanations for the acquired low-oxygen tolerance include the efficient synthesis of proteins involved in stress tolerance before the drastic energy crisis begins, and protein synthesis is impaired. Protein turnover has also been observed to be reduced in germinating lettuce seeds under anoxia (Pradet and Raymond, 1983).

Similar to the increased anoxia tolerance induced by hypoxic acclimation, mild heat treatments (38 °C, 1.5 h) have also been shown to induce anoxia tolerance in Arabidopsis seedlings. An anoxic pre-treatment on the other hand, followed by heat stress did not result in any improved tolerance to heat stress (Banti *et al.*, 2008). Analyses of global transcriptional and translational changes under hypoxia and anoxia in rice and Arabidopsis clearly demonstrated the induction of a large number of HSP and other heat-related genes, which are also often induced under conditions that involve oxidative stress (Loreti *et al.*, 2005; Swindell, 2006). Moreover, Loreti *et al.* (2005) showed that sucrose supplementation to Arabidopsis seedlings enhances their tolerance to anoxia. This ameliorative effect does not account for the higher substrate availability for glycolysis, since it also occurred in *ADH* knockout mutants (Banti *et al.*, 2008). Sucrose-related increased tolerance, on the other hand, may be associated with increased HSP induction under anoxia, since sucrose enhances the anoxia-inducibility of HSP-encoding genes (Banti *et al.*, 2008). In higher plants five major families of HSPs/chaperones have been described (Nover and Scharf, 1997): the HSP70 family, the chaperonins (HSP60), the HSP90 family, the HSP100 family, and the low molecular mass (12–40 kDa) small HSP (sHSP) family. HSPs act as molecular chaperones (Lindquist, 1986), assisting protein folding, assembly, translocation and degradation. A comparison of transcriptome profiling under different low-oxygen conditions showed that after a few hours of anoxia, a number of HSPs belonging to different families were induced. Two hours of hypoxia, on the other hand, were only able to upregulate small HSPs, and only later (9 h) were members of the HSP90 and 100 induced. sHSPs are believed to prevent, in an ATP independent manner, protein aggregation (Lee *et al.*, 1997; Löw *et al.*, 2000; Smýkal *et al.*, 2000) caused by heat stress but also during developmental transitions such as during chromoplast biogenesis (Lawrence *et al.*, 1997), flower development, (Dafny-Yelin *et al.*, 2008), embryogenesis (Barcala *et al.*, 2008), and germination (Banti *et al.*, 2008). The highly conserved HSP100/ClpB chaperone system, on the other hand, is involved in reverting protein aggregation. However this mechanism depends on ATP availability, required for substrate binding, together with its hydrolysis to ADP, to release the disaggregated polypeptides (Balogi *et al.*, 2008; Bosl *et al.*, 2005). Based on distinct functions and different expression patterns under low-oxygen conditions, it is tempting to speculate that the early induction of small HSP is aimed at preventing protein aggregation. Subsequent induction of HSP100/90 may be required to solubilize protein aggregates when the sHSP preventive activity is not sufficient. In animal systems, protein aggregation occurs under anoxic conditions (Clegg, 2007) and can be caused by oxidative stress, by a direct reaction with ROS or indirectly by lipid peroxidation that further affects proteins. Amino acid oxidation results in the addition of carbonyl groups that alter protein conformation, increasing hydrophobicity and enhancing non-specific protein–protein interactions (Davies, 1995). Besides acting as molecular chaperone, sHSPs also exhibit interactions with lipids in prokaryotes, plants, and mammalian systems that modulate membrane properties (Nakamoto and Vígh, 2007; Tsvetkova *et al.*, 2002). This could suggest a positive effect of sHSP on membrane integrity maintenance during an energy crisis, together with a further protection of lipid peroxidation from re-oxygenation. Interestingly, sHSP translation rates remain either unchanged or even increased upon re-oxygenation with respect to hypoxia, while most of the protein involved in anaerobic metabolism is rapidly dissociated from polysomes (Branco-Price *et al.*, 2008).

IX. CONCLUSIONS

Oxygen sensing and signal transduction in plants is undoubtedly much less explored when compared to other eukaryotes. However, advances in analysis techniques together with the increasing interest in applications of low-oxygen treatments in crop science and technology have recently contributed considerably to a better understanding of the molecular mechanisms underlying plant adaptation to anaerobiosis. Exponential increase in genome sequences for crop plants is also extremely beneficial, allowing the comparison of tolerant and intolerant species. Translational biology to formulate new hypotheses seems extremely promising, as demonstrated by frequent comparisons and overlaps with other responses to different conditions, in plants as in other eukaryotes. The emerging crosstalk between reactive oxygen- and nitrogen-species and signaling pathways especially deserves further investigation. However, the more traditional signaling components such as the role of oscillations in calcium concentrations and phosphorylation cascades also require better understanding. Future research will probably focus on the cross talk mechanisms between the three transcriptionally active organelles, nucleus, chloroplast, and mitochondria, since the last two are linked with oxygen, one in terms of its production and the other its consumption.

Although much attention has been paid to the mechanisms behind gene activation under hypoxia and anoxia, there has been practically no focus on how general transcriptional repression is achieved. Despite the relatively small amount of knowledge gained, compared with other stress conditions, very encouraging results have already been provided regarding applications for the marker assisted selection of flood tolerant rice cultivars (Xu *et al.*, 2006). Distinguishing between tissue and cell specific responses and adaptations in different species would probably help to understand what strategies really affect individual fitness and thus, which mechanism is worth transferring in to breeding programs.

Definitions of the acronyms referring to genes or proteins cited in the text

Acronym	Full name	Function
ACO	ACC oxidase	Catalyze ethylene synthesis
ACS	ACC synthase	Catalyze synthesis of ACC
ADH	Alcohol dehydrogenase	Catalyze the interconversion between ethanol and acetaldehyde
AMY1A	α-Amylase isoform 1A	Catalyze breakdown of starch
AMY3D	α-Amylase isoform 3D	Catalyze breakdown of starch
AOX	Alternative oxydase	Catalyze electron transfer to oxygen
AP2-ERF	Apetala2-ethylene responsive factor	Binding to DNA sequences
bHLH	Basic helix loop helix	Binding to DNA sequences
COX	Cytochrome *c* oxidase	Catalyze electron transfer to oxygen
CREB	cAMP response element binding	Transcriptional co-activating protein
CRR1	Copper response regulator 1	Locus containing elements that mediate copper deficiency response in *Chlamydomonas reinhardtii*
CycOs1	Cyclin of *Oryza sativa* isoform 1	Regulates cell cycle progression
EDS1	Enhanced disease Susceptability 1	Component of the R-gene mediated disease resistance
eIF4E	Eukaryotic translation initiation factor 4E	Initiates protein synthesis from a mRNA molecule in eukaryots
ETR2	Ethylene receptor 2	Involved in ethylene sensing and signal transduction
GDH	Glutamate dehydrogenase	Catalyzes the conversion of glutamate to α-ketoglutarate
GHB	γ-Hydroxybutyrate dehydrogenase	Catalyze the conversion of γ-hydroxybutirate to succinate semialdehyde
HB1	Non-symbiotic heomoglobin 1	Catalyze scavenging of NO
HIF	Hypoxia inducible factor	Mediates the transcriptional reprogramming of the hypoxic response in animal cells
HSPs	Heat shock proteins	Prevent or dismantle protein aggregates caused by denaturation (heat or oxidation)

KIN10	SNF1 kinase homolog 10	Conserved energy sensor, controlling convergent reprogramming of transcription in response to seemingly unrelated stress conditions that deplete sugar and energy supplies
KIN11	SNF1 kinase homolog 11	Conserved energy sensor, controlling convergent reprogramming of transcription in response to seemingly unrelated stress conditions that deplete sugar and energy supplies
LDH	Lactate dehydrogenase	Catalyzes interconversion of lactate to pyruvate
LSD1	Lesion simulating	Top of form
	Disease 1	Negatively regulates a plant cell death pathway
		Bottom of form
MDHAR	Monodehydroascorbato reductase	Catalyze the regeneration of reduced MDHA
MKK11	Map kinase kinase 11	Involved in signal transduction through phosphorylation cascades
MKK9	Map Kinase Kinase 9	Involved in signal transduction through phosphorylation cascades
MPK3	MAP protein kinase 3	Involved in signal transduction through phosphorylation cascades
MPK6	MAP protein kinase 6	Involved in signal transduction through phosphorylation cascades
MYB2	Myb transcription factor 2	Mediates the transcriptional response to abscissic acid
MYBS1	Myb transcription factor acting on the Sugar response complex isoform 1	Mediates transcriptional reprogramming caused by gibberellins and sucrose starvation in rice
NIP2;1	Nodulin 26 intrinsic protein	Lactate transporter
OFD1	2-OG-Fe(II) dioxygenase	Accelerates Sre1N degradation in the presence of oxygen
OsERL1	*Oryza sativa* ethylene receptor 2-like 1	Involved in ethylene sensing and signal transduction
p300		Transcriptional co-activating protein
PAD4	Phytoalexin deficient 4	Mediates salicylic acid signaling
PAS	PER-ARNT-SIM	Protein domain that functions as sensory module for oxygen tension, redox potential or light intensities
PDC	Pyruvate decarboxylase	Catalyze conversion of pyruvate to acetaldehyde
PDH	Pyruvate dehydrogenase	Catalyzes conversion of pyruvate to acetyl-CoA
PFK	Phosphofructokinase	Phosphorylates fructose 6-P
PHD	Prolyl-4-hydroxylase	Catalyze the oxygen dependent degradation of HIFα
PIF3	Phytochrome interacting factor 3	Mediates gene transcriptional regulation by light

(*continued*)

(*continued*)

Acronym	Full name	Function
PIF4	Phytochrome interacting factor 4	Mediates gene transcriptional regulation by light
pVHL	von Hippel–Lindau tumor suppressor protein	Component of the protein complex responsible for ubiquitination and degradation of HIF
ROP10	Rho of Plants isoform 10	Involved in modulation of ADH expression in response to hypoxia in *Arabidopsis thaliana*
SH1	Sucrose synthase 1	Catalyze sucrose catabolism in maize (isoform 1)
SLR1	Slender rice 1	Repress transcriptional activity of bHLH transcription factors
SLRL1	Slender rice 1 like isoform 1	Repress transcriptional activity of bHLH transcription factors
SNF1	Sucrose non-fermenting 1	Essential for release from glucose repression in eukaryotes
SnRKs	Snf1-related protein kinases	Involved in signal transduction through phosphorylation cascades
SRE1	Sterol regulatory element binding protein homolog 1	Activates large part of the transcriptional response to hypoxia in yeast
SSADH	Succinic semialdehyde dehydrogenase	Catalyzes the oxidation of succinate semialdehyde to succinate
SUB1	Submergence 1	Locus responsible for inhibition of flooding stimulated elongation and determining submergence tolerance in rice
SUS1	Sucrose synthase 1	Catalyze conversion of sucrose to fructose and UDP-glucose (isoform 4 of *Arabidopsis thaliana*)
SUS4	Sucrose synthase 4	Catalyze conversion of sucrose to fructose and UDP-glucose (isoform 4 of *Arabidopsis thaliana*)
TaMYB1	Myb transcription factor 1 of *Triticum aestivum*	Mediates transcriptional reprogramming caused by drought stress
ZAT10	Zinc finger protein of *Arabidopsis thaliana*	Mediates the transcriptional reprogramming caused by oxydative stress
ZAT12	Zinc finger protein of *Arabidopsis thaliana*	Mediates the transcriptional reprogramming caused by oxydative stress

REFERENCES

Abat, J. K., Mattoo, A. K. and Deswal, R. (2008). *S*-nitrosylated proteins of a medicinal CAM plant *Kalanchoe pinnata*; ribulose-1,5-bisphosphate carboxylase/oxygenase activity targeted for inhibition. *FEBS Journal* **275**, 2862–2872.

Abe, H., Urao, T., Ito, T., Seki, M., Shinozaki, K. and Yamaguchi-Shinozaki, K. (2003). Arabidopsis AtMYC2 (bHLH) and AtMYB2 (MYB) function as transcriptional activators in abscisic acid signaling. *Plant Cell* **15**, 63–78.

Acker, T., Fandrey, J. and Acker, H. (2006). The good, the bad and the ugly in oxygen-sensing: ROS, cytochromes and prolyl-hydroxylases. *Cardiovascular Research* **71**, 195–207.

Affourtit, C., Krab, K. and Moore, A. L. (2001). Control of plant mitochondrial respiration. *Biochimica et Biophysica Acta (BBA): Bioenergetics* **1504**, 58–69.

Aharoni, A., Dixit, S., Jetter, R., Thoenesa, E., Van Arkel, G. and Pereira, A. (2004). The SHINE clade of AP2 domain transcription factors activates wax biosynthesis, alters cuticle properties, and confers drought tolerance when overexpressed in Arabidopsis. *Plant Cell* **16**, 2463–2480.

Amirsadeghi, S., Robson, C. A., McDonald, A. E. and Vanlerberghe, G. C. (2006). Changes in plant mitochondrial electron transport alter cellular levels of reactive oxygen species and susceptibility to cell death signaling molecules. *Plant Cell Physiology* **47**, 1509–1519.

Appelhoff, R. J., Tian, Y.-M., Raval, R. R., Turley, H., Harris, A. L., Pugh, C. W., Ratcliffe, P. J. and Gleadle, J. M. (2004). Differential function of the prolyl hydroxylases PHD1, PHD2, and PHD3 in the regulation of hypoxia-inducible factor. *Journal of Biological Chemistry* **279**, 38458–38465.

Appleby, C. A., Bogus, D., Dennis, E. S. and Peacock, W. J. (1988). A role for haemoglobin in all plant roots? *Plant, Cell and Environment* **11**, 359–367.

Armstrong, W. (1979). Aeration in higher plants. *Advances in Botanical Research* **7**, 225–332.

Armstrong, W. and Beckett, P. M. (1987). Internal aeration and the development of stelar anoxia in submerged roots. A multishelled mathematical model combining axial diffusion of oxygen in the cortex with radial losses to the stele, the wall layers and the rhizosphere. *New Phytologist* **105**, 221–245.

Armstrong, W., Healy, M. T. and Lythe, S. (1983). Oxygen diffusion in pea. II. The oxygen status of the primary root as affected by growth, the production of laterals and radial oxygen loss. *New Phytologist* **94**, 549–559.

Aurisano, N., Bertani, A. and Reggiani, R. (1995). Involvement of calcium and calmodulin in protein and amino acid metabolism in rice roots under anoxia. *Plant Cell Physiology* **36**, 1525–1529.

Baena-Gonzalez, E., Rolland, F., Thevelein, J. M. and Sheen, J. (2007). A central integrator of transcription networks in plant stress and energy signalling. *Nature* **448**, 938–942.

Bailey-Serres, J. and Dawe, R. K. (1996). Both 5′ and 3′ sequences of maize ADH1 mRNA are required for enhanced translation under low-oxygen conditions. *Plant Physiology* **112**, 685–695.

Bailey-Serres, J. and Voesenek, L. (2008). Flooding stress: Acclimations and genetic diversity. *Annual Review of Plant Biology* **59**, 313–339.

Baird, S. D., Turcotte, M., Korneluk, R. G. and Holcik, M. (2006). Searching for IRES. *RNA* **12**, 1755–1785.

Balogi, Z., Cheregi, O., Giese, K. C., Juhasz, K., Vierling, E., Vass, I., Vigh, L. and Horvath, I. (2008). A mutant small heat shock protein with increased thylakoid association provides an elevated resistance against uv-b damage in *Synechocystis* 6803. *Journal of Biological Chemistry* **283**, 22983–22991.

Banga, M., Bögemann, G. M., Blom, W. P. M. and Voesenek, L. A. C. J. (1997). Flooding resistance of Rumex species strongly depends on their response to ethylene: Rapid shoot elongation or foliar senescence. *Physiologia Plantarum* **99**, 415–422.

Banti, V., Loreti, E., Novi, G., Santaniello, A., Alpi, A. and Perata, P. (2008). Heat acclimation and cross-tolerance against anoxia in *Arabidopsis*. *Plant, Cell and Environment* **31**, 1029–1037.

Barcala, M., García, A., Cubas, P., Almoguera, C., Jordano, J., Fenoll, C. and Escobar, C. (2008). Distinct heat-shock element arrangements that mediate the heat shock, but not the late-embryogenesis induction of small heat-shock proteins, correlate with promoter activation in root-knot nematode feeding cells. *Plant Molecular Biology* **66**, 151–164.

Barone, M. C., Darley-Usmar, V. M. and Brookes, P. S. (2003). Reversible inhibition of Cytochrome *c* oxidase by peroxynitrite proceeds through ascorbate-dependent generation of nitric oxide. *Journal of Biological Chemistry* **278**, 27520–27524.

Baxter-Burrell, A., Yang, Z., Springer, P. S. and Bailey-Serres, J. (2002). RopGAP4-dependent Rop GTPase rheostat control of *Arabidopsis* oxygen deprivation tolerance. *Science* **296**, 2026–2028.

Benschop, J. J., Jackson, M. B., Gahl, K., Vreeburg, R. A. M., Croker, S. J., Peeters, A. J. M. and Voesenek, L. A. C. J. (2005). Contrasting interactions between ethylene and abscisic acid in *Rumex* species differing in submergence tolerance. *The Plant Journal* **44**, 756–768.

Berry, L. J. and Norris, W. E. (1949). Studies on onion root respiration. I. Velocity of oxygen consumption in different segments of roots at different temperatures as a function of partial pressure of oxygen. *Biochimica et Biophysica Acta* **3**, 593–606.

Bertani, A., Brambilla, I. and Menegus, F. (1981). Effect of anaerobiosis on carbohydrate content in rice roots. *Biochemie Und Physiologie Der Pflanzen* **176**, 835–840.

Biemelt, S., Keetman, U. and Albrecht, G. (1998). Re-aeration following hypoxia or anoxia leads to activation of the antioxidative defense system in roots of wheat seedlings. *Plant Physiology* **116**, 651–658.

Bieniawska, Z., Barratt, D. H. P., Garlick, A. P., Thole, V., Kruger, N. J., Martin, C., Zrenner, R. and Smith, A. M. (2007). Analysis of the sucrose synthase gene family in *Arabidopsis*. *The Plant Journal* **49**, 810–828.

Blokhina, O. B., Chirkova, T. V. and Fagerstedt, K. V. (2001). Anoxic stress leads to hydrogen peroxide formation in plant cells. *Journal of Experimental Botany* **52**, 1179–1190.

Bologa, K. L., Fernie, A. R., Leisse, A., Ehlers Loureiro, M. and Geigenberger, P. (2003). A bypass of sucrose synthase leads to low internal oxygen and impaired metabolic performance in growing potato tubers. *Plant Physiology* **132**, 2058–2072.

Borisjuk, L., Macherel, D., Benamar, A., Wobus, U. and Rolletschek, H. (2007). Low oxygen sensing and balancing in plant seeds: A role for nitric oxide. *New Phytologist* **176**, 813–823.

Bosl, B., Grimminger, V. and Walter, S. (2005). Substrate binding to the molecular chaperone HSP104 and its regulation by nucleotides. *Journal of Biological Chemistry* **280**, 38170–38176.

Bouché, N., Fait, A., Bouchez, D., Moller, S. G. and Fromm, H. (2003). Mitochondrial succinic-semialdehyde dehydrogenase of the gamma-aminobutyrate shunt is required to restrict levels of reactive oxygen intermediates in plants. *Proceedings of the National Academy of Sciences of the United States of America* **100**, 6843–6848.

Bouny, J. M. and Saglio, P. H. (1996). Glycolytic flux and hexokinase activities in anoxic maize root tips acclimated by hypoxic pretreatment. *Plant Physiology* **111**, 187–194.

Branco-Price, C., Kawaguchi, R., Ferreira, R. B. and Bailey-Serres, J. (2005). Genome-wide analysis of transcript abundance and translation in Arabidopsis seedlings subjected to oxygen deprivation. *Annals of Botany* **96**, 647–660.

Branco-Price, C., Kaiser, K. A., Jang, C. J. H., Larive, C. K. and Bailey-Serres, J. (2008). Selective mRNA translation coordinates energetic and metabolic adjustments to cellular oxygen deprivation and reoxygenation in *Arabidopsis thaliana*. *The Plant Journal* **56**, 743–755. doi:10.1111/j.1365-313X.2008.03642.x.

Brauer, D., Conner, D. and Tu, S. (1992). Effects of pH on proton transport by vacuolar pumps from maize roots. *Physiologia Plantarum* **86**, 63–70.

Breitkreuz, K. E., Allan, W. L., Van Cauwenberghe, O. R., Jakobs, C., Talibi, D., Andre, B. and Shelp, B. J. (2003). A novel γ-hydroxybutyrate dehydrogenase: Identification and expression of an *Arabidopsis* cDNA and potential role under oxygen deficiency. *Journal of Biological Chemistry* **278**, 41552–41556.

Bucher, M., Brädle, R. and Khuhlemeier, C. (1994). Ethanolic fermentation in transgenic tobacco expressing Zymomonas mobilis pyruvate decarboxylase. *EMBO Journal* **13**, 2755–2763.

Campbell, R. and Drew, M. C. (1983). Electron microscopy of gas space (aerenchyma) formation in adventitious roots of *Zea mays* L. subjected to oxygen shortage. *Planta* **157**, 350–357.

Carystinos, G. D., Macdonald, H. R., Monroy, A. F., Dhindsa, R. S. and Poole, R. J. (1995). Vacuolar H^+-translocating pyrophosphatase is induced by anoxia or chilling in seedlings of rice. *Plant Physiology* **108**, 641–649.

Chaffey, N. (2007). Esau's Plant anatomy, meristems, cells, and tissues of the plant body: Their structure, function, and development. *Annals of Botany* **99**, 785–786.

Chen, X., Goodwin, M., Boroff, V. L., Liu, X. and Jenks, M. A. (2003). Cloning and characterization of the WAX2 gene of Arabidopsis involved in cuticle membrane and wax Pproduction. *Plant Cell* **15**, 1170–1185.

Chen, P.-W., Chiang, C.-M., Tseng, T.-H. and Yu, S.-M. (2006). Interaction between rice MYBGA and the gibberellin response element controls tissue-specific sugar sensitivity of α-amylase genes. *Plant Cell* **18**, 2326–2340.

Cho, H. T. and Kende, H. (1997). Expansins and internodal growth of deepwater rice. *Plant Physiology* **113**, 1145–1151.

Cho, J.-I., Ryoo, N., Ko, S., Lee, S.-K., Lee, J., Jung, K.-H., Lee, Y.-H., Bhoo, S., Winderickx, J., An, G., Hahn, T.-R. and Jeon, J.-S. (2006). Structure, expression, and functional analysis of the hexokinase gene family in rice (*Oryza sativa* L.). *Planta* **224**, 598–611.

Choi, W. G. and Roberts, D. M. (2007). Arabidopsis NIP2;1, a major intrinsic protein transporter of lactic acid Induced by anoxic stress. *Journal of Biological Chemistry* **282**, 24209–24218.

Christopher, M. E. and Good, A. G. (1996). Characterization of hypoxically inducible lactate dehydrogenase in maize. *Plant Physiology* **112**, 1015–1022.

Chung, H.-J. and Ferl, R. J. (1999). Arabidopsis alcohol dehydrogenase expression in both shoots and roots is conditioned by root growth environment. *Plant Physiology* **121**, 429–436.

Clegg, J. S. (2007). Protein stability in *Artemia* embryos during prolonged anoxia. *Biology Bulletin* **212**, 74–81.

Clifton, R., Lister, R., Parker, K., Sappl, P., Elhafez, D., Millar, A., Day, D. and Whelan, J. (2005). Stress-induced co-expression of alternative respiratory chain components in *Arabidopsis thaliana*. *Plant Molecular Biology* **58**, 193–212.

Colcombet, J. and Hirt, H. (2008). *Arabidopsis* MAPKs: A complex signalling network involved in multiple biological processes. *The Biochemical Journal* **413**, 217–226.

Colmer, T. D. (2003). Long-distance transport of gases in plants: A perspective on internal aeration and radial oxygen loss from roots. *Plant, Cell and Environment* **26**, 17–36.

Colmer, T. and Pederse, O. (2008). Underwater photosynthesis and respiration in leaves of submerged wetland plants: Gas films improve CO_2 and O_2 exchange. *New Phytologist* **177**, 918–926.

Colmer, T., Gibberd, M., Wiengweera, A. and Tinh, T. (1998). The barrier to radial oxygen loss from roots of rice (*Oryza sativa* L.) is induced by growth in stagnant solution. *Journal of Experimental Botany* **49**, 1431–1436.

Colmer, T. D., Huang, S. and Greenway, H. (2001). Evidence for down-regulation of ethanolic fermentation and K+ effluxes in the coleoptile of rice seedlings during prolonged anoxia. *Journal of experimental Botany* **52**, 1507–1517.

Contento, A. L., Kim, S.-J. and Bassham, D. C. (2004). Transcriptome profiling of the response of *Arabidopsis* suspension culture cells to Suc starvation. *Plant Physiology* **135**, 2330–2347.

Cox, M. C. H., Millenaar, F. F., Van Berkel, Y. E. M. D. J., Peeters, A. J. M. and Voesenek, L. A. C. J. (2003). Plant Movement. Submergence-induced petiole elongation in *Rumex palustris* depends on hyponastic growth. *Plant Physiology* **132**, 282–291.

Couee, I., Defontaine, S., Carde, J. P. and Pradet, A. (1992). Effects on anoxia on mitocondrial biogenesis in rice shoots—modification of in organelle translation characteristics. *Plant Physiology* **98**, 411–421.

Dafny-Yelin, M., Tzfira, T., Vainstein, A. and Adam, Z. (2008). Non-redundant functions of sHSP-CIs in acquired thermotolerance and their role in early seed development in *Arabidopsis*. *Plant Molecular Biology* **67**, 363–373.

D'Angelo, G., Duplan, E., Boyer, N., Vigne, P. and Frelin, C. (2003). Hypoxia up-regulates prolyl hydroxylase activity: A feedback mechanism that limits HIF-1 responses during reoxygenation. *Journal of Biological Chemistry* **278**, 38183–38187.

Das, A., Nanda, B. B., Sarkar, R. K. and Lodh, S. B. (2000). Effect of complete submergence on the activity of starch phosphorylase enzyme in rice (*Oryza Sariva* L.) leaves. *Jornal of Plant Biochemistry and Biotechnology* **9**, 41–43.

Davies, D. D. D. (1980). Anaerobic metabolism and the production of organic acids. *The biochemistry of plants* **2**, 581–611.

Davies, K. J. (1995). Oxidative stress: The paradox of aerobic life. *In* "Free Radicals and Oxidative Stress: Environment, Drugs, and Food Additives" (C. Rice-Evans, B. Halliwell and G. G. Lunt, eds.), pp. 7–18. Portaland Press, London.

Davies, W. J. and Zhang, J. (1991). Root signals and the regulation of growth and development of roots in drying soil. *Annual Reviews on Plant Physiology and Plant Molecular Biology* **42**, 55–76.

Davletova, S., Schlauch, K., Coutu, J. and Mittler, R. (2005). The zinc-finger protein ZAT12 plays a central role in reactive oxygen and abiotic stress signaling in *Arabidopsis*. *Plant Physiology* **139**, 847–856.

De Klerk, G.-J., Van Der Krieken, W. and De Jong, J. C. (1999). Review the formation of adventitious roots: New concepts, new possibilities. *In Vitro Cellular and Developmental Biology: Plant* **35**, 189–199.

De Lucas, M., Daviere, J. M., Rodriguez-Falcon, M., Pontin, M., Iglesis-Pedraz, J. M., Lorrain, S., Fankhauser, C., Blasquez, M. A., Titarenko, E. and Prat, S. (2008). A molecular framework for light and gibberellin control of cell elongation. *Nature* **451**, 480–484.

De Oliveira, H. C., Wulff, A., Saviani, E. E. and Salgado, I. (2008). Nitric oxide degradation by potato tuber mitochondria: Evidence for the involvement of external NAD(P)H dehydrogenases. *Biochimica et Biophysica Acta (BBA): Bioenergetics* **1777**, 470–476.

Dolferus, R., Jacobs, M., Peacock, W. J. and Dennis, E. S. (1994). Differential interactions of promoter elements in stress responses of the *Arabidopsis* adh gene. *Plant Physiology* **105**, 1075–1087.

Dolferus, R., Wolansky, M., Rebecka Carroll, R., Yo Miyashita, Y., Kathleen Ismond, K. and Good, A. (2008). Functional analysis of lactate dehydrogenase during hypoxic stress in *Arabidopsis*. *Functional Plant Biology* **35**, 131–140.

Dordas, C., Hasinoff, B. B., Igamberdiev, A. U., Manac, H. N., Rivoal, J. and Hill, R. D. (2003). Expression of a stress-induced hemoglobin affects NO levels produced by alfalfa root cultures under hypoxic stress. *The Plant Journal* **35**, 763–770.

Drew, M. C. (1997). Oxygen deficiency and root metabolism: Injury and acclimation under hypoxia and anoxia. *Annual Review of Plant Physiology and Plant Molecular Biology* **48**, 223–250.

Drew, M. C., Jackson, M. B. and Giffard, S. (1979). Ethylene-promoted adventitious rooting and development of cortical air spaces (aerenchyma) in roots may be adaptive responses to flooding in *Zea mays* L. *Planta* **147**, 83–88.

Drew, M. C., He, C.-J. and Morgan, P. W. (2000). Programmed cell death and aerenchyma formation in roots. *Trends in Plant Science* **5**, 123–127.

Ellis, M. H. and Setter, T. L. (1999). Hypoxia induces anoxia tolerance in completely submerged rice seedlings. *Journal of Plant Physiology* **154**, 219–230.

Ellis, M. H., Dennis, E. S. and James Peacock, W. (1999). Arabidopsis roots and shoots have different mechanisms for hypoxic stress tolerance. *Plant Physiology* **119**, 57–64.

Eriksson, M., Moseley, J. L., Tottey, S., Del Campo, J. A., Quinn, J., Kim, Y. and Merchant, S. (2004). Genetic dissection of nutritional copper signaling in *Chlamydomonas* distinguishes regulatory and target genes. *Genetics* **168**, 795–807.

Evans, D. E. (2004). Aerenchyma formation. *New Phytologist* **161**, 35–49.

Fähling, M. (2008). Cellular oxygen sensing, signalling and how to survive translational arrest in hypoxia. *Acta Physiologica* **195**, 205–230. doi:10.1111/j.1748-1716.2008.01894.x.

Faiz-ur-Rahman, A. T. M., Trewavas, A. J. and Davies, D. D. (1974). The Pasteur effect in carrot root tissue. *Planta* **118**, 195–210.

Fan, T. W.-M., Higashi, R. M., Frenkiel, T. A. and Lane, A. N. (1997). Anaerobic nitrate and ammonium metabolism in flood-tolerant rice coleoptiles. *Journal of Experimental Botany* **48**, 1655–1666.

Felle, H. H. (2001). pH: Signal and messenger in plant cells. *Plant Biology* **3**, 577–591.

Felle, H. H. (2005). pH Regulation in anoxic plants. *Annals of Botany* **96**, 519–532.

Felle, H. H. (2006). Apoplastic pH during low-oxygen stress in barley. *Annals of Botany* **98**, 1085–1093.

Fox, T. C. and Kennedy, R. A. (1991). Mitochondrial enzymes in aerobically and anaerobically germinated seedlings of *Echinichloa* and rice. *Planta* **184**, 510–514.

Friedberg, S. H. and Davidson, D. (1971). Cell population studies in developing root primordia. *Annals of Botany* **35**, 523–533.

Fukao, T. and Bailey-Serres, J. (2004). Plant responses to hypoxia—is survival a balancing act? *Trends in Plant Science* **9**, 449–456.

Fukao, T. and Bailey-Serres, J. (2008). Submergence tolerance conferred by Sub1A is mediated by SLR1 and SLRL1 restriction of gibberellin responses in rice. *Proceedings of the National Academy of Science of the United States of America* **105**, 16814–16819.

Fukao, T., Xu, K., Ronald, P. C. and Bailey-Serres, J. (2006). A variable cluster of ethylene response factor-like genes regulates metabolic and developmental acclimation responses to submergence in rice. *Plant Cell* **18**, 2021–2034.

Garrocho-Villegas, V. and Arredondo-Peter, R. (2008). Molecular cloning and characterization of a moss (*Ceratodon purpureus*) Nonsymbiotic Hemoglobin provides insight into the early evolution of plant nonsymbiotic hemoglobins. *Molecular Biology and Evolution* **25**, 1482–1487.

Garthwaite, A. J., Bothmer, R. V. and Colmer, T. D. (2003). Diversity in root aeration traits associated with waterlogging tolerance in the genus *Hordeum*. *Functional Plant Biology* **30**, 875–889.

Geigenberger, P. (2003). Response of plant metabolism to too little oxygen. *Current Opinion in Plant Biology* **6**, 247–256.

Germain, V., Raymond, P. and Ricard, B. (1997a). Differential expression of two tomato lactate dehydrogenase genes in response to oxygen deficit. *Plant Molecular Biology* **35**, 711–721.

Germain, V., Ricard, B., Raymond, P. and Saglio, P. H. (1997b). The role of sugars, hexokinase, and sucrose synthase in the determination of hypoxically induced tolerance to anoxia in tomato roots. *Plant Physiology* **114**, 167–175.

Gibbs, J. and Greenway, H. (2003). Mechanisms of anoxia tolerance in plants. I. Growth, survival and anaerobic catabolism. *Functional Plant Biology* **30**, 1–47.

Gibbs, J., Morrell, S., Valdez, A., Setter, T. L. and Greenway, H. (2000). Regulation of alcoholic fermentation in coleoptiles of two rice cultivars differing in tolerance to anoxia. *Journal of Experimental Botany* **51**, 785–796.

Goldstein, J. L., Debose-Boyd, R. A. and Brown, M. S. (2006). Protein sensors for membrane sterols. *Cell* **124**, 35–46.

Gonzali, S., Loret, E., Novi, G., Poggi, A., Alpi, A. and Perata, P. (2005). The use of microarrays to study the anaerobic response in *Arabidopsis*. *Annals of Botany* **96**, 661–668.

Gout, E., Boisson, A.-M., Aubert, S., Douce, R. and Bligny, R. (2001). Origin of the Cytoplasmic pH changes during anaerobic stress in higher plant cells. Carbon-13 and phosphorous-31 nuclear magnetic resonance studies. *Plant Physiology* **125**, 912–925.

Gray, W. M. (2004). Hormonal regulation of plant growth and development. *PloS Biology* **2**, 1270–1273.

Greenway, H. and Gibbs, J. (2003). Mechanisms of anoxia tolerance in plants. II. Energy requirements for maintenance and energy distribution to essential processes. *Functional Plant Biology* **30**, 999–1036.

Greenwood, D. J. (1967). Studies on oxygen transport in mustard seedlings (*Sinapsis alba* L.). *New Phytologist* **66**, 597–606.

Greer, S., Wen, M., Bird, D., Wu, X., Samuels, L., Kunst, L. and Jetter, R. (2007). The Cytochrome P450 enzyme CYP96A15 is the midchain alkane hydroxylase responsible for formation of secondary alcohols and ketones in stem cuticular wax of *Arabidopsis*. *Plant Physiology* **145**, 653–667.

Grennan, A. K. (2007). Protein *S*-nitrosylation: Potential targets and roles in signal transduction. *Plant Physiology* **144**, 1237–1239.

Guglielminetti, L., Perata, P. and Alpi, A. (1995). Effect of anoxia on carbohydrate metabolism in rice seedlings. *Plant Physiology* **108**, 735–741.

Guglielminetti, L., Loreti, E., Perata, P. and Alpi, A. (1999). Sucrose synthesis in cereal grains under oxygen deprivation. *Journal of Plant Research* **112**, 353–359.

Guglielminetti, L., Morita, A., Yamaguchi, J., Loreti, E., Perata, P. and Alpi, A. (2006). Differential expression of two fructokinases in *Oryza sativa* seedlings grown under aerobic and anaerobic conditions. *Journal of Plant Research* **119**, 351–356.

Gunawardena, A. H. L. A. N., Pearce, D. M., Jackson, M. B., Hawes, C. R. and Evans, D. E. (2001). Characterisation of programmed cell death during aerenchyma formation induced by ethylene or hypoxia in roots of maize (*Zea mays* L.). *Planta* **212**, 205–214.

Gunter, T. E., Gunter, K. K., Sheu, S. S. and Gavin, C. E. (1994). Mitochondrial calcium transport: Physiological and pathological relevance. *American Journal of Physiology: Cell Physiology* **267**, 313–339.

Gupta, K. J., Stoimenova, M. and Kaiser, W. M. (2005). In higher plants, only root mitochondria, but not leaf mitochondria reduce nitrite to NO, *in vitro* and *in situ*. *Journal of Experimental Botany* **56**, 2601–2609.

He, C. J., Morgan, P. W. and Drew, M. C. (1996). Transduction of an ethylene signal is required for cell death and lysis in the root cortex of maize during aerenchyma formation induced by hypoxia. *Plant Physiology* **112**, 463–472.

He, D.-Y., Yazaky, Y., Nishizaway, Y., Takay, R., Yamada, K., Saka, K., Shibuya, Y. and Minami, E. (1998). Gene activation by cytoplasmic acidification in suspension-cultured rice cells in response to the potent elicitor, N-acetylchitoheptaose. *Molecular Plant-Microbe Interactions* **11**, 1167–1174.

Hieta, R. and Myllyharju, J. (2002). Cloning and characterization of a low molecular weight prolyl 4-hydroxylase from *Arabidopsis thaliana*. Effective hydroxylation of proline-rich, collagen-like, and hypoxia-inducible transcription factor α-like peptides. *Journal of Biological Chemistry* **277**, 23965–23971.

Hoeren, F. U., Dolferus, R., Wu, Y., Peacock, W. J. and Dennis, E. S. (1998). Evidence for a role for AtMYB2 in the induction of the *Arabidopsis* alcohol dehydrogenase gene (ADH1). Low Oxygen. *Genetics* **149**, 479–490.

Horchani, F., Gallusci, P., Baldet, P., Cabasson, C., Maucourt, M., Rolin, D., Aschi-Smiti, S. and Raymond, P. (2008). Prolonged root hypoxia induces ammonium accumulation and decreases the nutritional quality of tomato fruits. *Journal of Plant Physiology* **165**, 1352–1359.

Huber, S. C. and Hardin, S. C. (2004). Numerous posttranslational modifications provide opportunities for the intricate regulation of metabolic enzymes at multiple levels. *Current Opinion in Plant Biology* **7**, 318–322.

Hughes, B. T. and Espenshade, J. (2008). Oxygen-regulated degradation of fission yeast SREBP by Ofd1, a prolyl hydroxylase family member. *The EMBO Journal* **27**, 1491–1501.

Igamberdiev, A. U. and Hill, R. D. (2008). Plant mitochondrial function during anaerobiosis. *Annals of Botany* **103**, 259–268. doi:10.1093/aob/mcn100.

Igamberdiev, A. U. and Kleczkowski, L. A. (2003). Membrane potential, adenylate levels and Mg^{2+} are interconnected via adenylate kinase equilibrium in plant cells. *Biochimica et Biophysica Acta (BBA): Bioenergetics* **1607**, 111–119.

Igamberdiev, A., Seregélyes, C., Manac'h, N. and Hill, R. (2004). NADH-dependent metabolism of nitric oxide in alfalfa root cultures expressing barley hemoglobin. *Planta* **219**, 95–102.

Igamberdiev, A., Bykova, N. and Hill, R. (2006). Nitric oxide scavenging by barley hemoglobin is facilitated by a monodehydroascorbate reductase-mediated ascorbate reduction of methemoglobin. *Planta* **223**, 1033–1040.

Ismond, K. P., Dolferus, R., De Pauw, M., Dennis, E. S. and Good, A. G. (2003). Enhanced low oxygen survival in *Arabidopsis* through increased metabolic flux in the fermentative pathway. *The American Society for Plant Biologists* **132**, 1292–1302.

Jackson, M. B. (2008). Ethylene-promoted elongation: An adaptation to submergence stress. *Annals of Botany* **101**, 229–248.

Jackson, M. B. and Colmer, T. D. (2005). Response and adaptation by plants to flooding stress—Preface. *Annals of Botany* **96**, 501–505.

Jones-Rhoades, M. W. and Bartel, D. P. (2004). Computational identification of plant microRNAs and their targets, including a stress-induced miRNA. *Molecular Cell* **14**, 787–799.

Jung, J., Won, S., Suh, S., Kim, H., Wing, R., Jeong, Y., Hwang, I. and Kim, M. (2007). The barley ERF-type transcription factor HvRAF confers enhanced pathogen resistance and salt tolerance in *Arabidopsis*. *Planta* **225**, 575–588.

Justin, S. H. F. W. and Armstrong, W. (1987). The anatomical characteristics of roots and plant response to soil flooding. *New Phytologist* **106**, 465–495.

Kaiser, W. M. and Brendle-Behnisch, E. (1995). Acid-base-modulation of nitrate reductase in leaf tissues. *Planta* **196**, 1–6.

Kanellis, A. K., Solomos, T. and Roubelakis-Angelakis, K. A. (1991). Suppression of cellulase and polygalacturonase and induction of alcohol dehydrogenase isoenzymes in avocado fruit mesocarp subjected to low oxygen stress. *Plant Physiology* **96**, 269–274.

Kawaguchi, R. and Bailey-Serres, J. (2005). mRNA sequence features that contribute to translational regulation in *Arabidopsis*. *Nucleic Acids Research* **33**, 955–965.

Kawaguchi, R., Girke, T., Bray, E. A. and Bailey-Serres, J. (2004). Differential mRNA translation contributes to gene regulation under non-stress and dehydration stress conditions in *Arabidopsis thaliana*. *The Plant Journal* **38**, 823–839.

Khan, A. A., Thakur, R., Akbar, M., Hillerislambers, D. and Seshu, D. V. (1987). Relationship of ethylene production to elongation in deepwater rice. *Crop Science* **27**, 1188–1196.

Kimpara, T., Aohara, T., Soga, K., Wakabayashi, K., Hoson, T., Tsumuraya, Y. and Kotake, T. (2008). β-1,3: 1,4-Glucan synthase activity in rice seedlings under water. *Annals of Botany* **102**, 221–226.

Klok, E. J., Wilson, I. W., Wilson, D., Chapman, S. C., Ewing, R. M., Somerville, S. C., Peacock, W. J., Dolferus, R. and Dennis, E. S. (2002). Expression profile analysis of the low-oxygen response in *Arabidopsis* root cultures. *Plant Cell* **14**, 2481–2494.

Komar, A. A. and Hatzoglou, M. (2005). Internal ribosome entry sites in cellular mRNAs: Mystery of their existence. *Journal of Biological Chemistry* **280**, 23425–23428.

Kropat, J., Tottey, S., Birkenbihl, R. P., Depáge, N., Huijser, P. and Merchant, S. (2005). A regulator of nutritional copper signaling in Chlamydomonas is an SBP domain protein that recognizes the GTAC core of copper response element. *Proceedings of the National Academy of Sciences of the United States of America* **102**, 18730–18735.

Kunst, L. and Samuels, A. L. (2003). Biosynthesis and secretion of plant cuticular wax. *Progress on Lipid Research* **42**, 51–80.

Lamport, E. A. (1965). The protein component of primary cell wall. *Advances in Botanical Research* **2**, 151–218.

Lasanthi-Kudahettige, R., Magneschi, L., Loreti, E., Gonzali, S., Licausi, F., Novi, G., Beretta, O., Vitulli, F., Alpi, A. and Perata, P. (2007). Transcript profiling of the anoxic rice coleoptile. *Plant Physiology* **144**, 218–231.

Laskowski, M., Biller, S., Stanley, K., Kajstura, T. and Prusty, R. (2006). Expression profiling of auxin-treated *Arabidopsis* roots: Toward a molecular analysis of lateral root emergence. *Plant Cell Physiology* **47**, 788–792.

Lawrence, S. D., Cline, K. and Moore, G. A. (1997). Chromoplast development in ripening tomato fruit: Identification of cDNAs for chromoplast-targeted proteins and characterization of a cDNA encoding a plastid-localized low-molecular-weight heat shock protein. *Plant Molecular Biology* **33**, 483–492.

Lee, Y. and Kende, H. (2002). Expression of α-expansin and expansin-like genes in deepwater rice. *Plant Physiology* **130**, 1396–1405.

Lee, G. J., R. A. M., Saibil, H. R. and Vierling, E. (1997). A small heat shock protein stably binds heat-denatured model substrates and can maintain a substrate in a folding-competent state. *The EMBO Journal* **16**, 659–671.

Lee, T. G., Jang, C. S., Kim, J. Y., Kim, D. S., Park, J. H., Kim, D. Y. and Seo, Y. W. (2007). A Myb transcription factor TaMyb1 from wheat roots is expressed during hypoxia: Roles in response to the oxygen concentration in root environment and abiotic stresses. *Physiologia Plantarum* **129**, 375–385.

Li, X., Cobb, C. E. and May, J. M. (2002). Mitochondrial recycling of ascorbic acid from dehydroascorbic acid: Dependence on the electron transport chain. *Archives of Biochemistry and Biophysics* **403**, 103–110.

Libourel, I. G. L., Van Bodegom, P. M., Fricker, M. D. and Ratcliffe, R. G. (2006). Nitrite reduces cytoplasmic acidosis under anoxia. *Plant Physiology* **142**, 1710–1717.

Lindermayr, C., Saalbach, G., Bahnweg, G. and Durner, J. (2006). Differential inhibition of Arabidopsis methionine adenosyltransferases by protein *S*-nitrosylation. *Journal of Biological Chemistry* **281**, 4285–4291.

Lindquist, S. (1986). The heat-shock response. *Annual Review of Biochemistry* **55**, 1151–1191.

Liu, T.-Y. and Castelfranco, P. (1970). The biosynthesis of ethyl-β-glucoside in extracts of pea seedlings. *Plant Physiology* **45**, 424–428.

Locatelli, F., Bracale, M., Magaraggia, F., Faoro, F., Manzocchi, L. A. and Coraggio, I. (2000). The product of the rice myb7 unspliced mRNA dimerizes with the maize leucine zipper Opaque2 and stimulates its activity in a transient expression assay. *Journal of Biological Chemistry* **275**, 17619–17625.

Lorbiecke, R. and Sauter, M. (1998). Induction of cell growth and cell division in the intercalary meristem of submerged deepwater rice (*Oryza sativa* L.). *Planta* **204**, 140–145.

Loreti, E., Yamaguchi, J., Alpi, A. and Perata, P. (2003). Sugar modulation of α-amylase genes under anoxia. *Annals of Botany* **91**, 143–148.

Loreti, E., Poggi, A., Novi, G., Alpi, A. and Perata, P. (2005). A genome-wide analysis of the effects of sucrose on gene expression in arabidopsis seedlings under anoxia. *Plant Physiology* **137**, 1130–1138.

Löw, D., Brändle, K., Nover, L. and Forreiter, C. (2000). Cytosolic heat-stress proteins Hsp17.7 class I and Hsp17.3 class II of tomato act as molecular chaperones *in vivo*. *Planta* **211**, 575–582.

Lu, C.-A., Ho, T.-H. D., Ho, S.-L. and Yu, S.-M. (2002). Three novel myb proteins with one DNA binding repeat mediate sugar and hormone regulation of α-amylase gene expression. *Plant Cell* **14**, 1963–1980.

Lu, C.-A., Lin, C.-C., Lee, K.-W., Chen, J.-L., Huang, L.-F., Ho, S.-L., Liu, H.-J., Hsing, Y.-I. and Yu, S.-M. (2007). The SnRK1A protein kinase plays a key role in sugar signaling during germination and seedling growth of rice. *Plant Cell* **19**, 2484–2499.

Macejak, D. G. and Sarnow, P. (1991). Internal initiation of translation mediated by the 5′ leader of a cellular mRNA. *Nature* **353**, 90–94.

MacLeod, R. D. and McLachlan, S. M. (1975). The development of large primordia in *Vicia faba* L.: Some cytological and anatomical changes. *Protoplasma* **85**, 291–304.

Magneschi, L. and Perata, P. (2009). Rice germination and seedling growth in the absence of oxygen. *Annals of Botany* **103**, 181–196.

Malamy, J. and Benfey, P. (1997). Organization and cell differentiation in lateral roots of *Arabidopsis thaliana*. *Development* **124**, 33–44.

Mardanova, E. S., Zamchuk, L. A. and Ravin, N. V. (2007). The 5′ untranslated region of the maize Alcohol Dehydrogenase gene provides efficient translation of mRNA in plants under stress conditions. *Molecular Biology* **41**, 914–919.

Mardanova, E. S., Zamchuk, L. A., Skulachev, M. V. and Ravin, N. V. (2008). The 5′ untranslated region of the maize alcohol dehydrogenase gene contains an internal ribosome entry site. *Gene* **420**, 11–16.

Margis, R. and Margis-Pinheiro, M. (2003). Phytocalpains: Orthologous calcium-dependent cysteine proteinases. *Trends in Plant Science* **8**, 58–62.

Mariano, A. B., Valente, C., Cadena, S. M. S. C., Rocha, M. E. M., De Oliveira, M. B. M. and Carnieri, E. G. S. (2005). Sensitivities of the alternative respiratory components of potato tuber mitochondria to thiol reagents and Ca^{2+}. *Plant Physiology and Biochemistry* **43**, 61–67.

Marillia, E.-F., Micallef, B. J., Micallef, M., Weninger, A., Pedersen, K. K., Zou, J. and Taylor, D. C. (2003). Biochemical and physiological studies of *Arabidopsis thaliana* transgenic lines with repressed expression of the mitochondrial pyruvate dehydrogenase kinase1. *Journal of Experimental Botany* **54**, 259–270.

Mateo, A., Muhlenbock, P., Rusterucci, C., Chang, C. C.-C., Miszalski, Z., Karpinska, B., Parker, J. E., Mullineaux, P. M. and Karpinski, S. (2004). Lesion simulating disease 1 is required for acclimation to conditions that promote excess excitation energy. *Plant Physiology* **136**, 2818–2830.

Mattana, M., Vannini, C., Espen, L., Bracale, M., Genga, A., Marsoni, M., Iriti, M., Bonazza, V., Romagnoli, F., Baldoni, E., Coraggio, I. and Locatelli, F. (2007). The rice Mybleu transcription factor increases tolerance to oxygen deprivation in *Arabidopsis* plants. *Physiologia Plantarum* **131**, 106–121.

McDonald, M. P., Galwey, N. W. and Colmer, T. D. (2001). Waterlogging tolerance in the tribe *Triticeae*: The adventitious roots of *Critesion marinum* have a relatively high porosity and a barrier to radial oxygen loss. *Plant, Cell and Environment* **24**, 585–596.

McDonald, M. P., Galwey, N. W. and Colmer, T. D. (2002). Similarity and diversity in adventitious root anatomy as related to root aeration among a range of wetland and dryland grass species. *Plant, Cell and Environment* **25**, 441–451.

Melis, A. and Happe, T. (2004). Trails of green alga hydrogen research—From Hans Gaffron to new frontiers. Discoveries in Photosynthesis. *Photosynthesis Research* **80**, 401–409.

Menegus, F., Cattaruzza, L., Mattana, M., Beffagna, N. and Ragg, E. (1991). Response to anoxia in rice and wheat seedlings: Changes in the ph of intracellular compartments, glucose-6-phosphate level, and metabolic rate. *Plant Physiology* **95**, 760–767.

Menegus, F., Cattaruzza, L., Molinari, H. N. and Ragg, E. (1993). Rice and wheat seedlings as plant models of high and low tolerance to anoxia. *In* "Surviving Hypoxia: Mechanisms of Adaptation and Control". (P. W. Hochachka, P. L. Lutz, T. Sick, M. Rosenthal and G. van der Tillart, eds.), pp. 53–64. CRC press, Boca raton.

Mertens, E., Larondelle, Y. and Hers, H.-G. (1990). Induction of pyrophosphate: fructose 6-phosphate 1-phosphotransferase by anoxia in rice seedlings. *Plant Physiology* **93**, 584–587.

Millar, A. H., Eubel, H., Jänsch, L., Kruft, V., Heazlewood, J. L. and Braun, H.-P. (2004). Mitochondrial Cytochrome *c* oxidase and succinate dehydrogenase complexes contain plant specific subunits. *Plant Molecular Biology* **56**, 77–90.

Mohanty, B., Wilson, P. M. and AP Rees, T. (1993). Effects of anoxia on growth and carbohydrate metabolism in suspension cultures of soybean and rice. *Phytochemistry* **34**, 75–82.

Mohanty, B., Krishnan, S. P. T., Swarup, S. and Bajic, V. B. (2005). Detection and preliminary analysis of motifs in promoters of anaerobically induced genes of different plant species. *Annals of Botany* **96**, 669–681.

Mommer, L., Pons, T. L., Wolters-Arts, M., Venema, J. H. and Visser, E. J. W. (2005). Submergence-induced morphological, anatomical, and biochemical responses in a terrestrial species affect gas diffusion resistance and photosynthetic performance. *Plant Physiology* **139**, 497–508.

Mommer, L., Lenssen, J. M. P., Huber, H., Visser, E. J. W. and DE Kroon, H. (2006). Ecophysiological determinants of plant performance under flooding: A comparative study of seven plant families. *Journal of Ecology* **94**, 1117–1129.

Moncousin, C. and Gaspar, T. (1983). Peroxidase as a marker for rooting improvement of *Cynara scolymus* L. cultured *in vitro*. *Biochemie und Physiologie von Pflanzen* **178**, 263–271.

Monk, L. S., Ceawford, R. M. M. and Brandle, R. (1984). Fermentation rates and ethanol accumulation in relation to flooding tolerance in rhizomes of monocotyledonous species. *Journal of Experimental Botany* **35**, 738–745.

Moons, A. (2008). Transcriptional profiling of the PDR gene family in rice roots in response to plant growth regulators, redox perturbations and weak organic acid stresses. *Planta* **229**, 53–71. doi:10.1007/s00425-008-0810-5.

Moons, A., Valcke, R., Van, M. and Montagu (1998). Low-oxygen stress and water deficit induce cytosolic pyruvate orthophosphate dikinase (PPDK) expression in roots of rice, a C3 plant. *The Plant Journal* **15**, 89–98.

Morita, A., Umemura, T.-A., Kuroyanagi, M., Futsuhara, Y., Perata, P. and Yamaguchi, J. (1998). Functional dissection of a sugar-repressed α-amylase gene (RAmy1α) promoter in rice embryos. *FEBS Letters* **423**, 81–85.

Moseley, J., Quinn, J., Mats, E. and Merchant, S. (2000). The *Crd1* gene encodes a putative di-iron enzyme required for photosystem I accumulation in copper deficiency and hypoxia in *Chlamydomonas*. *The EMBO Journal* **19**, 2139–2151.

Muhlenbock, P., Plaszczyca, M., Plaszczyca, M., Mellerowicz, E. and Karpinski, S. (2007). lysigenous aerenchyma formation in *Arabidopsis* is controlled by lesion simulating disease1. *Plant Cell* **19**, 3819–3830.

Mus, F., Dubini, A., Seibert, M., Posewitz, M. C. and Grossman, A. R. (2007). Anaerobic acclimation in *Chlamydomonas reinhardtii*: Anoxic gene expression, hydrogenase induction, and metabolic pathways. *Journal of Biological Chemistry* **282**, 25475–25486.

Mustroph, A., Albrecht, G., Hajirezaei, M., Grimm, B. and Biemelt, S. (2005). Low levels of pyrophosphate in transgenic potato plants expressing *E. coli* pyrophosphatase lead to decreased vitality under oxygen deficiency. *Annals of Botany* **96**, 717–726.

Mustroph, A., Boamfa, E., Laarhoven, L., Harren, F., Albrecht, G. and Grimm, B. (2006). Organ-specific analysis of the anaerobic primary metabolism in rice and wheat seedlings. I: Dark ethanol production is dominated by the shoots. *Planta* **225**, 103–114.

Nakamoto, H. and Vígh, L. (2007). The small heat shock proteins and their clients. *Cellular and Molecular Life Sciences* **64**, 294–306.

Neal, M. J. and Girton, R. E. (1955). The Pasteur effect in Maize. *American Journal of Botany* **42**, 733–737.

Neinhuis, C. and Barthlott, W. (1997). Characterisation and distribution of water-repellent, self-cleaning plant surfaces. *Annals of Botany* **79**, 667–677.

Neuteboom, L. W., Veth-Tello, L. M., Clijdesdale, O. R., Hooykaas, P. J. J. and Van Der Zaal, B. J. (1999). A novel subtilisin-like protease gene from *Arabidopsis thaliana* is expressed at sites of lateral root emergence. *DNA Research* **6**, 13–19.

Nicolai, M., Roncato, M. A., Canoy, A. S., Rouquie, D., Sarda, X., Freyssinet, G. and Robaglia, C. (2006). Large-scale analysis of mRNA translation states during sucrose starvation in *Arabidopsis* cells identifies cell proliferation and chromatin structure as targets of translational control. *Plant Physiology* **141**, 663–673.

Nomura, T., Kono, Y. and Akazawa, T. (1969). Enzymic mechanism of starch breakdown in germinating rice seeds II. Scutellum as the site of sucrose synthesis. *Plant Physiology* **44**, 765–769.

Nover, L. and Scharf, K. D. (1997). Heat stress proteins and transcription factors. *Cellular and Molecular Life Sciences (CMLS)* **53**, 80–103.

Ochsenbein, C., Przybyla, D., Danon, A., Landgraf, F., Göbel, C., Imboden, A. and Apel, K. (2006). The role of *EDS1* (enhanced disease susceptibility) during singlet oxygen-mediated stress responses of Arabidopsis. *The Plant Journal* **47**, 445–456.

Parker, J. E., Holub, E. B., Frost, L. N., Falk, A., Gunn, N. D. and Daniels, M. J. (1996). Characterization of eds1, a mutation in *Arabidopsis* suppressing resistance to peronospora parasitica specified by several different rpp genes. *Plant Cell* **8**, 2033–2046.

Pasentsis, K., Falara, V., Pateraki, I., Gerasopoulos, D. and Kanellis, A. K. (2007). Identification and expression profiling of low oxygen regulated genes from *Citrus flavedo* tissues using RT-PCR differential display. *Journal of Experimental Botany* **58**, 2203–2216.

Pederse, T., Rich, S. M. and Colmer, T. (2009). Surviving floods: Leaf gas films improve O_2 and CO_2 exchange, root aeration, and growth of completely submerged rice. *Plant Journal*, doi:10.1111/j.1365-313X.2008.03769.x.

Peng, H.-P., Chan, C.-S., Shih, M.-C. and Yang, S. F. (2001). Signaling events in the hypoxic induction of alcohol dehydrogenase gene in *Arabidopsis*. *Plant Physiology* **126**, 742–749.

Peng, H.-P., Lin, T.-Y., Wang, N.-N. and Shih, M.-C. (2005). Differential expression of genes encoding 1-aminocyclopropane-1-carboxylate synthase in *Arabidopsis* during hypoxia. *Plant Molecular Biology* **58**, 15–25.

Perata, P. and Alpi, A. (1991). Ethanol-induced injuries to carrot cells: The role of acetaldehyde. *Plant Physiology* **95**, 748–752.

Perata, P. and Alpi, A. (1993). Plant responses to anaerobiosis. *Plant Science* **93**, 1–17.

Perata, P., Alpi, A. and Loschiavo, F. (1984). Physiological responses of cereal seedlings to ethanol. *Journal of Plant Physiology* **126**, 181–188.

Perata, P., Vernieri, P., Armellini, D., Bugnoli, M., Tognoni, F. and Alpi, A. (1992). Immunological detection of acetaldehyde-protein adducts in ethanol-treated carrot cells. *Plant Physiology* **98**, 913–918.

Perata, P., Matsukura, C., Vernieri, P. and Yamaguchi, J. (1997). Sugar repression of a gibberellin-dependent signaling pathway in barley embryos. *Plant Cell* **9**, 2197–2208.

Perazzolli, M., Dominici, P., Romero-Puertas, M. C., Zago, E., Zeier, J., Sonoda, M., Lamb, C. and Delledonne, M. (2004). *Arabidopsis* nonsymbiotic hemoglobin AHb1 modulates nitric oxide bioactivity. *Plant Cell* **16**, 2785–2794.

Peschke, V. M. and Sachs, M. M. (1994). Characterization and expression of transcripts induced by oxygen deprivation in maize (*Zea mays* L.). *Plant Physiology* **104**, 387–394.

Pierik, R., Van Aken, J. M. and Voesenek, L. A. C. J. (2008). Is elongation-induced leaf emergence beneficial for submerged *Rumex* species? *Annals of Botany* **103**, 353–357. doi:10.1093/aob/mcn143.

Planchet, E. and Kaiser, W. M. (2006). Nitric oxide (NO) detection by DAF fluorescence and chemiluminescence: A comparison using abiotic and biotic NO sources. *Journal of Experimental Botany* **57**, 3043–3055.

Plaxton, W. C. and Podestá, F. E. (2006). The functional organization and control of plant respiration. *Critical Reviews in Plant Sciences* **25**, 159–198.

Pope, D. G. (1977). Relationships between hydroxyproline-containing proteins secreted into the cell wall and medium by suspension-cultured *Acer pseudoplatanus* cells. *Plant Physiology* **59**, 894–900.

Pradet, A. and Raymond, P. (1983). Adenine-nucleotide rations adenylate energy–charge in energy-metabolsim. *Annual Review of Plant Physiology and Plant Molecular Biology* **34**, 199–224.

Pronk, J. T., Steesma, Y. H. and Van Dijken, J. P. (1996). Pyruvate metabolism in *Saccharomyces cerevisiae*. *Yeast* **16**, 1607–1633.

Quinn, J. M. and Merchant, S. (1995). Two copper-responsive elements associated with the *Chlamydomonas* Cyc6 gene function as targets for transcriptional activators. *Plant Cell* **7**, 623–638.

Raskin, I. and Kende, H. (1983). How does deep water rice solve its aeration problem? *Plant Physiology* **72**, 447–454.

Reggiani, R., Cantu, C. A., Brambilla, I. and Bertani, A. (1988). Accumulation and interconversion of amino acids in rice roots under anoxia. *Plant Cell Physiology* **29**, 981–987.

Reggiani, R., Hochkoeppler, A. and Bertani, A. (1989). Polyamines and anaerobic elongation of rice coleoptile. *Plant Cell Physiology* **30**, 893–898.

Reggiani, R., Giussani, P. and Bertani, A. (1990). Relationship between the accumulation of putrescine and the tolerance to oxygen-deficit stress in *Gramineae* seedlings. *Plant Cell Physiology* **31**, 489–494.

Ricard, B., Rivoal, J., Spiteri, A. and Pradet, A. (1991). Anaerobic stress induces the transcription and translation of sucrose synthase in rice. *Plant Physiology* **95**, 669–674.

Riewe, D., Grosman, L., Zauber, H., Wucke, C., Fernie, A. R. and Geigenberger, P. (2008). Metabolic and developmental adaptations of growing potato tubers in response to specific manipulations of the adenylate energy status. *Plant Physiology* **146**, 1579–1598.

Rivoal, J. and Hanson, A. D. (1994). Metabolic control of anaerobic glycolisis: Overexpression of lactate dehydrogenase in transgenic tomato roots supports the Davies–Roberts hypothesis and points to a critical role for lactate secretion. *Plant Physiology* **106**, 1179–1185.

Rivoal, J., Ricard, B. and Pradet, A. (1991). Lactate dehydrogenase in *Oryza sativa* L. seedlings and roots: Identification and partial characterization. *Plant Physiology* **95**, 682–686.

Roberts, J. K., Callis, J., Jardetzky, O., Walbot, V. and Freeling, M. (1984). Cytoplasmic acidosis as a determinant of flooding intolerance in plants.

Proceedings of the National Academy of Sciences of the United States of America **81**, 6029–6033.

Roberts, J. K. M., Chang, K., Webster, C., Callis, J. and Walbot, V. (1989). Dependence of ethanolic fermentation, cytoplasmic pH regulation, and viability on the activity of alcohol dehydrogenase in hypoxic maize root tips. *Plant Physiology* **89**, 1275–1278.

Rossel, J. B., Wilson, P. B., Hussain, D., Woo, N. S., Gordon, M. J., Mewett, O. P., Howell, K. A., Whelan, J., Kazan, K. and Pogson, B. J. (2007). Systemic and intracellular responses to photooxidative stress in *Arabidopsis*. *Plant Cell* **19**, 4091–4110.

Rowland, O., Zheng, H., Hepworth, S. R., Lam, P., Jetter, R. and Kunst, L. (2006). CER4 encodes an alcohol-forming fatty acyl-coenzyme A reductase involved in cuticular wax production in *Arabidopsis*. *Plant Physiology* **142**, 866–877.

Russell, D. A. and Sachs, M. M. (1989). Differential expression and sequence analysis of the maize glyceraldehyde-3-phosphate dehydrogenase gene family. *Plant Cell* **1**, 793–803.

Saglio, P. H., Drew, M. C. and Pradet, A. (1988). Metabolic acclimation to anoxia induced by low (2–4 kPa partial pressure) pretreatment (hypoxia) in root tips of. *Zea mays Plant Physiology* **86**, 61–66.

Saika, H., Okamoto, M., Miyoshi, K., Kushiro, T., Shinoda, S., Jikumaru, Y., Fujimoto, M., Arikawa, T., Takahashi, H., Ando, M., Arimura, S.-I. Miyao, A. *et al.* (2007). Ethylene promotes submergence-induced expression of OsABA8ox1, a gene that encodes ABA 8′-hydroxylase in rice. *Plant Cell Physiology* **48**, 287–298.

Saint-Ges, V., Roby, C., Bligny, R., Pradet, A. and Douce, R. (1991). Kinetic studies of the variations of cytoplasmic pH, nucleotide triphosphates (^{31}P-NMR) and lactate during normoxic and anoxic transitions in maize root tips. *European Journal of Biochemistry* **200**, 477–482.

Samarakoon, A. B. and Horton, R. F. (1984). Petiole growth in *Ranunculus sceleratus* L.: Ethylene synthesis and submergence. *Annals of Botany* **54**, 263–270.

Santosa, I., Ram, P., Boamfa, E., Laarhoven, L., Reuss, J., Jackson, M. and Harren, F. (2007). Patterns of peroxidative ethane emission from submerged rice seedlings indicate that damage from reactive oxygen species takes place during submergence and is not necessarily a post-anoxic phenomenon. *Planta* **226**, 193–202.

Seago, J. L., JR, Marsch, L. C., Stevens, K. J., Soukup, A., Votrubova, O. and Enstone, D. E. (2005). A re-examination of the root cortex in wetland flowering plants with respect to aerenchyma. *Annals of Botany* **96**, 565–579.

Sedbrook, J. C., Kronebusch, P. J., Borisy, G. G., Trewavas, A. J. and Masson, P. H. (1996). Transgenic AEQUORIN reveals organ-specific cytosolic Ca^{2+} responses to anoxia in *Arabidopsis thaliana* seedling. *Plant Physiology* **111**, 243–257.

Semenza, G. L. (2007). Hypoxia-inducible factor 1 (HIF-1). Pathway. *Science STKE.* doi:10.1126/stke.4072007eg9.

Serpa, V., Vernal, J., Lamattina, L., Grotewold, E., Cassia, R. and Terenzi, H. (2007). Inhibition of AtMYB2 DNA-binding by nitric oxide involves cysteine *S*-nitrosylation. *Biochemical and Biophysical Research Communications* **361**, 1048–1053.

Setter, T. L. and Waters, I. (2003). Review of prospects for germplasm improvement for waterlogging tolerance in wheat, barley and oats. *Plant and Soil* **253**, 1–34.

Smirnoff, N. (1995). Antioxidant systems and plant response to the environment. *In* "Environment and plant metabolism: Flexibility and acclimation" (N. Smirnoff) pp. 217–244. BIOS Scientific Publishers, Oxford.

Smit, B. A., Neuman, D. S. and Stachowiak, M. L. (1990). Root hypoxia reduces leaf growth: Role of factors in the transpiration stream. *Plant Physiology* **92**, 1021–1028.

Smith, A. M., Zeeman, S. C. and Smith, S. M. (2005). Starch degradation. *Annual Review of Plant Biology* **56**, 73–98.

Smýkal, P., Mašín, J., Hrdý, I., Konopásek, I. and Dárský, V. (2000). Chaperone activity of tobacco HSP18, a small heat-shock protein, is inhibited by ATP. *The Plant Journal* **23**, 703–713.

Steffens, B., Wang, J. and Sauter, M. (2006). Interactions between ethylene, gibberellin and abscisic acid regulate emergence and growth rate of adventitious roots in deepwater rice. *Planta* **223**, 604–612.

Stitt, M. (1990). Fructose-2,6-bisphosphate as a regulatory molecule in plants. *Trends in Plant Science* **41**, 153–185.

Stoimenova, M., Libourel, I. G. L., Ratcliffe, R. G. and Kaiser, W. M. (2003). The role of nitrate reduction in the anoxic metabolism of roots II. Anoxic metabolism of tobacco roots with or without nitrate reductase activity. *Plant and Soil* **253**, 155–167.

Stoimenova, M., Igamberdiev, A. U., Gupta, K. J. and Hill, R. D. (2007). Nitrite-driven anaerobic ATP synthesis in barley and rice root mitochondria. *Planta* **226**, 465–474.

Subbaiah, C. C., Bush, D. S. and Sachs, M. M. (1994a). Elevation of cytosolic calcium precedes anoxic gene expression in maize suspension-cultured cells. *Plant Cell* **6**, 1747–1762.

Subbaiah, C. C., Zhang, J. and Sachs, M. M. (1994b). Involvement of intracellular calcium in anaerobic gene expression and survival of maize seedlings. *Plant Physiology* **105**, 369–376.

Subbaiah, C. C., Bush, D. S. and Sachs, M. M. (1998). Mitochondrial contribution to the anoxic Ca^{2+} signal in maize suspension-cultured cells. *Plant Physiology* **118**, 759–771.

Subbaiah, C. C., Kollipara, K. P. and Sachs, M. M. (2000). A Ca^{2+}-dependent cysteine protease is associated with anoxia-induced root tip death in maize. *Journal of Experimental Botany* **51**, 721–730.

Subbaiah, C. C., Palaniappan, A., Duncan, K., Rhoads, D. M., Huber, S. C. and Sachs, M. M. (2006). Mitochondrial localization and putative signaling function of sucrose synthase in maize. *Journal of Biological Chemistry* **281**, 15625–15635.

Sweetlove, L. J., Dunford, R., Ratcliffe, R. G. and Kruger, N. J. (2000). Lactate metabolism in potato tubers deficient in lactate dehydrogenase activity. *Plant, Cell and Environment* **23**, 873–881.

Swindell, W. R. (2006). The association among gene expression responses to nine abiotic stress treatments in *Arabidopsis thaliana*. *Genetics* **174**, 1811–1824.

Szarka, A., Horemans, N., Kovács, Z., Gróf, P., Mayer, M. and Bánhegyi, G. (2007). Dehydroascorbate reduction in plant mitochondria is coupled to the respiratory electron transfer chain. *Physiologia Plantarum* **129**, 225–232.

Tadege, M., Brändle, R. and Kuhlemeier, C. (1998). Anoxia tolerance in tobacco roots: Effect of overexpression of pyruvate decarboxylase. *The Plant Journal* **14**, 327–335.

Tari, I. and Nagy, M. (1996). Abscisic acid and Ethrel abolish the inhibition of adventitious root formation of paclobutrazol-treated bean primary leaf cuttings. *Biologia Plantarum* **38**, 369–375.

Taylor, E. R., Nie, X. Z., Macgregor, A. W. and Hill, R. D. (1994). A cereal haemoglobin gene is expressed in seed and root tissues under anaerobic conditions. *Plant Molecular Biology* **24**, 853–862.

Tiainen, P., Myllyharju, J. and Koivunen, P. (2005). Characterization of a second *Arabidopsis thaliana* prolyl 4-hydroxylase with distinct substrate specificity. *Journal of Biological Chemistry* **280**, 1142–1148.

Tischner, R., Planchet, E. and Kaiser, W. M. (2004). Mitochondrial electron transport as a source for nitric oxide in the unicellular green alga Chlorella sorokiniana. *FEBS Letters* **576**, 151–155.

Todd, B. L., Stewart, E. V., Burg, J. S., Hughes, A. L. and Espenshade, P. J. (2006). Sterol regulatory element binding protein is a principal regulator of anaerobic gene expression in fission yeast. *Mol. Cell. Biol.* **26**, 2817–2831.

Topa, M. A. and Cheeseman, J. M. (1994). Maintenance of ^{32}P uptake and transport in *Pinus serotina* seedlings under hypoxic growth conditions. *Physiologia Plantarum* **92**, 171–180.

Tournaire-Roux, C., Sutka, M., Javot, H. L. N., Gout, E., Gerbeau, P., Doan-Trung, L., Bligny, R. and Maurel, C. (2003). Cytosolic pH regulates root water transport during anoxic stress through gating of aquaporins. *Nature* **425**, 393–397.

Tsuji, H., Nakazono, M., Saisho, D., Tsutsumi, N. and Hirai, A. (2000). Transcript levels of the nuclear-encoded respiratory genes in rice decrease by oxygen deprivation: Evidence for involvement of calcium in expression of the alternative oxidase 1a gene. *FEBS Letters* **471**, 201–204.

Tsuji, H., Saika, H., Tsutsumi, N., Hirai, A. and Nakazono, M. (2006). Dynamic and reversible changes in histone H3-lys4 methylation and H3 acetylation occurring at submergence-inducible genes in rice. *Plant Cell Physiology* **47**, 995–1003.

Tsvetkova, N. M., Horváth, I., Török, Z., Wolkers, W. F., Balogi, Z., Shigapova, N., Crowe, L. M., Tablin, F., Vierling, E., Crowe, J. H. and Vígh, L. S. (2002). Small heat-shock proteins regulate membrane lipid polymorphism. *Proceedings of the National Academy of Sciences of the United States of America* **99**, 13504–13509.

Ueguchi-Tanaka, M., Masatoshi Nakajima, M., Etsuko Katoh, E., Hiroko Ohmiya, H., Kenji Asano, K., Shoko Saji, S., Xiang Hongyu, X., Motoyuki Ashikari, M., Hidemi Kitano, H., Isomaro Yamaguchi, I. and Matsuoka, M. (2007). Molecular interactions of a soluble gibberellin receptor, GID1, with a rice DELLA protein, SLR1, and gibberellin. *The Plant Cell* **19**, 2140–2155.

Van Dongen, J. T., Roeb, G. W., Dautzenberg, M., Froehlich, A., Vigeolas, H., Minchin, P. E. H. and Geigenberger, P. (2004). Phloem import and storage metabolism are highly coordinated by the low oxygen concentrations within developing wheat seeds. *Plant Physiology* **135**, 1809–1821.

Van Dongen, J. T., Frohlich, A., Ramirez-Aguilar, S. J., Schauer, N., Fernie, A. R., Erban, A., Kopka, J., Clark, J., Langer, A. and Geigenberger, P. (2008). Transcript and metabolite profiling of the adaptive response to mild decreases in oxygen concentration in the roots of *Arabidopsis* plants. *Annals of Botany* **103**, 269–280.

Vartapetian, B. B. (1982). Pasteur effect visualization by electron microscopy. *Naturwissenschaften* **69**, 99–99.

Vartapetian, B. B. and Jackson, M. B. (1997). Plant adaptations to anaerobic stress. *Annals of Botany* **79**, 3–20.

Vartapetian, B. B., Adreeva, I. N., Generazova, I. P., Polyakova, L. I., Maslova, I. P., Dolgikh, Y. I. and Stepenova, A. Y. (2003). Functional electron microscopy in studies of plant response and adaptation to anaerobic stress. *Annals of Botany* **91**, 155–172.

Vigeolas, H., Van Dongen, J. T., Waldeck, P., Huhn, D. and Geigenberger, P. (2003). Lipid storage metabolism is limited by the prevailing low oxygen concentrations within developing seeds of oilseed rape. *Plant Physiology* **133**, 2048–2060.

Visser, E. J. W., Colmer, T. D., Blom, C. W. P. M. and Voesenek, L. A. C. J. (2000). Changes in growth, porosity, and radial oxygen loss from adventitious roots of selected mono- and dicotyledonous wetland species with contrasting types of aerenchyma. *Plant Cell and Environment* **23**, 1237–1245.

Voesenek, L., Banga, M., Thier, R. H., Mudde, C. M., Harren, F., Barendse, G. and Blom, C. (1993). Submergence-induced ethylene synthesis, entrapment, and growth in two plant species with contrasting flooding resistances. *Plant Physiology* **103**, 783–791.

Voesenek, L., Colmer, T. D., Pierik, R., Millenaar, F. F. and Peeters, A. J. M. (2006). How plants cope with complete submergence. *New Phytologist* **170**, 213–226.

Vogel, J. T., Zarka, D. G., Van Buskirk, H. A., Fowler, S. G. and Thomashow, M. F. (2005). Roles of the CBF2 and ZAT12 transcription factors in configuring the low temperature transcriptome of *Arabidopsis*. *The Plant Journal* **41**, 195–211.

Vogel, S. (2006). Living in a physical world VIII. Gravity and life in water. *Journal of Bioscience* **31**, 309–322.

Vreeburg, R. A. M., Benschop, J. J., Peeters, A. J. M., Colmer, T. D., Ammerlaan, A. H. M., Staal, M., Elzenga, T. M., Staals, R. H. J., Darley, C. P., McQueen-Mason, S. J. and Voesenek, L. A. C. J. (2005). Ethylene regulates fast apoplastic acidification and expansin A transcription during submergence-induced petiole elongation in *Rumex palustris*. *The Plant Journal* **43**, 597–610.

Vriezen, W. H., Hulzink, R., Mariani, C. and Voesenek, L. A. C. J. (1999). 1-Aminocyclopropane-1-Carboxylate oxidase activity limits ethylene biosynthesis in *Rumex palustris* during submergence. *Plant Physiology* **121**, 189–196.

Wagner, P., Fürstner, R., Barthlott, W. and Neinhuis, C. (2003). Quantitative assessment to the structural basis of water repellency in natural and technical surfaces. *Journal of Experimental Botany* **385**, 1295–1303.

Walker, J. C., Howard, E. A., Dennis, E. S. and Peacock, W. J. (1987). DNA sequences required for anaerobic expression of the maize alcohol dehydrogenase 1 gene. *Proceedings of the National Academy of Sciences of the United States of America* **84**, 6624–6628.

Wang, C., Barry, J. K., Min, Z., Tordsen, G., Rao, A. G. and Olsen, O.-A. (2003). The calpain domain of the maize DEK1 protein contains the conserved catalytic triad and functions as a cysteine proteinase. *Journal of Biological Chemistry* **278**, 34467–34474.

Watanabe, C. K., Hachiya, T. I. T. and Noguchi, K. (2008). The lack of alternative oxidase at low temperature leads to a disruption of the balance in carbon and nitrogen metabolism, and to an up-regulation of antioxidant defence systems in *Arabidopsis thaliana* leaves. *Plant, Cell and Environment* **31**, 1190–1202.

Waters, I., Morrell, S., Greenway, H. and Colmer, T. D. (1991). Effects of Anoxia on wheat seedlings: II. Influence of O_2 supply prior to anoxia on tolerance to anoxia, alcoholic fermentation, and sugar levels. *Journal of Experimental Botany* **42**, 1437–1447.

Webster, C., Kim, C.-Y. and Roberts, J. K. M. (1991). Elongation and termination reactions of protein synthesis on maize root tip polyribosomes studied in a homologous cell-free system. *Plant Physiology* **96**, 418–425.

Wenger, R. H. (2002). Cellular adaptation to hypoxia: O_2-sensing protein hydroxylases, hypoxia-inducible transcription factors, and O_2-regulated gene expression. *The FASEB Journal* **16**, 1151–1162.

Xia, J. H. and Roberts, J. (1994). Improved cytoplasmic pH regulation, increased lactate efflux, and reduced cytoplasmic lactate levels are biochemical traits expressed in root tips of whole maize seedlings acclimated to a low-oxygen environment. *Plant Physiology* **105**, 651–657.

Xin, Z., Zhao, Y. and Zheng, Z.-L. (2005). Transcriptome analysis reveals specific modulation of abscisic acid signaling by Rop10 small GTPase in *Arabidopsis*. *Plant Physiology* **139**, 1350–1365.

Xu, K., Xu, X., Fukao, T., Canlas, P., Maghirang-Rodriguez, R., Heuer, S., Ismail, A. M., Bailey-Serres, J., Ronald, P. C. and Mackill, D. J. (2006). Sub1A is an ethylene-response-factor-like gene that confers submergence tolerance to rice. *Nature* **442**, 705–708.

Xu, Z.-S., Xia, L.-Q., Chen, M., Cheng, X.-G., Zhang, R.-Y., Li, L.-C., Zhao, Y.-X., Lu, Y., Ni, Z.-Y., Liu, L., Qiu, Z.-G. and Ma, Y.-Z. (2007). Isolation and molecular characterization of the *Triticum aestivum* L. ethylene-responsive factor 1 (TaERF1) that increases multiple stress tolerance. *Plant Molecular Biology* **65**, 719–732.

Xu, J., Li, Y., Wang, Y., Liu, H., Lei, L., Yang, H., Liu, G. and Ren, D. (2008). Activation of MAPK Kinase 9 induces ethylene and camalexin biosynthesis and enhances sensitivity to salt stress in *Arabidopsis*. *Journal of Biological Chemistry* **283**, 26996–27006.

Yoshida, K. and Komae, K. (2006). A rice family 9 glycoside hydrolase isozyme with broad substrate specificity for hemicelluloses in type II cell walls. *Plant Cell Physiology* **47**, 1541–1554.

Youm, J., Jeon, J., Choi, D., Yi, S., Joung, H. and Kim, H. (2008). Ectopic expression of pepper CaPF1 in potato enhances multiple stresses tolerance and delays initiation of *in vitro* tuberization. *Planta* **228**, 701–708.

Yu, D., Ranathunge, K., Huang, H., Pei, Z., Franke, R., Schreiber, L. and He, C. (2008). Wax crystal-sparse leaf1 encodes a β-ketoacyl CoA synthase involved in biosynthesis of cuticular waxes on rice leaf. *Planta* **228**, 675–685.

Zabalza, A., Van Dongen, J. T., Froehlich, A., Oliver, S., Faix, B., Kapuganti, J. G., Schmalzlin, E., Igal, M., Orcaray, L., Royuela, M. and Geigenberger, P. (2009). Regulation of respiration and fermentation to control the plant internal oxygen concentration. *Plant Physiology*. doi:10.1104/pp.108.129288.

Zhang, Q. and Greenway, H. (1994). Anoxia tolerance and anaerobic catabolism of aged beetroot storage tissues. *Journal of Experimental Botany* **45**, 567–575.

Zhang, Z., Wei, L., Zou, X., Tao, Y., Liu, Z. and Zheng, Y. (2008). Submergence-responsive microRNAs are potentially involved in the regulation of morphological and metabolic adaptations in maize root cells. *Annals of Botany* **102**, 509–519.

Zhou, Z., Vriezen, W., Van Caeneghem, W., Van Montagu, M. and Van der Straeten, D. (2001). Rapid induction of a novel ACC synthase gene in deepwater rice seedlings upon complete submergence. *Euphytica* **121**, 137–143.

Zuckermann, H., Harren, F., Reuss, J. and Parker, D. H. (1997). Dynamics of acetaldehyde production during anoxia and post-anoxia in red bell pepper studied by photoacoustic techniques. *Plant Physiology* **113**, 925–932.

Roles of Circadian Clock and Histone Methylation in the Control of Floral Repressors

RYM FEKIH, RIM NEFISSI, KANA MIYATA, HIROSHI EZURA AND TSUYOSHI MIZOGUCHI

Gene Research Center, University of Tsukuba, Tennodai 1-1-1, Tsukuba, Ibaraki 305-8572, Japan

Advances in Botanical Research, Vol. 50

0065-2296/09 $35.00
DOI: 10.1016/S0065-2296(08)00805-7

ABSTRACT

The flowering of *Arabidopsis* is controlled by several signaling pathways that converge on a small set of floral activator genes (e.g., *FT*, *SOC1*, and *LFY*) that function as pathway integrators. Both floral activators and repressors play key roles in controlling flowering time. Temporal balance between floral repressor and promoter activity both daily and seasonally is crucial in helping plants determine when to flower. This review summarizes recent progress on understanding interactions between floral repressors and activators. The possible roles of the circadian clock and histone methylation in the control of floral repressors are mainly discussed.

I. INTRODUCTION

Plants meet the challenge of environmental change with developmental change as an adaptation to their sessile lifestyle. According to Thomas *et al.* (2006), the ability to detect and respond to seasonal change confers a selective advantage on plants because it provides a means of anticipating, and consequently preventing, the adverse effects of a particular seasonal environment. A main developmental switch in the life cycle of a flowering plant is the transition from vegetative to reproductive phase. Timing the floral transition during the most favorable conditions is crucial in agriculture and horticulture to maximize reproductive success (Boss *et al.*, 2004).

The decision to flower is affected by many environmental and endogenous factors, and genetic analysis of a large number of *Arabidopsis* flowering-time mutants has led to the identification of several genetic pathways involved in controlling flowering time. These flowering pathways include the photoperiod pathway, gibberellic acid (GA) pathway, autonomous pathway, and vernalization pathway (Boss *et al.*, 2004). Considering the process of flowering as a default developmental program that must be suppressed early in the life cycle of the plant, Boss *et al.* (2004) and Komeda (2004) divided the floral pathways into those that enable the floral transition and those that promote it. Based on their model, the pathway that enables flowering regulates the expression of floral repressors. In contrast, the signals of the promoting pathway converge to activate the expression of the *floral pathway integrators*, or genes that cause the floral transition (Parcy, 2005). A major question is whether vegetative development is the result of the activation of a special vegetative program or the repression of the reproductive program. In *Arabidopsis*, few if any genes have been found that are exclusively expressed in the vegetative shoot; thus, repression of flowering is likely the main mechanism for maintaining the vegetative program (Sung *et al.*, 2003).

In *Arabidopsis*, *FLOWERING LOCUS C* (*FLC*) encoding a MADS-box protein is a major floral repressor. Knowledge of the roles and regulation of the floral repressors has increased greatly in the past 5 years. This is mainly due to recent findings that most of the regulators of *FLC* expression appear to play key roles in chromatin remodeling and form protein complexes (Exner and Hennig, 2008). Many other floral repressors have been identified and we have started to understand how gene expression and protein accumulation of the floral repressors together with activators are regulated under various environmental conditions.

General mechanisms on flowering time and roles of floral activators have already been summarized (Boss *et al.*, 2004; Jaeger *et al.*, 2006; Komeda, 2004; Mizoguchi *et al.*, 2006; Parcy, 2005; Putterill *et al.*, 2004). In this review, we aim to summarize recent progress on the regulation of *FLC* expression in the autonomous/vernalization pathways, highlighting the roles of histone methylation and demethylation.

Chromatin physically restricts the accessibility of the genome to regulatory proteins that function in transcription (Pfluger and Wagner, 2007). These proteins include DNA binding factors and co-activators and co-repressors for transcriptional regulation. The restriction is dynamic and alters during developmental processes and under different environmental conditions. Histone modifications such as acetylation, methylation, phosphorylation and ubiquitination, play major roles in the epigenetic control of chromatin to affect gene expression (Pfluger and Wagner, 2007).

Next, we describe the roles of circadian clock and photoperiod in the control of floral repressors such as the MADS-box proteins SHORT VEGETATIVE PHASE (SVP) and MADS AFFECTING FLOWERING 1–5 (MAF1–5) and the AP2-domain proteins SCHLAFMÜTZE (SMZ), SCHNARCHZAPFEN (SNZ) and TARGET OF EAT1 (TOE1). We then show floral regulation by many pairs of floral repressors and activators. Finally, we discuss the molecular mechanisms underlying floral reversion in different plant species based on the functions of the floral repressors.

II. REGULATION OF GENE EXPRESSION OF THE FLORAL REPRESSOR *FLOWERING LOCUS C* (*FLC*) BY HISTONE METHYLATION

FLC is one of the major floral repressors in Arabidopsis. In this section, roles of the *FLC* and its homologous genes *MAF1–5* in Arabidopsis are shown. There have been many reports on key players in the autonomous/vernalization pathways that affect expression levels of the *FLC* and *MAF1–5*. Recent findings on regulation of these MADS-box genes by histone methylation/demethylation are also summarized in this section.

A. FLORAL REPRESSOR GENE *FLC* AND ITS HOMOLOGOUS GENES IN *ARABIDOPSIS*

1. FLC

The floral repressor FLC is a MADS-box transcriptional factor considered as central regulator of the flowering-enabling pathway (Boss *et al.*, 2004) (Fig. 1). It is expressed predominantly in shoot and root apices (Michaels and Amasino, 2000) and quantitatively prevents flowering through the repression of the floral integrator genes *FLOWERING LOCUS T* (*FT*), *SUPPRESSOR OF OVEREXPRESSION OF CO 1.* (*SOC1*), and *LEAFY* (*LFY*; Nilsson *et al.*, 1998). The flowering-enabling pathways (vernalization and autonomous pathways) remove these repressors and allow activation of the floral integrator genes by the floral promotion pathways (photoperiod, light quality, hormones, and ambient temperature; Thomas *et al.*, 2006). *FLC* levels are then determinant in the final decision to flower. The expression of the *FLC* gene is maintained at high levels by the *FRIGIDA* (*FRI*) gene. *FRI* encodes a protein with two coiled coil domains. FRI promotes the accumulation of *FLC* mRNA (Michaels and Amasino, 1999) and functions as a floral

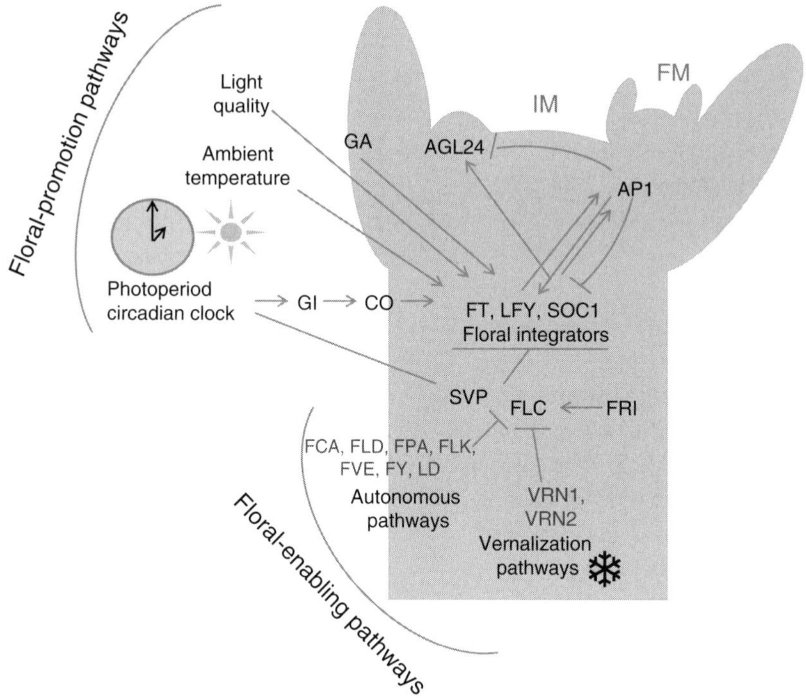

Fig. 1. Illustration of pathways controlling flowering time and floral meristem identity in *Arabidopsis*.

repressor. Prolonged exposure to cold temperatures (vernalization) strongly downregulates *FLC* levels and accelerates flowering (Michaels and Amasino, 1999; Sheldon *et al.*, 1999). According to Boss *et al.* (2004), many early-flowering accessions of *Arabidopsis* carry loss of function of *FRI* alleles. However, subsequent genetics studies that have compared different winter-annual and summer-annual accessions of *Arabidopsis* have revealed that variation at one or both of two loci, *FLC* and *FRI*, can account for a large portion of the winter-annual habit in *Arabidopsis* (Boss *et al.*, 2004). Thus, FLC and FRI synergistically delay flowering in winter-annual accessions of *Arabidopsis*, and loss-of-function mutations in either gene result in the loss of the late-flowering phenotype (Johanson *et al.*, 2000; Michaels and Amasino, 1999).

2. MADS AFFECTING FLOWERING Genes Family (MAFs)

Five genes, *MAF1–5*, encode proteins that are closely related to the floral repressor FLC (Ratcliffe *et al.*, 2001, 2003). *MAF1* gene was also identified as *FLOWERING LOCUS M* (*FLM*; Scortecci *et al.*, 2001). Overexpression of *MAF1* caused late-flowering phenotypes similar to those of the *FLC* over-expressor line (Ratcliffe *et al.*, 2001). In contrast to *FLC*, however, *MAF1* expression showed a less clear correlation with vernalization response. *FLC* expression was not affected by the overexpression of *MAF1*. Therefore, MAF1 appears to act downstream or independently of *FLC* transcription. Genetic analysis has suggested that MAF1 together with SVP may play key roles in the photoperiodic pathway rather than the autonomous/vernalization pathway (Scortecci *et al.*, 2003).

Four *FLC* paralogs (*MAF2–5*) are arranged in tandem on the bottom of *Arabidopsis* chromosome 5 (Ratcliffe *et al.*, 2003). Overexpression of *MAF2* in a Landsberg *erecta* (L*er*) background caused late flowering without affecting *FLC* expression. *maf2* mutant plants displayed a pronounced vernalization response when subjected to relatively short cold periods that were insufficient to elicit a strong flowering response in the wild type despite producing a large reduction in *FLC* levels. The effect of vernalization on the reduction of gene expression of *MAF2* was much weaker than that of *FLC*. These findings suggest that MAF2 can prevent premature vernalization in response to brief cold spells and that MAF2 and FLC may have both common and distinct roles in vernalization signaling.

The transcript levels of *MAF3*, and to some extent *MAF4*, appeared to be responsive to a 6 Weeks vernalization similar to those of *FLC* and *MAF1* (Ratcliffe *et al.*, 2003). However, the expression patterns of *MAF3* and *MAF4* differed from those of *FLC*. Transcript levels were not consistently higher in nonvernalized samples from each of the three late-flowering backgrounds (i.e., Pitztal [Pi-0] wild type, Stockholm [St-0] wild type, and *fca-9* mutant [Columbia; Col]) relative to nonvernalized Col wild-type plants. Overexpression of MAF3 or MAF4 produced alterations in flowering time, suggesting that these genes also act as floral regulators. However, the final gene in the cluster, MAF5, was upregulated by vernalization, suggesting that MAF5 could play an opposite role from FLC in vernalization response.

B. REPRESSION OF *FLC* EXPRESSION BY THE AUTONOMOUS/VERNALIZATION PATHWAYS

Genes in the autonomous and vernalization pathways play key roles in the flowering-enabling pathway (Marquardt *et al.*, 2006) (Fig. 1). The autonomous pathway comprises a group of seven genes (*FCA*, *FY*, *FLD*, *FVE*, *FPA*, *LD*, and *FLK*) that, when mutated, produce late-flowering phenotypes (Mouradov *et al.*, 2002). Like the vernalization pathway genes, the autonomous pathway genes converge on the major floral repressor *FLC* and promote flowering by repressing *FLC* expression (Bastow and Dean, 2003). However, recent experiments have suggested that different genes use different mechanisms in the autonomous pathway to repress *FLC* expression (Putterill *et al.*, 2004). The *FCA* and *FY* genes function together to control the degree of FLC mRNA levels and flowering time by regulating polyadenylation site selection (Quesada *et al.*, 2003).

The *FLD* gene is one of the seven identified members of the autonomous floral pathway. The *fld* mutant exhibits late-flowering phenotypes due to an increase in *FLC* expression (He *et al.*, 2003). In contrast to hypoacetylation of histones, which is associated with transcriptionally silent chromosomal regions, hyperacetylation of histones is associated with transcriptionally active genes (Reyes *et al.*, 2002). FLD is homologous to the human protein KIAA060 that is a component of histone deacetylase 1 and 2 (HDAC 1/2) and therefore has been proposed to regulate *FLC* by deacetylating histone H4 in *FLC* chromatin (He *et al.*, 2003) (Table I).

C. ROLES OF HISTONE METHYLATION AND DEMETHYLATION IN *FLC* EXPRESSION

Vernalization promotes flowering by reducing the activity of FLC, thus counteracting the effect of the *FRI* gene (Boss *et al.*, 2004). Although vernalization is an irreversible process, the plant continues to grow for several weeks before flowering. The question is, how do plants that have been vernalized remember this signal and flower months later? Somehow, the cells remember this signal as they divide; the vernalized state can be passed on through

TABLE I
Summary of floral regulators

Gene name	Accession No.	R/A	Domain	Function	Interactors
AGL24	At4g24540	A	MADS box	DNA binding	
CCA1	At2g46830		myb domain	DNA binding	LHY, ELF3
CO	At5g15840	A	Zinc finger (B-box)		
ELF3	At2g25930	R			CCA1, SVP
ELF6	At5g04240	R	Jmj N/C domain, C2H2 zinc finger	Histone demethylation	BES1
ELF7	At1g79730	R	Yeast PAF! Homolog	Histone demethylation	
ELF8/VIP6	At2g06210	R	Yeast CTR9 homolog	Histone demethylation	
FD	At4g35900	A	bZIP	DNA binding	FT
FDP	At2g17770		bZIP	DNA binding	FT, TFL1
FLC	At5g10140	R	MADS box	DNA binding	SVP
FLD	At3g10390	R	SWIRM, amine oxidease	Histone deacetylation	
FT	At1g65480	A	phosphatidylethanolamine binding domain?		FD
FVE	At2g19520		WD40 repeat		
GI	At1g22770	A			
LHP1/TFL2	At5g17690	R	HP1, chromo domain	DNA binding, methylated histone binding	
LHY	At1g01060		myb domain	DNA binding	CCA1
MAF1/FLM	At1g77080	R	MADS box	DNA binding	
MAF2	At5g65060	R	MADS box	DNA binding	
MAF3	At5g65060	R	MADS box	DNA binding	
MAF4	At5g65070	R	MADS box	DNA binding	
MAF5	At5g65080		MADS box	DNA binding	
PIE1	At3g12810		ISWI, SWI2/SNF2	ATP/DNA binding, helicase	SEF, APR6

(*continues*)

TABLE I (*continued*)

Gene name	Accession No.	R/A	Domain	Function	Interactors
REF6	At3g48430	A	Jmj N/C domain, C2H2 zinc finger	Histone demethylation	BES1
SDG8/EFS	At1g77300	R	SET domain	H3K36 methylation	
SDG26/ASHH1	At1g76710	A	SET domain	H3K36 methylation	
SMZ	At3g54990	R	AP2 domain	DNA binding	
SNP	At2g39250	R	AP2 domain	DNA binding	
SVZ	At2g22540	R	MADS box	DNA binding	FLC, ELF3
TFL1	At5g03840	R	Phosphatidylethanolamine binding domain?		FDP, FD
TOE1	At2g28550	R	AP2 domain	DNA binding	
VIN3	At5g61150		PHD finger	Histone methylation	VRN2, FIE, CLF, SWN
VIP4	At5g61150		Leo1-like protein, member of PAF1 complex	Histone methylation	
VRN1	At3g18990		B3 domain	Histone H3 methylation	
VRN2	At4g16845		PcG, zinc finger	Histone H3 methylation	
VRN5/VIL1	At3g24440		PHD finger	Histone methylation	

R/A indicate floral repressor or activator, respectively.

mitotic cell divisions even in the absence of cold, but it is lost after meiosis (Putterill *et al.*, 2004). This indicates that the repression of *FLC* expression is epigenetic (Schmitz and Amasino, 2007).

According to Thomas *et al.* (2006), this epigenetic silencing of *FLC* is mediated by *VERNALIZATION 1* (*VRN1*) and *VERNALIZATION 2* (*VRN2*) genes, which are involved in histone methylation and the formation of mitotically stable and transcriptionally silent heterochromatin. The *VRN2* gene encodes a nuclear-localized zinc-finger protein similar to the *Drosophila* Polycomb Group (PcG) protein SU(Z)12 (Gendall *et al.*, 2001). PcG proteins are components of complexes that repress gene expression by maintaining the chromatin in a state incompatible with transcription (Gendall *et al.*, 2001). VRN2 may function in a similar manner to other PcG proteins to maintain the vernalization-induced repression of *FLC* (Putterill *et al.*, 2004).

As shown in the previous section, both acetylation and methylation of histones play key roles in the epigenetic control of gene expression. Lysine residues (K) of histones (H) can be mono-, di-, or trimethylated, and this provides an ample magnitude of epigenetic information for transcription regulation. In fungi, SET2 is the sole methyltransferase responsible for the mono-, di-, and trimethylation of the 36th K of histone H3 (H3K36) (Strahl *et al.*, 2002). Two *Arabidopsis* SET2 homologs, SET domain group 8 (SDG8) and SDG26, which each methylated oligonucleosomes *in vitro*, were localized in the nucleus (Xu *et al.*, 2008). *sdg8* and *sdg26* mutants exhibited early- and late-flowering phenotypes, respectively. Several MADS-box flowering repressors were downregulated by *sdg8* but upregulated by *sdg26*, consistent with the flowering phenotypes. The *sdg8* but not the *sdg26* mutant plants showed a dramatically reduced level of both di- and trimethyl-H3K36 and an increased level of monomethyl-H3K36, suggesting that SDG8 may be specifically required for di- and trimethylation of H3K36 (Xu *et al.*, 2008). *SDG8* and *VERNALIZATION INDEPENDENCE 4* (*VIP4*), both of which encode proteins similar to components of the yeast PAF1 complex, acted independently and synergistically in transcription regulation (Xu *et al.*, 2008).

Plant homeodomain (PHD) finger-containing proteins, VERNALIZATION INSENSITIVE 3 (VIN3; Sung and Amasino, 2004) and VERNARIZATION 5 (VRN5; Greb *et al.*, 2007), are shown to form a heterodimer necessary for establishing the vernalization-induced chromatin modifications. These modifications include histone deacetylation and trimethylation of H3K27 that are required for the epigenetic silencing of *FLC* (Greb *et al.*, 2007). VIN3-LIKE 1 (VIL1)/VRN5 also plays important roles in not only vernalization but also photoperiod pathways by regulating expression of both FLC and FLM/MAF1 (Sung *et al.*, 2006).

EARLY FLOWERING 6 (ELF6) and its homolog RELATIVE OF EARLY FLOWERING 6 (REF6) are Jumonji N/C (JmjN/C) domain-containing proteins and regulate flowering time in *Arabidopsis* (Noh *et al.*, 2004). Many Jmj proteins have histone demethylase activity (Klose and Zhang, 2007). Consistent with this, chromatin immunoprecipitation (ChIP) experiments indicated that histone 3 lysine 9 (H3K9) methylation status was changed in *elf6* and *ref6* mutants (Yu *et al.*, 2008). A late-flowering phenotype of the *ref6* mutation was associated with an increased level of *FLC* expression. These results indicate that histone methylation by SET2-like proteins and histone demethylation by JmjN/C domain-containing protein(s) play important roles in controlling flowering *via* expression of *FLC*.

Although FLD was previously proposed to be involved in deacetylation of histone H4 as shown in Section I.B, it has been reported that FLD has significant homology to the human lysine-specific demethylase 1 (LSD1) (Jiang *et al.*, 2007). LSD1 is an integral component of several mammalian HDAC corepressor complexes (Humphrey *et al.*, 2001; Hakimi *et al.*, 2002). In these complexes, HDACs and LSD1 are thought to cooperate to remove acetyl and methyl histone modifications (Shi *et al.*, 2005; Lee *et al.*, 2006). FCA requires FLD for downregulation of *FLC* expression. *fca* mutations increase H3K4 dimethylation in the central region of *FLC*, supporting a close association of FCA and FLD in mediating H3K4 demethylation to control FLC expression (Liu *et al.*, 2007b).

Methylation not only of lysine (K) but also arginine (R) residues of histones plays a key role in chromatin regulation (Bedford and Richard, 2005; Wysocka *et al.*, 2006). Shk1 binding protein 1 (SKB1) in *Arabidopsis* catalyzed symmetric dimethylation of histone H4R3 (H4R3sme2; Wang *et al.*, 2007). Mutation in the *SKB1* gene increased *FLC* expression and delayed flowering time under both long-day (LD) and short-day (SD) conditions. SKB1 bound to the *FLC* promoter, and *skb1* mutation resulted in reduced H4R3sme2, especially in the promoter of *FLC* chromatin (Wang *et al.*, 2007). These findings indicate that methylation and demethylation of histones of arginine as well as lysine residues may play key roles in controlling flowering time.

III. REGULATION OF FLORAL REPRESSORS BY CIRCADIAN CLOCK OR PHOTOPERIOD

There is no doubt for the major roles of FLC and MAFs of *Arabidopsis* and their homologous genes in *Brassica* species in floral repression, but the *FLC/MAF* gene family does not appear to be conserved in all plant species. In this section, roles of other floral repressors such as *SVP* and its homologous genes encoding MADS-box proteins and *SMZ*, *SNZ* and *TOE1* encoding AP2-domain proteins are shown. Although the autonomous/vernalization pathways have important roles in the control of the *FLC* and *MAF* gene expression, protein accumulation of SVP protein and gene expression of *SMZ*, *SNZ*, and *TOE1* appear to be affected by circadian clock and photoperiods. Recent findings on these clock and photoperiod-dependent regulations are summarized in this section.

A. SHORT VEGETATIVE PHASE (SVP)

1. SVP gene

The *SVP* gene encodes a MADS-box protein acting as a floral repressor. Studies at the cellular level have shown a strong expression signal in young leaves and throughout the shoot apical meristem that disappears in the inflorescence apical meristem. During flower development, expression disappears prior to the emergence of sepals (Hartmann *et al.*, 2000). The *svp* mutant shows early-flowering phenotypes without displaying other obvious features (Hartmann *et al.*, 2000), and thus the *svp* mutant seems to pass more rapidly through the vegetative development stage. According to Hartmann *et al.* (2000), vernalization, which usually reduces flowering time in *Arabidopsis* wild type under both long-day (LD) and short-day (SD) conditions, could shorten the flowering time of *svp* mutants only when grown under SD conditions.

Besides its role in controlling flowering time, the *SVP* gene also functions as a modulator of meristem identity. Ectopic expression of the *SVP* gene inhibits floral meristem identity in *Arabidopsis*, causing floral abnormalities such as the conversion of sepals and petals to leaf-like structures (Brill and Watson, 2004; Masiero *et al.*, 2004) and floral reversion (Tooke *et al.*, 2005) through the production of inflorescence-like structures within the flowers (Brill and Watson, 2004).

2. Functional interaction of SVP gene with other MADS-box genes

A successful floral transition requires the activation and repression of appropriate flowering-time genes (i.e., activation of promoters and repression of suppressors) and specific activation of floral meristem identity genes that will subsequently activate downstream organ identity genes to build the flower (Boss *et al.*, 2004).

Genetic studies by Scortecci *et al.* (2003) showed that *svp* mutations overcame the late flowering conferred by overexpression of the *FLM*/*MAF1* gene, and *svp;flm* double mutants behaved like single mutants. Their study indicates that FLM/MAF1 and SVP appear to function in the same floral pathway, which interacts with the photoperiodic pathway. Moreover, as SVP mRNA levels were not affected by *flm* mutations or FLM/MAF1 overexpression (Scortecci *et al.*, 2003), and *flm* mutations suppressed the late-flowering effect of *SVP* overexpression, SVP and FLM act as co-regulated partners to control the floral transition.

AGAMOUS-like 24 (*AGL24*) belongs to the same MADS-box gene subfamily as *SVP* (Becker and Theißen, 2003). However, these two MADS-box genes regulate the floral transition in an antagonistic manner. Whereas the *SVP* gene represses flowering, *AGL24* activates it (Michaels *et al.*, 2003; Yu *et al.*, 2002) by promoting the inflorescence identity. Furthermore, this antagonism in the regulation of the floral transition does not prevent a synergistic regulation of floral meristem identity achieved by both MADS-box genes (Yu *et al.*, 2004). In that sense, both genes suppress *AG* by recruiting a co-repressor complex along with *APETALA1* (*AP1*; Gregis *et al.*, 2006).

A more recent study by Liu *et al.* (2007a) highlighted the master role of the MADS-box protein AP1 as a direct suppressor of the expression of flowering-time genes (notably *SVP*, *AGL24*, and *SOC1*) in floral meristems to prevent the continuation of the shoot developmental program. Liu *et al.* (2007a) showed that the misexpression of the *SVP* gene in floral meristems affects normal flower development, which confirms the function of the *SVP* gene as a flower meristem regulator. They also showed a new role of AP1 as a repressor of flowering-time genes; thus, AP1 plays a dual role in regulating floral meristem development by both activating and repressing different sets of genes.

3. *Common and distinct roles of SVP and SVP-like genes in plants*

MADS-box genes similar to *SVP* have been identified in species such as tomato (Mao *et al.*, 2000), *Antirrhinum* (Masiero *et al.*, 2004), *Solanum tuberosum* (Garcia-Maroto *et al.*, 2000), *Paulownia kawakamii* (Prakash and Kumar, 2002), rice (Duan *et al.*, 2006), *Triticum aestivum* (Kane *et al.*, 2005, 2007), barley (*Hordeum vulgare*; Trevaskis *et al.*, 2007), and *Eucalyptus grandis* (Brill and Watson, 2004).

The 35S promoter from the cauliflower mosaic virus (CaMV) is frequently used to drive the constitutive and ectopic expression of plant genes (Sanger *et al.*, 1990). In some species, ectopic expression of *SVP*-like genes under the 35S promoter in either a homologous or heterologous system causes a reversion to vegetative features after the floral transition. For example, when *StMADS16* from *S. tuberosum* was ectopically expressed in tobacco, transformants overexpressing the *StMADS16* gene exhibited vegetative features in the flowers (Garcia-Maroto *et al.*, 2000). *Antirrhinum INCOMPOSITA* (*INCO*) is an *SVP*-like gene that is also capable of repressing flowering when ectopically overexpressed (Masiero *et al.*, 2004). Floral reversion phenotypes appeared when *INCO* was overexpressed in an *Arabidopsis* background, suggesting a common role of this MADS-box gene in controlling floral meristem identity (Masiero *et al.*, 2004). In barley (*H. vulgare*), a monocot, ectopic expression of Barley MADS1 (BM1) or BM10 inhibits spike (flower structure in cereal) development and causes floral reversion (Trevaskis *et al.*, 2007), and ectopic expression of OsMADS22 in transgenic rice plants results in abnormal floral morphogenesis characterized by a disorganized palea, an elongated glume, and a two-floret spikelet (Sentoku *et al.*, 2005).

An SVP-group MADS-box protein in rice, OsMDP1 (Duan *et al.*, 2006), was shown to have a negative role in brassinosteroid (BR) signaling. Two years later, Lee *et al.* (2008) showed that *OsMADS22* and *OsMADS55*, two other SVP-group MADS-box genes in rice, work as negative regulators of BR responses. However, these MADS-box genes are not involved in photoperiod-dependent flowering regulation. Based on these observations, the role of the floral meristem modulator has been attributed to SVP-like MADS-box genes in some dicotyledonous species.

According to Sablowski (2007), determinancy in the *Arabidopsis* shoot is a property of the floral meristems, whereas the vegetative and inflorescence meristems are indeterminate. The activity of the vegetative and the inflorescence meristems is antagonized by flower-specific regulators.

B. ROLES OF CIRCADIAN CLOCK PROTEINS *LATE ELONGATED HYPOCOTYL* (*LHY*) AND *CIRCADIAN CLOCK ASSOCIATED 1* (*CCA1*) IN THE CONTROL OF SVP PROTEIN ACCUMULATION

1. Circadian clock proteins LHY and CCA1 in Arabidopsis

Circadian rhythms are driven by endogenous biological clocks that regulate many biochemical, physiological, and behavioral processes in a wide variety of organisms (Dunlap, 1999 and references therein). The circadian clocks are endogenous timekeepers that control many rhythmic processes (circadian rhythms) in organisms as they experience the 24 h cycle of day and night (Dunlap, 1999). Interactions between photoreceptors and the circadian clock are thought to allow plants to distinguish between different daylengths.

The basic molecular mechanisms underlying the generation of circadian rhythms have been investigated in model organisms such as *Synechococcus elongatus*, *Neurospora crassa*, *Drosophila melanogaster*, *Arabidopsis thaliana* and mouse (Mizoguchi *et al.*, 2006). In *Arabidopsis*, *LATE ELONGATED*

HYPOCOTYL (*LHY*; Schaffer *et al.*, 1998), *CIRCADIAN CLOCK ASSOCIATED 1* (*CCA1*; Wang and Tobin, 1998), *TIMING OF CAB EXPRESSION 1* (*TOC1*; Strayer *et al.*, 2000)/*PSEUDO RESPONSE-REGULATOR 1* (*PRR1*; Matsushika *et al.*, 2000), *PRR5*, *PRR7*, *PRR9*, *GIGANTEA* (*GI*; Fowler *et al.*, 1999; Park *et al.*, 1999), *EARLY FLOWERING 3* (*ELF3*; Hicks *et al.*, 2001), *EARLY FLOWERING 4* (*ELF4*; Doyle *et al.*, 2002), and several other genes have been shown to be important components of the circadian clock in *Arabidopsis*. Connection between circadian clock and photoperiodic flowering has been summarized in recent reviews (Imaizumi and Kay, 2006; Izawa, 2007; Mizoguchi *et al.*, 2006; Niinuma *et al.*, 2007).

2. Diurnal accumulation of SVP protein controlled by LHY and CCA1

Three floral activators such as GI, CONSTANS (CO), and FLOWERING LOCUS T (FT) play key roles in the photoperiodic flowering responses of the LD plant *Arabidopsis* (Mizoguchi *et al.*, 2005). The GI-CO-FT pathway is highly conserved in plants. In addition to repressing the floral transition under SD and LD conditions, the circadian clock proteins LHY and CCA1 accelerated flowering when the plants were grown under continuous light (LL; Fujiwara *et al.*, 2008; Mizoguchi *et al.*, 2002, 2005; Mizoguchi and Yoshida, 2009). LHY and CCA1 accelerated flowering in LL by promoting *FT* expression through a genetic pathway apparently independent of the canonical photoperiodic pathway involving GI and CO proteins (Fujiwara *et al.*, 2008). A genetic screen revealed that the late-flowering phenotype of the *lhy;cca1* double mutant under LL was suppressed through mutations in genes for two MADS-box transcription factors, SVP and FLC (Fujiwara *et al.*, 2008). Yeast two-hybrid analysis demonstrated an interaction between SVP and FLC, and genetic analysis indicated that these two proteins act as partially redundant repressors of flowering time (Fujiwara *et al.*, 2008). SVP protein showed a diurnal accumulation in wild-type plants under LD condition and the SVP protein level increased in *lhy;cca1* plants under LL (Fujiwara *et al.*, 2008). These findings suggest that LHY and CCA1 may accelerate flowering in part by reducing the abundance of SVP and thereby antagonizing its capacity to repress *FT* expression under LL (Fujiwara *et al.*, 2008).

Spatial as well as temporal controls of floral activator genes have been demonstrated to be crucial for the precise regulation of the photoperiodic flowering (Li *et al.*, 2008). Li *et al.* have shown that the flowering repressor SVP is controlled by the autonomous, thermosensory, and gibberelic acid (GA) pathways, and suppress directly expression of *SOC1* in the shoot apex and leaf. *FT* expression in the leaf is also modulated by SVP. SVP protein associates with the promoter regions of both *FT* and *SOC1* and FLC also binds these regions (Li *et al.*, 2008). These findings are consistent with our recent findings (Fujiwara *et al.*, 2008) and suggest that the interaction between SVP and FLC mediated by various flowering genetic pathways governs the integration of flowering signals. Li *et al.* provide mechanistic views on how FLC and SVP precisely regulate expression of two floral activator genes, *FT* and *SOC1*, in two different spaces, apical meristems and leaves (Li *et al.*, 2008).

3. *Possible role of ELF3 in the control of SVP protein accumulation by LHY and CCA1*

Accumulation of SVP protein in *lhy;cca1* appears to be responsible for the late-flowering phenotype of *lhy;cca1* under LL. It was, however, still not clear how the *lhy;cca1* affected the SVP protein level under LL, because the mRNA level of *SVP* was not greatly affected by *lhy;cca1* mutations (Fujiwara *et al.*, 2008), and protein–protein interactions between SVP and LHY or CCA1 were not detected by the yeast two-hybrid analysis (Fujiwara *et al.*, 2008). Finding a possible missing link between LHY/CCA1 and SVP/FLC has been a next important step for us to understand how SVP and FLC delayed flowering more strongly in *lhy;cca1* mutants than wild-type plants under LL. We have recently identified several *elf3* (Zagotta *et al.*, 1996) mutations as suppressors of *lhy;cca1* under LL (Yoshida *et al.*, in press). Overexpression of *ELF3* increased the SVP protein level and *elf3* mutation altered diurnal pattern of the SVP protein accumulation (Yoshida *et al.*, in press). Protein–protein interactions were detected between CCA1 and ELF3 and between ELF3 and SVP by yeast two-hybrid analysis (Yoshida *et al.*, in press). These results suggest that ELF3 may have an important role as a mediator between LHY/CCA1 and SVP/FLC in the control of flowering time. The evening-expressed protein ELF3 is a potential positive regulator of the morning-clock genes *CCA1* and *LHY* (Alabadi *et al.*, 2001; Doyle *et al.*, 2002; Hicks *et al.*, 2001). Further screening and analysis of suppressor mutations of the late-flowering phenotype of *lhy;cca1* and of the early-flowering phenotype of *elf3* are underway (Nefissi, Natsui and Mizoguchi, unpublished). These would be helpful to understand molecular mechanisms underlying the regulation of SVP protein accumulation by clock proteins LHY and CCA1 in *Arabidopsis*.

C. PHOTOPERIODIC CONTROL OF GENE EXPRESSION FOR APETALA 2 (AP2)-DOMAIN PROTEINS SCHLAFMÜTZE (SMZ), SCHNARCHZAPFEN (SNZ) AND TARGET OF EAT1 (*TOE1*)

Arabidopsis is a facultative LD plant and flowers much earlier under LD than SD conditions. The photoperiodic inductive condition not only upregulates floral activator genes such as GI and CO but also downregulates the floral repressor genes SMZ and SNZ. These two paralogous genes encode transcription factors with the AP2 domain (Schmid *et al.*, 2003).

MicroRNAs are involved in a variety of biological processes including flowering-time regulation. Of these microRNAs, miR172 promotes flowering primarily by posttranscriptionally repressing a set of *AP2*-like genes, including *TOE1*, *TOE2*, and *TOE3* (Aukerman and Sakai, 2003). Overexpression of *TOE1* caused late flowering, whereas miR172-overexpressing plants exhibited early flowering under both LD and SD conditions. A global analysis of *Arabidopsis* gene expression identified SMZ and SNZ as additional miR172 targets (Schmid *et al.*, 2003). miR172 abundance is regulated by photoperiod *via* GI-mediated miRNA processing (Jung *et al.*, 2007). miR172 promotes photoperiodic flowering through a CO-independent genetic pathway. GI-mediated photoperiodic flowering appears to be governed by the coordinated interaction of two distinct genetic pathways: one mediated *via* CO and the other mediated *via* miR172 and its targets (Jung *et al.*, 2007). The CO-independent promotion of flowering controlled by GI has been proposed by independent research (Mizoguchi *et al.*, 2005). The early flowering 3 (*elf3*) mutation also accelerates flowering in a CO-independent process (Kim *et al.*, 2005).

IV. FLORAL REVERSION

Floral reversion includes flower and inflorescence reversion. Even though these processes are thought to be unusual, these are useful to understand a molecular mechanism underlying a switch from floral development back to vegetative development. Floral reversion appears to be caused by altered balance of activity of floral activators and repressors. First, examples of pairs of floral activator and repressor are shown in this section. Then our knowledge on floral reversion in *Arabidopsis* and other plant species is summarized to map the floral regulators on the complicated floral signaling networks.

A. FLORAL REPRESSION AND ACTIVATION IS CONTROLLED BY PAIRS OF FLORAL REGULATORS

As shown in the previous sections, flowering time appears to be finely tuned by pairs of positive and negative floral regulators. The first set identified was the floral repressor TERMINAL FLOWER 1 (TFL1) and the activator FT (Kardailsky *et al.*, 1999; Kobayashi *et al.*, 1999). Both of these proteins share significant homology to phosphatidylethanolamine-binding proteins. FT, a candidate of florigen, has been shown to be a mobile protein from leaf to apex (Corbesier *et al.*, 2007; Jaeger and Wigge, 2007; Tamaki *et al.*, 2007) and requires the bZIP-type transcription factor FD to function as a floral activator (Abe *et al.*, 2005; Wigge *et al.*, 2005). *FD* but not *FT* is expressed in the apical meristem. FT protein produced in leaves may move to the apical meristem and interact with FD to control genes required for flowering and organ identity (Abe *et al.*, 2005). FDP is a closely related homolog of FD. Protein–protein interactions exist not only between FDP and FT but also between FDP and TFL1 (Wigge *et al.*, 2005). FT–FD and TFL1–FDP may function as a floral activator and repressor, respectively, and these pairs of floral regulators may play opposite roles in controlling flowering.

The second set of positive and negative floral regulators reported was a pair of MADS-box transcription factors, the floral repressor SVP and the floral activator AGL24 (Hartmann *et al.*, 2000; Michaels *et al.*, 2003; Yu *et al.*, 2002). Both of these proteins interact with the floral activators AP1 and SOC1. Competition for interaction with these floral activators may be one explanation for the opposite roles of SVP and AGL24 in controlling flowering time.

Two pairs of proteins, ELF6–REF6 and SDG8–SDG26, appear to be involved in controlling methylation/demethylation of histones, as shown in the previous section (Xu *et al.*, 2008; Yu *et al.*, 2008). ELF6 and REF6 encode JmjN/C domain-containing proteins and may have histone demethylase activity.

B. FLORAL REVERSION IN *ARABIDOPSIS*

In *Arabidopsis*, the vegetative phase is characterized by the formation of the rosette shoot, as the vegetative shoot does not undergo internode elongation (Sung *et al.*, 2003). However, stem internodes start to elongate (bolting) after the transition from vegetative to reproductive state. This first phase transition is regulated by a large number of flowering-time genes. The second phase change involves the development of the main inflorescence meristem. The floral meristems that originate from inflorescence meristems are determinates, in contrast to the inflorescence meristems, which are indeterminates with the production of secondary inflorescences subtended by cauline leaves, or flowers (Yu *et al.*, 2004). Floral meristem identity genes such as *LFY* and *AP1* are required to confer floral identity on newly arising meristems; in the absence of these genes, floral meristems are partially or completely replaced by shoot meristems (Parcy *et al.*, 2002) (Fig. 2). An antagonistic interaction between the shoot identity gene *TFL1* and the flower meristem identity genes controls the final fate of these meristems. Nevertheless, the simple establishment of floral meristem identity is not sufficient to guarantee floral identity (Yu *et al.*, 2004). Its maintenance is required after its establishment to avoid floral reversion (Anthony *et al.*, 1996; Battey and Lyndon, 1990; Tooke *et al.*, 2005).

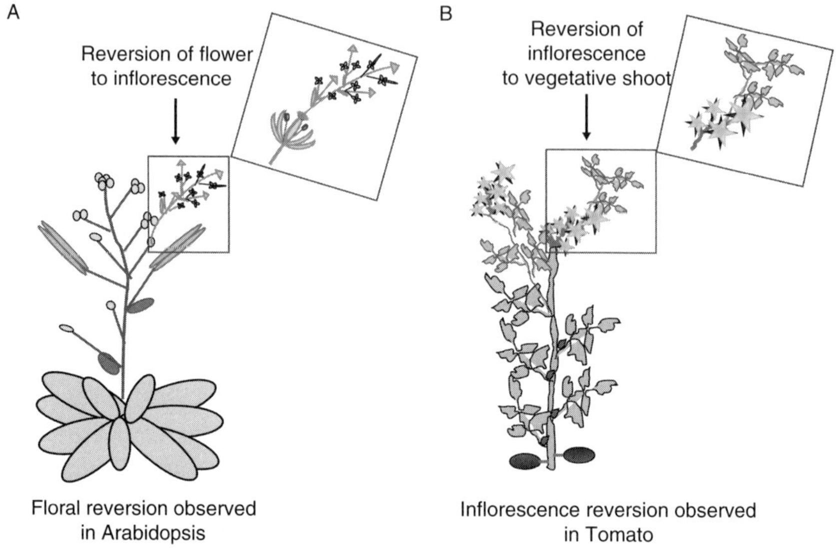

Fig. 2. Schematic illustration of the floral reversion phenotype in two different species *Arabidopsis* and Tomato.

Defining floral reversion as "a return to leaf production after a period of flower development," Tooke *et al.* (2005) distinguished two distinct types: (a) inflorescence reversion, in which vegetative development occurs after inflorescences have already developed; and (b) flower reversion, in which the form of the flower itself is altered as individual flowers stop producing floral organs and initiate vegetative organs instead (Fig. 2). In *Arabidopsis*, floral reversion has been described in terms of mutant genetic backgrounds such as *ap1*, *lfy*, and *ag* (Gregis *et al.*, 2006) and backgrounds overexpressing a large number of genes involved in specifying flower organ identity or flowering-time genes, notably MADS-box genes such as *AGL24*, *SVP*, and *SOC1* (Brill and Watson, 2004; Liu *et al.*, 2007; Masiero *et al.*, 2004; Trevaskis *et al.*, 2007).

Based on the genetic characterization of floral homeotic mutants of *A. thaliana* and *Antirrhinum majus*, the simple and elegant ABC model of flower development (Coen and Meyerowitz, 1991; Weigel and Meyerowitz, 1994) explains how the four types of floral organs (sepals, petals, stamens, and carpels) arranged in four concentric whorls are specified by A, B, and C classes of floral homeotic genes. Mutations in these floral homeotic genes result in substitutions or replacements of one organ type by another.

According to Sablowski (2007), the floral meristem in *Arabidopsis*, in contrast to indeterminate vegetative and inflorescence meristems, typically has a determinant growth habit. Consequently, in the case of floral reversion, a floral meristem will lose its determinancy in favor of indeterminacy. To prevent the floral meristem from becoming indeterminate, AG, which controls the development of stamens and carpels in the flower (Yanofsky *et al.*, 1990), acts by repressing the *WUSCHEL* (*WUS*) gene after a full set of floral organs is initiated. *WUS* and *SHOOT MERISTEMLESS* (*STM*) are two regulatory genes with central roles in shoot meristem development (Wang and Li, 2008). Among other functions, they are required to maintain the vegetative, inflorescence, and floral meristems (Wang and Li, 2008). During flower development, *WUS* is expressed in the organizing center of the floral meristem until the initiation of fourth whole organs (Mayer *et al.*, 1998). Also, the indeterminate floral phenotype in *ag* mutants is due to a possible negative regulation of the *WUS* gene by AG in a wild-type background; this was further confirmed when *WUS* expression remained in the center of *ag* mutant flowers (Lenhard *et al.*, 2001). These results lead to the establishment of a molecular mechanism whereby LFY and AP1 activate *AG* expression, which leads to the downregulation of *WUS* for the subsequent determinate floral meristem. This mechanism would enable floral reversion to take place.

In *Arabidopsis*, an example of floral reversion is provided by *ap1;cal* double mutants and *ap1;cal;ful* triple mutants in which the inflorescence shoot meristem produces primordia that behave like secondary inflorescence meristems (Ferrándiz *et al.*, 2000). Other examples include mutants that produce floral primordia that revert to an inflorescence shoot meristem after production of floral organs in different whorls; these mutants include *ag* mutants, *ap1;clv1* double mutants, and *lfy* heterozygote mutants grown under SD conditions (Clark *et al.*, 1993; Okamuro *et al.*, 1996; Parcy *et al.*, 2002). Also, floral reversion could be observed under the overexpression of the floral promoter AGL24 member of the MADS-box family of DNA-binding transcription factors (Yu *et al.*, 2002), suggesting a potential role of AGL24 as a promoter of inflorescence identity. The ectopic expression of the *SVP* gene (Brill and Watson, 2004) causes floral abnormalities and inhibits floral meristem identity, causing inflorescence-like structures to emerge within the flowers. According to the definition of Tooke *et al.* (2005), this latter phenotype is synonymous with floral reversion.

C. FLORAL REVERSION IN OTHER PLANT SPECIES

Impatiens balsamina (cv. Dwarf Bush Flowered) shows a typical flower reversion, as meristem activity can persist after gynoecium production (Took and Battey, 2003). An aberrant phenotype shows reversion of the terminal flower to leaf production when transferred to LD conditions, which limit flowering conditions for this species (Chiurugwi *et al.*, 2007).

The reproductive structure of tomato has long been considered to be a cyme (Sawhney and Greyson, 1972; Went, 1944), but further work done by Allen and Sussex (1996) indicated that it is a raceme (like the inflorescence) rather a cyme. These observations were confirmed by Welty *et al.* (2007) through the study of inflorescence development in two tomato species. Consequently, similar to *Arabidopsis*, the tomato inflorescence is regulated by two groups of identity genes: inflorescence and floral meristem identity genes. Although fewer mutants affect the flower or inflorescence identity of tomato compared to *Arabidopsis*, mutants such as *compound inflorescence* (*s*), *jointless* (*j*), and *falsiflora* (*fa*) alter the inflorescence architecture in different ways. *s* mutants produce branched inflorescences upon the reversion of the floral meristems to inflorescence meristems, which leads to a highly branched inflorescence (Quinet *et al.*, 2006), whereas *j*-mutant inflorescences revert to vegetative growth after having initiated two or three flowers (Mao *et al.*, 2000; Szymkowiak and Irish, 1999). *sft* and *fa* mutants are characterized by an inflorescence reverting to vegetative growth after the development of reproductive structure (Molinero-Rosales *et al.*, 2004), whereas *j* mutants are further distinguished by the lack of an abscission zone on the flower pedicels (Mao *et al.*, 2000).

V. PERSPECTIVES

Although SVP is proposed to be in the photoperiodic flowering pathway together with FLM/MAF1, the exact position of these floral repressors in this pathway is not known. *SVP* expression is not significantly affected by loss of function or overexpression of genes involved in the photoperiodic pathway (Fujiwara *et al.*, 2008). Biochemical analysis of SVP and FLM/MAF1 protein levels under different photoperiodic conditions and isolation of SVP- and FLM/MAF1-interacting proteins will help researchers to understand the molecular mechanisms underlying the possible roles of SVP in the photoperiodic pathway. ELF6 encoding a putative histone demethylase is also placed in the photoperiodic pathway based on genetics and analysis of *FLC* gene expressions. However, compared to those of its homolog REF6, the roles of ELF6 in controlling flowering time are much less understood. One attractive but still possible explanation is that SVP may interact with ELF6 directly or indirectly to play key roles in the photoperiodic pathway.

ACKNOWLEDGMENT

This work was supported in part by the Ministry of Economy, Trade, and Industry of Japan (METI) (H.E. and T.M.).

REFERENCES

Abe, M., Kobayashi, Y., Yamamoto, S., Daimon, Y., Yamaguchi, A., Ikeda, Y., Ichinoki, H., Notaguchi, M., Goto, K. and Araki, T. (2005). FD, a bZIP protein mediating signals from the floral pathway integrator FT at the shoot apex. *Science* **309**, 1052–1056.

Alabadi, D., Oyama, T., Yanovsky, M. J., Harmon, F. G., Mas, P. and Kay, S. A. (2001). Reciprocal regulation between TOC1 and LHY/CCA1 within the *Arabidopsis* circadian clock. *Science* **293**, 880–883.

Allen, K. D. and Sussex, I. M. (1996). Falsiflora and anantha control early stages of floral meristem development in tomato (*Lycopersicon esculentum* Mill.). *Planta* **200**, 254–264.

Anthony, R. G., James, P. E. and Jordan, B. R. (1996). Cauliflower (*Brassica oleracea* var. *botrytis* L.) curd development: The expression of meristem identity genes. *Journal of Experimental Botany* **47**, 181–188.

Aukerman, M. J. and Sakai, H. (2003). Regulation of flowering time and floral organ identity by a microRNA and its APETALA2-like target genes. *Plant Cell* **15**, 2730–2741.

Bastow, R. and Dean, C. (2003). Deciding when to flower. *Science* **302**, 1695–1697.

Battey, N. H. and Lyndon, R. F. (1990). Reversion of flowering. *Botanical Review* **56**, 163–189.

Becker, A. and Theißen, G. (2003). The major clades of MADS-box genes and their role in the development and evolution of flowering plants. *Molecular Phylogenetics and Evolution* **29**, 464–489.

Bedford, M. T. and Richard, S. (2005). Arginine methylation an emerging regulator of protein function. *Molecular Cell* **18**, 263–272.

Boss, P. K., Bastow, R. M., Mylne, J. S. and Dean, C. (2004). Multiple pathways in the decision to flower: Enabling, promoting, and resetting. *Plant Cell* **16**, S18–S31.

Brill, E. M. and Watson, J. M. (2004). Ectopic expression of a *Eucalyptus grandis SVP* orthologue alters the flowering time of *Arabidopsis thaliana*. *Functional Plant Biology* **31**, 217–224.

Chiurugwi, T., Pouteau, S., Nicholls, D., Tooke, F., Ordidge, M. and Battey, N. (2007). Floral meristem indeterminacy depends on flower position and is facilitated by acarpellate gynoecium development in *Impatiens balsamina*. *New Phytologist* **173**, 79–90.

Clark, S. E., Running, M. P. and Meyerowitz, E. M. (1993). *CLAVATA1*, a regulator of meristem and flower development in *Arabidopsis*. *Development* **119**, 397–418.

Coen, E. S. and Meyerowitz, E. M. (1991). The war of the whorls: Genetic interactions controlling flower development. *Nature* **353**, 31–37.

Corbesier, L., Vincent, C., Jang, S., Fornara, F., Fan, Q., Searle, I., Giakountis, A., Farrona, S., Gissot, L., Turnbull, C. and Coupland, G. (2007). FT protein movement contributes to long-distance signaling in floral induction of *Arabidopsis*. *Science* **316**, 1030–1033.

Doyle, M. R., Davis, S. J., Bastow, R. M., McWatters, H. G., Kozma-Bognar, L., Nagy, F., Millar, A. J. and Amasino, R. M. (2002). The *ELF4* gene controls circadian rhythms and flowering time in *Arabidopsis thaliana*. *Nature* **419**, 74–77.

Duan, K., Li, L., Hu, P., Xu, S. P., Xu, Z. H. and Xue, H. W. (2006). A brassinolide-suppressed rice MADS-box transcription factor,OsMDP1, has a negative regulatory role in BR signaling. *Plant Journal* **47**, 519–531.

Dunlap, J. C. (1999). Molecular bases for circadian clocks. *Cell* **96**, 271–290.

Exner, V. and Hennig, L. (2008). Chromatin rearrangements in development. *Current Opinion of Plant Biology* **11**, 64–69.

Ferrándiz, C., Gu, Q., Martienssen, R. and Yanofsky, M. F. (2000). Redundant regulation of meristem identity and plant architecture by *FRUITFULL*, *APETALA1* and *CAULIFLOWER*. *Development* **127**, 725–734.

Fowler, S., Lee, K., Onouchi, H., Samach, A., Richardson, K., Morris, B., Coupland, G. and Putterill, J. (1999). *GIGANTEA*: A circadian clock-controlled gene that regulates photoperiodic flowering in *Arabidopsis* and encodes a protein with several possible membrane-spanning domains. *EMBO Journal* **18**, 4679–4688.

Fujiwara, S., Oda, A., Yoshida, R., Niinuma, K., Miyata, K., Tomozoe, Y., Tajima, T., Nakagawa, M., Hayashi, K., Coupland, G. and Mizoguchi, T. (2008). Circadian clock proteins LHY and CCA1 regulate SVP protein accumulation to control flowering in *Arabidopsis*. *Plant Cell* **20**, 2960–2971.

Gendall, A. R., Levy, Y. Y., Wilson, A. and Dean, C. (2001). The *VERNALIZATION 2* gene mediates the epigenetic regulation of vernalization in *Arabidopsis*. *Cell* **107**, 525–535.

Greb, T., Mylne, J. S., Crevillen, P., Geraldo, N., An, H., Gendall, A. R. and Dean, C. (2007). The PHD finger protein VRN5 functions in the epigenetic silencing of *Arabidopsis FLC*. *Current Biology* **17**, 73–78.

Gregis, V., Sessa, A., Colombo, L. and Kater, M. M. (2006). AGL24, SHORT VEGETATIVE PHASE, and APETALA1 redundantly control AGAMOUS during early stages of flower development in Arabidopsis. *Plant Cell* **18**, 1373–1382.

Garcia-Maroto, F., Ortega, N., Lozano, R. and Carmona, M. J. (2000). Characterization of the potato MADS-box gene *STMADS16* and expression analysis in tobacco transgenic plants. *Plant Molecular Biology* **42**, 499–513.

Hakimi, M. A., Bochar, D. A., Chenoweth, J., Lane, W. S., Mandel, G. and Shiekhattar, R. (2002). A core-BRAF35 complex containing histone deacetylase mediates repression of neuronal-specific genes. *Proceedings of the National Academy of Sciences of the USA* **99**, 7420–7425.

Hartmann, U., Hohmann, S., Nettesheim, K., Wisman, E., Saedler, H. and Huijser, P. (2000). Molecular cloning of SVP: A negative regulator of the floral transition in *Arabidopsis*. *Plant Journal* **21**, 351–360.

He, Y., Michaels, S. and Amasino, R. (2003). Regulation of flowering time by histone acetylation in *Arabidopsis*. *Science* **302**, 1751–1754.

Hicks, K. A., Alberston, T. M. and Wagner, D. R. (2001). *Early Flowering3* encodes a novel protein that regulates circadian clock function and flowering in *Arabidopsis*. *Plant Cell* **13**, 1281–1292.

Humphrey, G. W., Wang, Y., Russanova, V. R., Hirai, T., Qin, J., Nakatani, Y. and Howard, B. H. (2001). Stable histone deacetylase complexes distinguished by the presence of SANT domain proteins CoREST/kiaa0071 and Mta-L1. *Journal of Biological Chemistry* **276**, 6817–6824.

Imaizumi, T. and Kay, S. A. (2006). Photoperiodic control of flowering: Not only by coincidence. *Trends in Plant Science* **11**, 550–558.

Izawa, T. (2007). Daylength measurements by rice plants in photoperiodic short-day flowering. *International Review of Cytology* **256**, 191–222.

Jaeger, K. E. and Wigge, P. A. (2007). FT protein acts as a long-range signal in *Arabidopsis*. *Current Biology* **17**, 1050–1054.

Jaeger, K. E., Graf, A. and Wigge, P. A. (2006). The control of flowering in time and space. *Journal of Experimental Botany* **57**, 3415–3418.

Jiang, D., Yang, W., He, Y. and Amasino, R. M. (2007). Arabidopsis relatives of the human lysine-specific Demethylase1 repress the expression of *FWA* and *FLOWERING LOCUS C* and thus promote the floral transition. *Plant Cell* **19**, 2975–2987.

Johanson, U., West, J., Lister, C., Michaels, S., Amasino, R. and Dean, C. (2000). Molecular analysis of FRIGIDA, a major determinant of natural variation in *Arabidopsis* flowering time. *Science* **290**, 344–347.

Jung, J. H., Seo, Y. H., Seo, P. J., Reyes, J. L., Yun, J., Chua, N. H. and Parka, C. M. (2007). The GIGANTEA-regulated MicroRNA172 mediates photoperiodic flowering independent of CONSTANS in *Arabidopsis*. *Plant Cell* **19**, 2736–2748.

Kane, N. A., Danyluk, J., Tardif, G., Ouellet, F., Laliberte, J. F., Limin, A. E., Fowler, D. B. and Sarhan, F. (2005). TaVRT-2, a member of the StMADS-11 clade of flowering repressors, is regulated by vernalization and photoperiod in wheat. *Plant Physiology* **138**, 2354–2363.

Kane, N. A., Agharbaoui, Z., Diallo, A. O., Adam, H., Tominaga, Y., Ouellet, F. and Sarhan, F. (2007). TaVRT2 represses transcription of the wheat vernalization gene *TaVRN1*. *Plant Journal* **51**, 670–680.

Kardailsky, I., Shukla, V. K., Ahn, J. H., Dagenais, N., Christensen, S. K., Nguyen, J. T., Chory, J., Harrison, M. J. and Weigel, D. (1999). Activation tagging of the floral inducer *FT*. *Science* **286**, 1962–1965.

Kim, S. Y., He, Y., Jacob, Y., Noh, Y. S., Michaels, S. and Amasino, R. (2005). Establishment of the vernalization-responsive, winter-annual habit in *Arabidopsis* requires a putative histone H3 methyl transferase. *Plant Cell* **17**, 3301–3310.

Klose, R. J. and Zhang, Y. (2007). Regulation of histone methylation by demethylimination and demethylation. *Nature Reviews Molecular Cell Biology* **8**, 307–318.

Kobayashi, Y., Kaya, H., Goto, K., Iwabuchi, M. and Araki, T. (1999). A pair of related genes with antagonistic roles in mediating flowering signals. *Science* **286**, 1960–1962.

Komeda, Y. (2004). Genetic regulation of time to flower in *Arabidopsis thaliana*. *Annual Review of Plant Biology* **55**, 521–535.

Lee M. G., Wynder C., Schmidt D. M., McCafferty D. G., and Shiekhattar R. (2006). Histone H3 Lysine 4 Demethylation Is a Target of Nonselective Antidepressive Medications. *Chemistry and Biology* **13**, 563–567.

Lee, S., Choi, S. C. and An, G. (2008). Rice SVP-group MADS-box proteins, OsMADS22 and OsMADS55, are negative regulators of brassinosteroid responses. *Plant Journal* **54**, 93–105.

Lenhard, M., Bohnert, A., Jurgens, G. and Laux, T. (2001). Termination of stem cell maintenance in *Arabidopsis* floral meristems by interactions between WUSCHEL and AGAMOUS. *Cell* **105**, 805–814.

Li, D., Liu, C., Shen, L., Wu, Y., Chen, H., Robertson, M., Helliwell, C. A., Ito, T., Meyerowitz, E. and Yu, H. (2008). A repressor complex governs the integration of flowering signals in *Arabidopsis*. *Developmental Cell* **15**, 110–120.

Liu, C., Zhou, J., Bracha-Drori, K., Yalovsky, S., Ito, T. and Yu, H. (2007a). Specification of *Arabidopsis* floral meristem identity by repression of flowering time genes. *Development* **134**, 1901–1910.

Liu, F., Quesada, V., Crevillén, P., Bäurle, I., Swiezewski, S. and Dean, C. (2007b). The Arabidopsis RNA-binding protein FCA requires a lysine-specific demethylase 1 homolog to downregulate FLC. *Molecular Cell* **28**, 398–407.

Mao, L., Begum, D., Chuang, H. W., Budiman, M. A., Symkowiak, E. J., Irish, E. E. and Wing, R. A. (2000). *JOINTLESS* is a MADS-box gene controlling tomato flower abscission zone development. *Nature* **206**, 910–913.

Marquardt, S., Boss, P. K., Hadfield, J. and Dean, C. (2006). Additional targets of the *Arabidopsis* autonomous pathway members, FCA and FY. *Journal of Experimental Botany* **57**, 3379–3386.

Masiero, S., Li, M. A., Will, I., Hartmann, U., Saedler, H., Huijser, P., Schwarz-Sommer, Z. and Sommer, H. (2004). *INCOMPOSITA*: A MADS-box gene controlling prophyll development and floral meristem identity in *Antirrhinum*. *Development* **13**, 5981–5990.

Matsushika, A., Makino, S., Kojima, M. and Mizuno, T. (2000). Circadian waves of expression of the APRR1/TOC1 family of pseudo-response regulators in *Arabidopsis thaliana*: Insight into the plant circadian clock. *Plant Cell and Physiology* **41**, 1002–1012.

Mayer, K. F., Schoof, H., Haecker, A., Lenhard, M., Jurgens, G. and Laux, T. (1998). Role of WUSCHEL in regulating stem cell fate in the *Arabidopsis* shoot meristem. *Cell* **95**, 805–815.

Michaels, S. D. and Amasino, R. M. (1999). *FLOWERING LOCUS C* encodes a novel MADS domain protein that acts as a repressor of flowering. *Plant Cell* **11**, 949–956.

Michaels, S. D. and Amasino, R. M. (2000). Memories of winter: Vernalization and the competence to flower. *Plant Cell and Environment* **23**, 1145–1153.

Michaels, S. D., Ditta, G., Gustafson-Brown, C., Pelaz, S., Yanofsky, M. and Amasino, R. M. (2003). AGL24 acts as a promoter of flowering in *Arabidopsis* and is positively regulated by vernalization. *Plant Journal* **33**, 867–874.

Mizoguchi, T. and Yoshida, Y. (2009). Punctual coordination: Switching on and off for flowering during a day. *Plant Signaling and Behavior* **4**, 113–115.

Mizoguchi, T., Wheatley, K., Hanzawa, Y., Wright, L., Mizoguchi, M., Song, H. R., Carré, I. A. and Coupland, G. (2002). *LHY* and *CCA1* are partially redundant genes required to maintain circadian rhythms in *Arabidopsis*. *Developmental Cell* **2**, 629–641.

Mizoguchi, T., Wright, L., Fujiwara, S., Cremer, F., Lee, K., Onouchi, H., Mouradov, A., Fowler, S., Kamada, H., Putterill, J. and Coupland, G. (2005). Distinct roles of *GIGANTEA* in promoting flowering and regulating circadian rhythms in Arabidopsis. *Plant Cell* **17**, 2255–2270.

Mizoguchi, T., Putterill, J. and Ohkoshi, Y. (2006). Kinase and phosphatase: The cog and spring of the circadian clock. *International Review of Cytology* **250**, 47–72.

Molinero-Rosales, N., Latorre, A., Jamilena, M. and Lozano, R. (2004). *SINGLE FLOWER TRUSS* regulates the transition and maintenance of flowering in tomato. *Planta* **218**, 427–434.

Mouradov, A., Cremer, F. and Coupland, G. (2002). Control of flowering time: Interacting pathways as a basis for diversity. *Plant Cell* **14**, 111–130.

Niinuma, K., Nakagawa, M., Calvino, M. and Mizoguchi, T. (2007). Dance of plants with the circadian clock. *Plant Biotechnology* **24**, 87–97.

Nilsson, O., Lee, I., Blazquez, M. A. and Weigel, D. (1998). Flowering time genes modulate the response to LEAFY activity. *Genetics* **150**, 403–410.

Noh, B., Lee, S. H., Kim, H. J., Yi, G., Shin, E. A., Lee, M., Jung, K. J., Doyle, M. R., Amasino, R. M. and Noh, Y. S. (2004). Divergent roles of a pair of homologous jumonji/zinc-finger-class transcription factor proteins in the regulation of Arabidopsis flowering time. *Plant Cell* **16**, 2601–2613.

Okamuro, J. K., Den Boer, B. G., Lotys-Prass, C., Szeto, W. and Jofuku, K. D. (1996). Flowers into shoots: Photo and hormonal control of a meristem identity switch in *Arabidopsis*. *Proceedings of the National Academy of Science of the USA* **93**, 13831–13836.

Parcy, F. (2005). Flowering: A time for integration. *International Journal of Developmental Biology* **49**, 585–593.

Parcy, F., Bomblies, K. and Weigel, D. (2002). Interaction of *LEAFY*, *AGAMOUS* and *TERMINAL FLOWER1* in maintaining floral meristem identity in *Arabidopsis*. *Development* **129**, 2519–2527.

Park, D. H., Somers, D. E., Kim, Y. S., Choy, Y. H., Lim, H. K., Soh, M. S., Kim, H. J., Kay, S. A. and Nam, H. G. (1999). Control of circadian rhythms and photoperiodic flowering by the *Arabidopsis GIGANTEA* gene. *Science* **285**, 1579–1582.

Pfluger, J. and Wagner, D. (2007). Histone modifications and dynamic regulation of genome accessibility in plants. *Current Opinion in Plant Biology* **10**, 645–652.

Prakash, A. P. and Kumar, P. P. (2002). *PkMADS1* is a novel MADS box gene regulating adventitious shoot induction and vegetative shoot development in *Paulownia kawakamii*. *Plant Journal* **29**, 141–151.

Putterill, J., Laurie, R. and Macknight, R. (2004). It's time to flower: The genetic control of flowering time. *Bioessays* **26**, 363–373.

Quesada, V., Macknight, R., Dean, C. and Simpson, G. G. (2003). Autoregulation of *FCA* pre-mRNA processing controls *Arabidopsis* flowering time. *EMBO Journal* **22**, 3142–3152.

Quinet, M., Dubois, C., Goffin, M. C., Chao, J., Dielen, V., Batoko, H., Boutry, M. and Kinet, J. M. (2006). Characterization of tomato (*Solanum lycopersicum* L.) mutants affected in their flowering time and in the morphogenesis of their reproductive structure. *Journal of Experimental Botany* **57**, 1381–1390.

Ratcliffe, O. J., Nadzan, G. C., Reuber, T. L. and Riechmann, J. L. (2001). Regulation of flowering in *Arabidopsis* by an FLC homologue. *Plant Physiology* **126**, 122–132.

Ratcliffe, O. J., Kumimoto, R. W., Wong, B. J. and Riechmann, J. L. (2003). Analysis of the Arabidopsis MADS AFFECTING FLOWERING gene family: MAF2 prevents vernalization by short periods of cold. *Plant Cell* **15**, 1159–1169.

Reyes, J. C., Hennig, L. and Gruissem, W. (2002). Chromatin-remodeling and memory factors. New regulators of plant development. *Plant Physiology* **130**, 1090–1101.

Sablowski, R. (2007). Flowering and determinacy in *Arabidopsis*. *Journal of Experimental Botany* **58**, 899–907.

Sanger, M., Daubert, S. and Goodman, R. M. (1990). Characteristics of a strong promoter from figwort mosaic virus: Comparison with the analogous 35S promoter from cauliflower mosaic virus and the regulated mannopine synthase promoter. *Plant Molecular Biology* **14**, 433–443.

Sawhney, V. K. and Greyson, R. I. (1972). On the initiation of the inflorescence and floral organs in tomato (*Lycopersicon esculentum*). *Canadian Journal of Botany* **50**, 1493–1495.

Schaffer, R., Ramsay, N., Samach, A., Corden, S., Putterill, J., Carré, I. A. and Coupland, G. (1998). The *late elongated hypocotyl* mutation of *Arabidopsis* disrupts circadian rhythms and the photoperiodic control of flowering. *Cell* **93**, 1219–1229.

Schmid, M., Uhlenhaut, N. H., Godard, F., Demar, M., Bressan, R., Weigel, D. and Lohmann, J. U. (2003). Dissection of floral induction pathways using global expression analysis. *Development* **130**, 6001–6012.

Schmitz, R. J. and Amasino, R. M. (2007). Vernalization: A model for investigating epigenetics and eukaryotic gene regulation in plants. *Biochimica et Biophysica Acta* **1769**, 269–275.

Scortecci, K. C., Michaels, S. D. and Amasino, R. M. (2001). Identification of a MADS-box gene, *FLOWERING LOCUS M*, that represses flowering. *Plant Journal* **26**, 229–236.

Scortecci, K., Michaels, S. D. and Amasino, R. M. (2003). Genetic interactions between *FLM* and other flowering-time genes in *Arabidopsis thaliana*. *Plant Molecular Biology* **52**, 915–922.

Sentoku, N., Kato, H., Kitano, H. and Imai, R. (2005). *OsMADS22*, an STMADS11-like MADS-box gene of rice, is expressed in non-vegetative tissues and its ectopic expression induces spikelet meristem indeterminacy. *Molecular Genetics and Genomics* **273**, 1–9.

Sheldon, C. C., Burn, J. E., Perez, P. P., Metzger, J., Edwards, J. A., Peacock, W. J. and Dennis, E. S. (1999). The *FLF* MADS box gene: A repressor of flowering in *Arabidopsis* regulated by vernalization and methylation. *Plant Cell* **11**, 445–458.

Shi, Y. J., Matson, C., Lan, F., Iwase, S., Baba, T. and Shi, Y. (2005). Regulation of LSD1 histone demethylase activity by its associated factors. *Molecular Cell* **19**, 857–864.

Strahl, B. D., Grant, P. A., Briggs, S. D., Sun, Z. W., Bone, J. R., Caldwell, J. A., Mollah, S., Cook, R. G., Shabanowitz, J., Hunt, D. F. and Allis, C. D. (2002). Set2 is a nucleosomal histone H3-selective methyltransferase that mediates transcriptional repression. *Molecular and Cellular Biology* **22**, 1298–1306.

Strayer, C., Oyama, T., Schultz, T. F., Raman, R., Somers, D. E., Más, P., Panda, S., Kreps, J. A. and Kay, S. A. (2000). Cloning of the *Arabidopsis* clock gene *TOC1*, an autoregulatory response regulator homolog. *Science* **289**, 768–771.

Sung, S. and Amasino, R. M. (2004). Vernalization in *Arabidopsis thaliana* is mediated by the PHD finger protein VIN3. *Nature* **427**, 159–164.

Sung, S., Schmitz, R. J. and Amasino, R. M. (2006). A PHD finger protein involved in both the vernalization and photoperiod pathways in *Arabidopsis*. *Genes and Development* **20**, 3244–3248.

Sung, Z. R., Chen, L., Moon, Y. H. and Lertpiriyapong, K. (2003). Mechanisms of floral repression in *Arabidopsis*. *Current Opinion of Plant Biology* **6**, 29–35.

Szymkowiak, E. J. and Irish, E. E. (1999). Interactions between *jointless* and wild-type tomato tissues during development of the pedicel abscission zone and the inflorescence meristem. *Plant Cell* **11**, 159–175.

Tamaki, S., Matsuo, S., Wong, H. L., Yokoi, S. and Shimamoto, K. (2007). Hd3a protein is a mobile flowering signal in rice. *Science* **316**, 1033–1036.

Thomas, B., Carre, I. and Jackson, S. (2006). "Photoperiodism and Flowering. The Molecular Biology and Biotechnology of Flowering," Second edition, pp. 3–25. CAB International, New York.

Tooke F., and Battey N. (2003). Transley Review Models of shoot apical meristem function. *New Phytologist* **159**, 37–52.

Tooke, F., Ordidge, M., Chiurugwi, T. and Battey, N. (2005). Mechanisms and function of flower and inflorescence reversion. *Journal of Experimental Botany* **56**, 2587–2599.

Trevaskis, B., Tadege, M., Hemming, M. N., Peacock, W. J., Dennis, E. S. and Sheldon, C. (2007). Short vegetative phase-like MADS box genes inhibit floral meristem identity in barley. *Plant Physiology* **143**, 225–235.

Wang, Y. and Li, J. (2008). Molecular basis of plant architecture. *Annual Review of Plant Biology* **59**, 253–279.

Wang, Z. Y. and Tobin, E. M. (1998). Constitutive expression of the *CIRCADIAN CLOCK ASSOCIATED 1* (*CCA1*) gene disrupts circadian rhythms and suppresses its own expression. *Cell* **93**, 1207–1217.

Wang, X., Zhang, Y., Ma, Q., Zhang, Z., Xue, Y., Bao, S. and Chong, K. (2007). SKB1-mediated symmetric dimethylation of histone H4R3 controls flowering time in Arabidopsis. *EMBO Journal* **26**, 1934–1941.

Weigel, D. and Meyerowitz, E. M. (1994). The ABCs of floral homeotic genes. *Cell* **78**, 203–209.

Welty, N., Radovich, C., Meulia, T. and Van der Knaap, E. (2007). Inflorescence development in two tomato species. *Canadian Journal of Botany* **85**, 111–118.

Went, F. W. (1944). Morphological observations on the tomato plant. *Bulletin of the Torrey Botanical Club* **71**, 77–92.

Wigge, P. A., Kim, M. C., Jaeger, K. E., Busch, W., Schmid, M., Lohmann, J. U. and Weigel, D. (2005). Integration of spatial and temporal information during floral induction in *Arabidopsis*. *Science* **309**, 1056–1059.

Wysocka, J., Allis, C. D. and Coonrod, S. (2006). Histone arginine methylation and its dynamic regulation. *Frontiers in Bioscience* **11**, 344–355.

Xu, L., Zhao, Z., Dong, A., Soubigou-Taconnat, L., Renou, J. P., Steinmetz, A. and Shen, W. H. (2008). Di- and tri- but not monomethylation on histone H3 Lysine 36 marks active transcription of genes involved in flowering time regulation and other processes in *Arabidopsis thaliana*. *Molecular and Cellular Biology* **28**, 1348–1360.

Yanofsky, M. F., Ma, H., Bowman, J. L., Drews, G. N., Feldmann, K. A. and Meyerowitz, E. M. (1990). The protein encoded by the *Arabidopsis* homeotic gene *agamous* resembles transcription factors. *Nature* **346**, 35–39.

Yoshida, R., Fekih, R., Fujiwara, S., Oda, A., Miyata, K., Tomozoe, Y., Nakagawa, M., Niinuma, K., Hayashi, K., Ezura, H., Coupland, G. and Mizoguchi, T. (in press) Possible role of EARLY FLOWERING 3 (ELF3) in the clock-dependent floral regulation by SHORT VEGETATIVE PHASE (SVP) in *Arabidopsis*. *New Phytologist*.

Yu, H., Xu, Y., Tan, E. L. and Kumar, P. P. (2002). AGAMOUS-LIKE 24, a dosage-dependent mediator of the flowering signals. *Proceedings of the National Academy of Science of the USA* **99**, 16336–16341.

Yu, H., Ito, T., Wellmer, F. and Meyerowitz, E. M. (2004). Repression of *AGAMOUS-LIKE 24* is a crucial step in promoting flower development. *Nature Genetic* **36**, 157–161.

Yu, X., Li, L., Li, L., Guo, M., Chory, J. and Yin, Y. (2008). Modulation of brassinosteroid-regulated gene expression by Jumonji domain-containing proteins ELF6 and REF6 in *Arabidopsis*. *Proceedings of the National Academy of Science of the USA* **105**, 7618–7623.

Zagotta, M. T., Hicks, K. A., Jacobs, C. I., Young, J. C., Hangarter, R. P. and Meeks-Wagner, D. R. (1996). The *Arabidopsis ELF3* gene regulates vegetative photomorphogenesis and the photoperiodic induction of flowering. *Plant Journal* **10**, 691–702.

AUTHOR INDEX

Numbers in **bold** refer to pages on which full references are listed.

C

SUBJECT INDEX